A Passion for Antiquities

P CVRTILIVS PLAGA

A Passion for Antiquities

Ancient Art from the Collection of Barbara and Lawrence Fleischman

The J. Paul Getty Museum
in association with
The Cleveland Museum of Art

17985 Pacific Coast Highway
Malibu, California 90265-5799

Published on the occasion of an exhibition organized by The J. Paul Getty Museum and cosponsored by The Cleveland Museum of Art; the exhibition will be on view in Malibu from October 13, 1994 to January 15, 1995 and in Cleveland from February 15 to April 23, 1995.

At the J. Paul Getty Museum:
Christopher Hudson, Publisher
John Harris, Editor

Project staff:
Marion True and Kenneth Hamma, Editors
Cynthia Newman Bohn, Manuscript Editor
Katy Homans, Designer
Elizabeth Burke Kahn, Production Coordinator
Bruce White and Jerry Thompson, Photography
Allan Jokisaari, Map Designer

Typesetting by G&S Typesetters, Inc., Austin, Texas
Printed by Nissha Printing Co., Ltd., Kyoto, Japan

Cover:
Antefix in the Form of a Maenad and Silenos Dancing.
Catalogue number 92. Photo: Ellen Rosenbery

Title page:
The Fleischmans' living room, January 1994

Library of Congress Cataloguing-in-Publication Data
A Passion for antiquities : ancient art from the collection of
Barbara and Lawrence Fleischman.
p. cm.
"Published on the occasion of an exhibition held at the J. Paul Getty Museum, October 13, 1994–January 15, 1995 and at the Cleveland Museum of Art, February 15–April 9, 1995"—T.p. verso
Includes bibliographical references.
ISBN 0-89236-223-5
1. Art, Classical—Exhibitions. 2. Fleischman, Lawrence Arthur, 1925–   —Art collections—Exhibitions. 3. Fleischman, Barbara—Art collections—Exhibitions. 4. Art—Private collections—New York (N.Y.)—Exhibitions. 5. Art—New York (N.Y.)—Exhibitions.
I. J. Paul Getty Museum. II. Cleveland Museum of Art.
N5603.M36G486 1994
709´.38´07479493—dc20 94-2552
CIP

Contents

Foreword

As the great American museums attest, this country has a well-established tradition of knowledgeable private collecting. What we see in the galleries is largely a mosaic of once private collections. These donations, rather than well-publicized individual purchases, form the basis of our museums. The feats of individual collectors provoke our curiosity about what originally inspired them to start collecting, about the taste that guided them, and about the passion that drove them, often at great personal sacrifice. Private collections are shaped by many factors: opportunity, personal interest, influential advisors, and contemporary tastes. Unlike museum collections that generally try to provide the public with as complete and representative a view of an artistic period or medium as possible, the private collection knows no such restrictions. The only considerations for the collectors are, Do I like it? Can I afford it? Can I live with it?

Private collections permit us to enter for a moment into the minds of the collectors and to feel vicariously the joy of discovery and imagine the satisfaction of choosing the piece or pieces that gives us the greatest pleasure. Unlike the museum experience, which often seems to impose a standard of historical significance and chronology upon the visitor, a private collection invites us to enjoy the object and make our own judgments.

The more than three hundred ancient Greek, Etruscan, and Roman objects patiently and lovingly gathered by Barbara and Lawrence Fleischman over the past forty years represent the most refined kind of private collecting. The collection reflects the passionate interests and tastes of the two people who formed it. Because they were chosen for display in a private apartment, not a grand museum gallery, the objects are generally small and invite intimate study. Examining a large number of finely wrought objects in a temporary exhibition can be taxing, so for the purposes of this exhibition, we have chosen to show only two-thirds of the entire collection.

We are moved by the objects of daily life: a mirror that a South Italian beauty once used to arrange her hair, a glass perfume bottle that adorned a Roman matron's table, a humorous terracotta lamp that gave light to some ancient insomniac. We are impressed by the power that still radiates from the bronze and marble images of pagan gods, the confident serenity of Tyche, the majesty of Zeus, the seductive smile of Dionysos. And we are amused, as ancient theater comes alive again on the great South Italian vases and in the comic masks and figures of actors in clay and bronze.

The opportunity to exhibit this collection has a particular appeal for both the J. Paul Getty Museum and the Cleveland Museum of Art. Like the Getty Museum's collection of antiquities, it is limited to the arts of Greece, Etruria, and Rome, and like the collections of both museums, the guiding factor in the selection of these pieces has been their exceptional artistic quality, not their archaeological interest.

It is a rare privilege for us to present this exhibition. The generosity that made it possible is characteristic of the Fleischmans, who for many years have welcomed scholars and connoisseurs from all over the world into their home. It was not difficult to persuade them that the time had come to share their extraordinary achievement with a larger public audience and to live for a year deprived of their household gods.

In addition to the Fleischmans, we would like to thank Ariel Herrmann, curator of the Fleischman collection for the past four years, who undertook the original research on most of the objects in the collection and provided many of the catalogue entries, in addition to overseeing the photography and assisting in the preparation for the transportation of the collection; Bruce White, who provided the wonderful photographs of the objects; Jerry Thompson, who photographed the collection in situ; Charles Passela and his staff, particularly Rebecca Branham and Ellen Rosenbery, in Photographic Services at the Getty Museum, who prepared the photographic materials for publication; the Conservation and Preparation staffs of the Getty Museum, especially Jerry Podany, Wayne Haak, Bruce Metro, and Rita Gomez, and Bruce Christman, Chief Conservator of the Cleveland Museum of Art, who managed all aspects of the complex packing, shipping, and installation of these irreplaceable pieces; Cynthia Newman Bohn, who edited the catalogue, Katy Homans, who is responsible for its elegant design, and Elizabeth Burke Kahn of the Publication Services staff, who coordinated all aspects of its production.

John Walsh
Director, The J. Paul Getty Museum

Robert P. Bergman
Director, The Cleveland Museum of Art

Preface

Building a collection is an ongoing adventure and exhibiting it brings us a special pleasure. What makes this odyssey so stimulating is meeting so many talented people, all of whom enrich our lives. Planning this exhibition has been a joy, giving us the opportunity to work with co-curators Marion True and Arielle Kozloff, two extraordinarily gifted friends. We would like to thank the staff of the J. Paul Getty Museum, a community of helpful and able people headed by John Walsh, who have contributed greatly to making this a most happy experience. In particular we are grateful to Deborah Gribbon, Dorothy Osaki, Rita Gomez, and Wayne Haak.

We are also most appreciative of the staff of the Cleveland Museum of Art, and its director, Robert P. Bergman, for making possible the exhibition's stay in Cleveland.

Interpreting our acquisitions is exceedingly important and has been accomplished by the intelligent and insightful catalogue text written by Maxwell L. Anderson, Janet Grossman, Robert Guy, Kenneth Hamma, Sybille Haynes, Ariel Herrmann, John Herrmann and Annewies van den Hoek, Marit Jentoft-Nilsen, Arielle Kozloff, Leo Mildenberg, Andrew Oliver, Michelle Roland, Oliver Taplin, Elana Towne, Dale Trendall, Marion True, and Karol Wight, with additional help from Jack Ogden, whose reports on the jewelry in the collection served as the basis for the catalogue entries on those objects, and Anne Leinster, who assisted in the verification of the bibliographic citations.

Scholars from all over the world have served as our professors in the informal university that antiquity collecting creates, and we have benefited enormously from their deep and varied knowledge. We must mention Maxwell L. Anderson and Dietrich von Bothmer, who from early days encouraged our reactivated interest in the ancient world, and others, like Michael Padgett, Brian Cook, and Dyfri Williams, who continue to share their enthusiasm and knowledge with us.

As we continue to build this collection we must acknowledge the vital role so many others play in assisting us. The antiquity dealers combine their particular knowledge and enterprise to make these beautiful and unique objects available to us and we have learned much from them.

Bruce White, whose splendid photographs enhance this catalogue, has been a boon to us. Lisa Pilosi helps watch over the physical condition of our treasures, and William Stenders has ably assisted in mounting the objects for display at home.

Friendship runs like a thread through our many associations with the people who inhabit this fascinating world. In particular we want to mention our friends and fellow collectors Shelby White and Leon Levy, who, in connection with the exhibition, are so generously underwriting a symposium at the Cleveland Museum of Art as well as helping to sponsor several performances of ancient comedy at the Getty Museum. Also a continual inspiration to us is Christos Bastis, dean of New York antiquity collectors. These friendships, born out of our shared passion, add a special warmth and dimension to our life and indeed are the "icing on the cake."

Barbara and Lawrence Fleischman

Barbara and Lawrence Fleischman: Guardians of the Past

Marion True and Arielle Kozloff

In an age of prefabricated buildings, disposable furnishings, and transient values, it is impossible not to wonder how a young couple from Detroit became interested in the arts of the Classical world. For Barbara and Lawrence Fleischman, the answer is not complicated—for them, collecting is pure pleasure. Their interest began before they were married with their mutual fascination with history and the visual arts. Nurtured by experience and personal friendships, their commitment to ancient art developed into a shared passion that over the years has provided an important focus for their partnership.

Larry Fleischman's attraction to antiquity began during the Second World War, when he was a soldier stationed in France, where he visited the Roman ruins of Besançon. A French doctor observed Larry, then a young man of nineteen, gazing intently at the remains of the Roman theater and asked him what he found so absorbing. Impressed by the seriousness of the young American's answer, he invited him home for an evening which would become the basis of a lasting friendship. The doctor, who was well-educated in the classics, and his family provided stimulating company during the long months abroad and encouraged Larry's interest in ancient history.

England, too, played a role in the formation of the Fleischmans' development as collectors. During the war, the British Museum had put most of its great treasures in storage for safekeeping. However, one gallery, the Gallery of Edward VII, containing a selection of masterpieces from the collections, remained open to the public. When he was on leave in London Larry visited this gallery; he was fascinated by the variety of objects and charmed by the way that ancient artists had made functional objects into things of beauty. Before the gallery closed in the early 1950s, he returned with his bride, Barbara, who proved to be an easy convert to the world of ancient art.

Back in Detroit, the Fleischmans' love and support for the arts became an important force in their lives. They soon developed a close friendship with the great director of the Detroit Institute of Arts, Edgar P. Richardson, who was to be a major influence on the direction of their collecting activities. Like the Fleischmans, Richardson was fascinated by ancient art, and together they visited many of the important American collections, where Richardson would point out interesting and often overlooked artifacts. In particular, it was the pieces that were associated with daily life that excited the Fleischmans most. As Barbara recalls, whenever they looked at objects such as ancient bronze seat markers from a Roman theater or a strainer that had been used in a Greek kitchen, they felt a strong and immediate connection with the people who had owned these things in antiquity.

Barbara and Lawrence Fleischman,
January 1994.

Their support for the Detroit Institute of Arts led to Larry's election as president of the museum's Founders Society in 1962.

The Fleischmans purchased their first ancient work of art, a Roman bronze lamp in the shape of a thyrsos (cat. no. 147) in 1951. Still a treasured part of the collection, this piece embodied all of their developing interests. They loved it not only for its elegant shape and color but also for its functional nature—its association with the practical aspects of daily life. The fact that someone had once used this beautiful object had a special significance for them, and the human connection to the past that it evoked was to remain a guiding principle in their future collecting. In 1963, for similar reasons, they purchased several Greek vases from the collection of William Randolph Hearst as well as a small vessel of Roman millefiore glass from Jacob Hirsch around the same time.

During these years, another interest, again fostered by Edgar Richardson, began to absorb the Fleischmans, and this was American art. Richardson was one of the foremost scholars in the field of American art and under his guidance they became increasingly impressed by the amount and variety of art that had been created during the United States' two brief centuries of existence. Deciding to focus their attention on works of art created during the life of their own country, Larry and Barbara spent the next decade building a major collection of American painting and sculpture. Not satisfied to simply own the works of art, the Fleischmans wanted to know everything they could about the artists and the intellectual achievements of the various periods. As a corollary to their own search for information, Larry at this time cofounded the Archives of American Art, an organization devoted to research and scholarship in the field, which he continues to support.

Ironically, it was the Fleischmans' collection of American art that took them to Greece for the first time. In 1959 their paintings went on tour to a number of European cities, including Athens, and Larry was invited to speak about American art in Greece. During this visit, the couple drove together to Delphi and there had an experience that would affect the rest of their lives: they saw the Delphi Charioteer. The emotional impact that monumental bronze figure made upon them both was instrumental in renewing their attraction to ancient art.

Fortunately, this revived interest came at an opportune moment, for it was not many years later that the Fleischmans found it necessary to relinquish the pleasure of collecting American art. During the years of forming their great collection, Larry had become a leading expert and lecturer in the field. And, as is often the case in such circumstances, his personal interest introduced him to a new profession. In 1966 he became a partner in the Kennedy Galleries, a private art gallery

The Fleischmans' living room, January 1994.

in New York devoted to American art; he would become full owner in 1983. He and Barbara moved their family from Detroit to Manhattan and set about creating a new life for themselves. Realizing that they could no longer collect in an area in which they would be competing with their clients, they sought another outlet for their passion. Several factors conspired to influence this decision in favor of antiquities.

The first was an extraordinary invitation from the Vatican to assist in the formation of a collection of modern religious art. Shortly after his coronation, Pope Paul VI had examined the holdings of the Vatican Museum. Although these were exceptionally strong in the arts of the past, the spiritual art of the twentieth century was not represented. The pope recognized that this situation must be corrected if the Vatican collections were to maintain their relevance to the modern church and its members around the world, and he decided to seek the advice of experts. Being himself fond of the work of Ben Shahn, Pope Paul asked his staff to contact Shahn's dealer, the Kennedy Galleries. Larry was called to Rome to consult with the Vatican staff about contemporary religious art, and with his guidance, the nucleus of the twentieth-century collection took shape. This close association with the Vatican over a number of years was to have a profound effect on both Barbara and Larry, while their frequent trips to Rome deepened their love of ancient history and refined their tastes in art.

During this period, the Fleischmans also made the acquaintance of Dietrich von Bothmer, Chairman of the Greek and Roman Department at the Metropolitan Museum of Art in New York. One of the greatest scholars of Greek vase painting, Bothmer had learned that Larry and

The Fleischmans' dining room, January 1994.

Barbara owned some vases from the William Randolph Hearst collection. As Bothmer himself had purchased sixty-five vases from the Hearst estate for the Metropolitan Museum, he had a particular interest in the collection and asked if he could come to see their pieces. With his typical honesty, he judged them to be genuine but of little aesthetic value. He advised the Fleischmans to sell most of these pieces and replace them with examples of greater quality if they wished to collect seriously. Recognizing the wisdom of this advice, Larry and Barbara began to dispose of the early purchases and to refine the quality of their ancient art collection.

But perhaps the single most important influence on the Fleischmans at this time was Maxwell Anderson, then Assistant Curator of Greek and Roman Art under Bothmer at the Metropolitan Museum, and now Director of the Michael C. Carlos Museum at Emory University in Atlanta, Georgia. Delighted by the Fleischmans' burgeoning interest in antiquities, he introduced them to dealers who specialized in ancient art and encouraged them to support the Metropolitan Museum's exhibitions of ancient art, most notably "The Amasis Painter and His World." With characteristic commitment, the Fleischmans became enthusiastic participants in the activities of the Greek and Roman Department.

Through their association with the Metropolitan Museum, Barbara and Larry met two other passionate collectors of antiquities, Shelby White and Leon Levy, who joined Christos Bastis and others when Larry and Dietrich von Bothmer formed a support group called the Philodoroi for the Metropolitan's Greek and Roman Department. When the moment came for the retirement of Bothmer, who had served as a source of encouragement and advice to both couples, the Fleischmans and the Levys together proposed to create a Distinguished Chair for Research in Bothmer's honor. It was at the urging of the Fleischmans that Shelby White and Leon Levy agreed to present their splendid collection to the public in exhibition "The Glories of the Past," which opened at the Metropolitan Museum in September 1990. The Fleischmans sponsored the symposium that was held on January 12, 1991, in honor of the exhibition.

One special passion of Barbara Fleischman, now shared wholeheartedly by Larry, is a love of the theater. On her graduation from college, she had actually gone to New York to work as an actress. As she recalls, it was much to her parents' relief that she returned to Detroit to marry Larry. Her pleasure in the theater was never forgotten, however, but simply diverted into other channels, which would affect the couple's commitment to collecting and collections. Within their own collection, the number of objects related to the theater or representing theatrical subjects is remarkably large, and many are of exceptional importance. This fact has led many theater historians and classicists with a special interest in ancient drama to visit the collection, providing Larry and Barbara with access to the special world of scholarship on ancient theater.

To stay abreast of the art world and contemporary theater, the Fleischmans maintain an apartment in London, where they have found themselves drawn back to the British Museum, where they spent so many meaningful hours early in their marriage. Now committed to the world of antiquities, Barbara and Larry rediscovered the pleasures of the British Museum's great sculpture galleries and the exceptional collections of Greek and Roman bronzes and Greek vases. Brian Cook, recently retired Keeper of the Greek and Roman Collections, encouraged the Fleischmans' interests and with their assistance inaugurated a support group for the department, the Caryatids. When the museum undertook a new restoration and installation of the frieze from the Temple of Apollo at Bassai, the Fleischmans underwrote both the cost of the work and the scholarly symposium that celebrated the completion of the project. They took a special pleasure in contributing to the conservation, reassessment, and protection of this important sculptural complex, hoping, as Larry says, in some small way to repay the British Museum for the pleasure that it has given them over the years. Their participation and support for the activities of British Museum's Greek and Roman department continue under the new Keeper, Dyfri Williams.

The commitment to conservation and presentation demonstrated

The Fleischmans' living room, January 1994.

by the Fleischmans' support for the Bassai frieze project is equally evident in their own collection. Since most of their objects have already survived two millennia or more, they are determined that these pieces will be passed on, thoroughly researched and in the best possible condition, in the endless chain of preservation. "We are the temporary custodians," both are quick to remind the visitor, but they are concerned and devoted guardians. A highly respected scholar of ancient sculpture, Ariel Herrmann, serves as the curator of the collection, carefully researching and documenting each new addition to the collection. Trained conservators have cleaned and mounted the individual objects, advising the Fleischmans on climate control and special environments for sensitive materials. All publications and changes in attribution or identification provided by visiting specialists are efficiently noted and retained in their growing catalogue files.

For the Fleischmans, collecting is not an act of accumulating trophies or private treasures. It is also neither complicated nor scientific. They collect on the basis of instantaneous emotional response to the object's aesthetic appeal and its historic interest for them. Later, they reflect on where it fits within the context of their collection. Although seemingly simple, such an approach demands great confidence and discernment, skills of connoisseurship acquired over decades of experience and exposure to works of art. Larry in particular has had the unusual opportunity to collect from three perspectives, as a museum president, as a dealer, and as a private collector, and he sums up the differences as follows: "When you are collecting for an institution, you are always influenced by what the collection needs; in commerce, you are motivated by what sells; but in forming a personal collection, you know that you will have to live with the object twenty-four hours a day, so you buy only what you react to most positively."

As the Fleischmans know well, living surrounded by works of art is quite a different experience from visiting an exhibition. Changing their appearance with the light of day or the mood of viewer, the objects provide a constant source of inspiration and discovery. Touring a guest through the collection, Barbara is sometimes perplexed that a statuette or vase has moved, but she explains with a smile that Larry often gets up and rearranges things in the night, creating new juxtapositions and installations that give a dynamic quality to the collection. And though the different textures of materials represented among the objects in the collection—gold, bronze, terracotta, marble, and glass—were gathered unintentionally, they tend to enhance one another by subtle contrasts.

Above all, however, it is the universality of the ancient art objects that has captured the Fleischmans' devotion. These objects, which have survived centuries of change and evolution, reflect many of the same desires and emotions that still color our lives. The yearning to raise the functional object to something beautiful is evident in every medium. Vases of metal and terracotta were decorated with lively compositions of human and mythical creatures or inlaid with precious metal and colored glass paste; the walls of rooms were expanded with painted fantasies; the waterspouts and antefixes of buildings were transformed into animals or figural groups. In death, the Roman silversmith wanted to be remembered for his craft, a Macedonian hunter for his youth and beauty.

For Barbara and Larry, their shared love for the arts of Greece, Rome, and Etruria has provided a powerful focus for their partnership. Happily, the vitality and enthusiasm with which they pursue their interest in ancient art has also infected their children, Rebecca, Arthur, and Martha, who, together with their granddaughter Jennifer, have encouraged and supported their parents' interests. At home or abroad, there is always a new museum to visit, a scholarly symposium to attend, or a gallery to explore, and antiquities have brought the Fleischmans into contact with a wealth of friends who share their commitment. Indeed, the enjoyment that they have derived from the experience of forming this collection is reflected in this exhibition, as each object, large or small, reflects the pleasure of its discovery and selection.

Chronological Overview of the Fleischman Collection

Karol Wight

The objects in this collection were produced over a span of about 3,000 years, from the flourishing of the early Bronze Age civilizations in the region of the Aegean Sea to the Roman Imperial domination of the areas we know today as France and Germany, and are among the finest examples of the arts produced by these numerous and varied cultures.

The earliest piece in the Fleischman collection, the magnificent head of an idol (cat. no. 6), was produced by a culture known as the Cycladic (circa 2900–2000 B.C.), because its peoples inhabited the Cyclades Islands in the central Aegean Sea. The archipelago was called the Cyclades by the Greeks because the islands encircle Delos, the sacred birthplace of the god Apollo and the goddess Artemis. The islands, especially Paros and Naxos, are rich in marble and provided a ready source of material for the highly skilled Cycladic sculptors. Numerous marble statuettes of standing or reclining figures have survived from this period, most of which are small in size. Some of the statuettes may have come from burials and served a funereal or religious function. The Cycladic culture also produced numerous vessels, in both terracotta and marble, many with precisely fitted lids.

Following the Cycladic period, the dominant maritime culture in the Aegean was that of the Minoans (circa 1800–1200 B.C.), whose center of power was the island of Crete. The culture is named after the mythical King Minos, son of the god Zeus and Europa, daughter of the King of Tyre. The Minoan political system was based on a monarchy, and the wealthy lived in sprawling palaces that were brightly painted with wall frescoes. The ceramic vessels produced at this time are both functional and fanciful. The Minoans' great love of nature is exemplified by the sea and animal creatures decorating their vessels. The Minoan culture went into decline about 1400 B.C., perhaps as a result of the general disruption in the region after the eruption of the volcanic island of Thera (modern-day Santorini).

The Mycenaean culture (circa 1800–1200 B.C.) existed contemporaneously with the Minoan and was also ruled by a monarchy. The culture is named after the site of the palace of Mycenae in the northeastern part of the Greek Peloponnesos. In contrast to the open palace structure of the Minoans, however, the Mycenaeans needed heavily fortified palaces for protection. The emphasis on a warrior society is reflected in Mycenaean art, which in addition to scenes of fish and animals influenced by the Minoans (see cat. nos. 2–5), often has images of armed warriors marching into combat. The *Iliad* and the *Odyssey,* epic poems composed by Homer during the eighth century B.C., record the legends of the great Greek warriors who lived at the end of the Mycenaean period. Led by King Agamemnon, these warriors sailed to Troy to recover the beautiful Helen, and the Greek armies spent a decade battling the Trojans below the walls of the city. Helen and her brothers, Kastor and Polydeukes (the Dioskouroi), were the offspring of Zeus and Leda, wife of the king of Sparta.

The Mycenaean culture was eventually overwhelmed by invasions of the Dorians, a people who inhabited the northern fringes of the Mycenaean world, and until about 900 B.C., artistic production was limited to smaller objects, primarily ceramics. The ensuing Geometric Period (circa 900–700 B.C.) is named after the geometric patterning found on many of the ceramic vessels produced in this span of years. In addition to ceramics, the technique of casting small bronze objects was refined; this developing technology is represented in the Fleischman collection by small pendants and statuettes (see cat. nos. 7–12).

As the societies of the Greek mainland stabilized in the eighth and seventh centuries B.C., the various communities, generally ruled by a monarch or a small group of privileged individuals, evolved into the city-states that dominated the later periods. This stability permitted greater artistic and architectural production, and large stone temples with programmatic sculptural decoration began to be built in the centers of worship that were sacred to all Greeks, such as Delphi and Olympia. The Archaic period (circa 700–480 B.C.) saw the development of the ceramic firing techniques that produced first the black-figured and later the red-figured vessels of Athens, which were widely exported throughout the Mediterranean world (see cat. nos. 33–42). The painters and potters responsible for the manufacture of these vessels took great pride in their work, and in contrast to the anonymity of earlier artists, many recorded their names on their masterpieces. In their subject matter, these vessels provide a detailed reflection of ancient life and religion. Although much of the large-scale wall painting described by ancient authors is now lost, the vases give us some idea of what techniques and styles were used by the muralists. The techniques of bronze casting were further refined, resulting in increasingly larger and finer bronze statuary. Life-size and larger marble statues were also produced, and numerous votive offerings of *korai* (draped maidens) and *kouroi* (nude youths) have been found at religious sanctuaries. Small-scale sculptures in bronze and terracotta, however, continued to be more frequently used as votives. Along with their larger counterparts, such offerings were left by believers in places sacred to the gods as a means of giving thanks or of pleading a cause.

Regional styles developed in the wealthier centers of production, and some artists, prompted by their local fame, began to record their names on their sculptures and other works of art. The arts of other cultures influenced Greek style and iconography at this time. Fantastic monsters such as griffins (see cat. no. 13) and sirens (see cat. no. 9) were introduced into the Greek artistic repertoire during the eighth and seventh centuries B.C., probably from the Persian (Achaemenid) empire via the colonies established by Greek cities along the western coast of modern-day Turkey.

The Classical Period (circa 480–323 B.C.) followed the successive invasions of the great Persian armies in 490 and 480 B.C. The Greek triumphs at the battles of Marathon (490 B.C.) and Salamis (480 B.C.), events recorded by the historian Herodotus, led to a tremendous sense of self-confidence among the Greeks, resulting in a period of great architectural and artistic production. The regional stability provided by the political alliances for mutual defense formed by the Greek city-states after the Persian invasions led to greater prosperity and made it possible to undertake ambitious architectural projects. It was also a time that witnessed the flourishing of literature at Athens, when such figures as Aeschylus, Sophocles, Euripides, and Aristophanes wrote their numerous dramatic works. In the later Classical Period, the philosopher Socrates pursued his discourses with the aristocratic youth of the city, including his best-known student, Plato.

Among the most notable of the public works of this period was the rebuilding of the Athenian Acropolis, which had been destroyed during its occupation by Persian invaders in 490 B.C. During what has been called the Golden Age of Athens, the sacred site was completely reconstructed under the direction of the Athenian statesman Pericles. The sculptural decoration of the buildings exhibits the major changes in artistic style, particularly in the Greek conception of the human form, which had evolved from the stiff and mannered representation of the Archaic period into a more naturalistic figural type.

As in the Archaic period, all aspects of life, myth, and religion are depicted in the works of art created at this time. One of the best-known Classical sculptors was Polykleitos, who, in about 450–440 B.C., wrote the *Kanon* (rule), a treatise that laid out rules (based upon numerical ratios) for the depiction of bodily proportions in works of art. To embody his theory, he constructed a bronze statue of a nude youth bearing a spear, the Doryphoros, which has survived from antiquity only in the form of Roman copies. The figure achieves a perfect balance between youth and manhood, slenderness and muscularity, rest and movement. Another of his works that exemplified this same system of proportions was the Diadoumenos (fillet-binder), a youth tying a ribbon, or diadem of victory, around his head (see cat. no. 180).

As recorded by the historian Thucydides, the last decades of the fifth century B.C. witnessed the Peloponnesian Wars, with continued clashes between rival city-states Athens and Sparta, and the final defeat of Athens in 404 B.C. The first half of the fourth century B.C. saw the steady rise of the Macedonian state in northern Greece; by mid-century all of Greece was under the control of Philip II, father of Alexander the Great. After Philip's assassination in 336 B.C., Alexander ascended the Macedonian throne and shortly afterwards led the armies of Macedonia and Greece into Persia. As the borders of the Greek world expanded with the conquests of Alexander's armies, artists increased their interest in images of other ethnic peoples and in representing farcical and grotesque figures (see cat. no. 109). The splendid finds from subterranean painted chamber tombs in Macedonia, especially the royal tombs of Vergina, indicate that it was a time of great wealth as the treasures of the Persian empire flowed into Greece.

The sculpture produced in the late Classical period saw the continued evolution of the human figure into an increasingly more expressive and complicated form. It was during this period that the first female nude was carved, the statue of Aphrodite made for the Knidians by the sculptor Praxiteles in the 340s B.C. The artist Lysippos, renowned as the court sculptor of Alexander, created the official portraits of the ruler (see cat. no. 106). Lysippos changed the system of bodily proportions established by Polykleitos. His works have slimmer proportions, with the heads smaller in relation to the overall height of the body, and often have more expressive and individualized facial features in contrast to the dignified and reserved expressions used during the Classical period. Portraiture became a major art form at this time, preserving the features of some of the great statesmen and literary figures of the day (Socrates, Plato, Pericles, etc.), in a still somewhat idealized but nonetheless recognizable manner.

After Alexander's untimely death in Babylon in 323 B.C., his empire was divided among his generals and what is known as the Hellenistic period began (circa 323–30 B.C.). Two of the dynastic lines established by these generals were the Ptolemies of Egypt and the Attalids of Pergamon; both dynasties sponsored the production of art and literature. The library at Alexandria, destroyed by fire in about 47 B.C., was famous throughout the Greek world as a center of learning, and the building and sculptural programs of the city of Pergamon, including the Great Altar of Zeus (now housed in the Pergamon Museum in Berlin), influenced artistic styles for centuries.

In the eighth and seventh centuries B.C., Greek cities had established western colonies on the Italic peninsula, primarily along the coast of southern Italy and Sicily. The artists of the South Italian colonies, originally immigrants from Greece, continued using and developing the same stylistic vocabulary as the artists of the Greek mainland. The figured vases produced in the various regions, many of which are monumental in scale, imitate the techniques of painted Athenian pottery. The iconography of these vessels, however, goes beyond the standard Greek repertoire, and in many instances, seems to reflect the local popularity of Greek drama (see the essay by Oliver Taplin below), and perhaps even specific theatrical productions (see cat. nos. 56–60, 62–64).

In what is now the Tuscan area of northern Italy, the lands were inhabited by a people known as the Etruscans, after whom the region is named. Etruscan artists were renowned for their bronze working, and numerous vessels and statuettes found in their tombs survive to attest to their great skill (see cat. nos. 68–71). The Etruscans had trade routes throughout the Mediterranean, and their goods were distributed over a wide area. Although they produced their own original types of painted ceramics (cat. nos. 86–90), they also appreciated the painted vases of Athens, and the majority of Athenian vessels that have survived have been discovered in Etruscan tombs. Often these tombs were chambered, cut directly into the soft volcanic stone, then plastered and painted with colorful representations of a variety of scenes from everyday life, the banquet for the deceased, and the afterlife. Because the local stone was unfit for sculpting, the Etruscans, as well as the Greeks of South Italy, became masters at manufacturing large-scale terracotta plaques (see cat. no. 91), sculpture, and vases (see cat. nos. 86–87). Much of the sculpture was used to ornament their buildings and temples (see cat. no. 92).

Founded in 753 B.C., the city of Rome was initially dominated by Etruscan kings, but after their expulsion in 510 B.C., the Romans established a republican form of government, and the city gradually grew in size and strength. By about 275 B.C., it had emerged as the dominant force on the Italic peninsula. The Roman empire gradually spread eastward, and by 100 B.C. almost all of the Hellenistic kingdoms were under its control. This eastward expansion culminated in the defeat, by Octavian and his allies, of Antony and Cleopatra, the last Ptolemaic ruler, at the battle of Actium in 31 B.C. The Romans' austere and virtuous view of life during the Republican period (circa 500–27 B.C.) is reflected in their early portraiture (see cat. no. 177).

In spite of their military control of the region, the Romans had a great respect and appreciation for Greek art, literature, and philosophy. This Roman love of things Greek caused them to remove many of the most famous works of art from the great Greek sanctuaries for transport to Rome, and even extended to the making of copies of famous sculptures for the decoration of their public places and homes. The Mahdia shipwreck of 100 B.C., found off the coast of Tunisia, and the Antikythera shipwreck of 75–50 B.C., discovered off the southern coast of the Greek Peloponnesos, both carried expensive works of art as their cargo, attesting to the widespread trade and shipment of art throughout the Mediterranean at this time. This transference of originals and copies of Greek art to Rome during the period of expansion had an enormous impact on the indigenous Roman style.

The Roman Imperial period began when the senate conferred the titles of Princeps (first citizen) and Augustus upon Octavian in 27 B.C. Augustus founded the Julio-Claudian dynasty, which ended with the death of Nero in 68 A.D. Augustus' biographer, Suetonius, records his boast that he found Rome built of brick and left her clothed in marble. The ambitious program of public works undertaken during his lifetime confirms this claim. Continuing a tradition begun by Julius Caesar, Augustus expanded and modified the Forum Romanum, the central and most sacred area of the city, completed the Forum Iulium begun by his adoptive father, and built the Forum Augustum adjacent to it. Theaters, public baths, and temples, including the Pantheon, were constructed throughout the city for the benefit of the citizens. Augustus was also a patron of the literary arts, encouraging the work of the poets Virgil, Ovid, and Horace, as well as the historian Livy.

The eruption of Mt. Vesuvius in A.D. 79 sealed the cities of Pompeii and Herculaneum in volcanic ash and mud, preserving a microcosm of Roman life. The excavations of such buildings as the Villa of the Mysteries in Pompeii and the Villa dei Papyri in Herculaneum have yielded numerous examples of interior decoration in the form of frescoed wall paintings (see cat. nos. 125–127), household furnishings (see cat. nos. 147, 154), and utensils (cat. no. 131), enhancing our understanding of the everyday life of the Romans in the last years of the Roman Republic and the first decades of the Empire.

With the continued expansion of the Roman empire northward under the emperors Trajan (reigned A.D. 98–117) and Hadrian (reigned A.D. 117–138) into what is now modern Europe, the artistic and architectural styles of the provinces merged with those of their conquerors (see cat. nos. 161, 162). This stylistic blending is most apparent in figural sculpture, in the representations of the indigenous gods in the guise of deities of the Roman pantheon (see cat. no. 162). Some provincial centers became leaders in the manufacture of certain products, such as the glass vessels made in the workshops of Cologne (ancient Colonia Claudia Ara Agrippinensium) (see cat. no. 174), and the bronzes with inlaid enamel fashioned in Gaul (see cat. nos. 150, 165). The provincial capitals built under the leadership of the Roman emperors established a solid foundation that lasted for centuries, even after the boundaries of the empire began to shrink under encroaching tribal migrations from eastern Europe during the fifth century A.D.

The Beauty of the Ugly: Reflections of Comedy in the Fleischman Collection

Oliver Taplin

Theater must be one of the most successful coups of cultural promotion in human history. The Athenians more or less invented the *theatron*—a time and place set aside for spectators to sit and watch players acting out narratives—and established its two main forms, tragedy and comedy (much though these have interplayed since then). In less than 250 years theater went from being virtually the creation of one city among the hundreds of the Greek world to being an important part of the shared cultural life of all Greeks from Georgia to Syria to Libya to Calabria, not to mention the Greek mainland and Asia Minor. This pre-eminence was to last for the following six hundred years or so of Greco-Roman antiquity. Throughout this period, the theater would leave its mark in the visual and decorative arts as well as in actual performance. In the following pages I shall attempt, very summarily, to trace its geographical and perceptual spread by characterizing the state of its development in 500, 400, and 300 B.C.

The New Theater (500 B.C.)

Athens—the conurbation along with the large surrounding country of Attica—had grown greatly in importance during the sixth century B.C. New centralized festivals were set up to reflect this concentration of power and influence, and one of these was the "Great," or "City," Dionysia. A competition between three playwrights of *tragoidia* was made the centerpiece of this novel occasion, which occupied the lives of a large proportion of all the citizens (i.e. freeborn males) for several days each spring. The date traditionally given for the performance of first tragedies is 534, but some scholars have recently argued for 508.[1] The later date would link tragedy directly with the introduction of the new power-for-the-people constitution, which was to develop over the next half-century into what was arguably the most radical participatory democracy the world has ever known. Thus in 500 tragedy was still very new, and in fact Aeschylus was to compete for the first time the next year. It may even have had only one actor at this stage, but it must already have had its most important formal characteristics, such as the alternation of the actor's speech with choral song and the wearing of masks by all characters.

As for comedy, it was not introduced into the festival until the 480s. We know from vase paintings that curiously costumed choruses, such as men riding on ostriches or dolphins, were already performing at Athens,[2] and that there were various licensed occasions for ribald dancing and joking. But it was not until, following the same basic model as tragedy, masked speaking actors were combined with the chorus and the performance was made a competition at the Great Dionysia that comedy, in the sense that would have significance for the future, was born.

The Passing of the Golden Age in Athens (400 B.C.)

By the end of the fifth century B.C. democratic Athens had passed the prime of its golden age, and yet, though it was staggering from some dire knocks, its great days were not over. The cultural, political, and economic rise of Athens was powered by her central role in the repulse of the Persian Empire in 490 and 480, which led to the establishment of a kind of "empire" of her own, whose members gave Athens revenue and loyalty in return for goodwill and support. The Great Dionysia, occurring just when the seas became safely navigable in the spring, was a gathering-time for the representatives of the allies, who no doubt went to see this much spoken-of institution of the theater. And the Athenian theater rose to meet the occasion, pouring energy and expense into costumes and scene painting, into the skills of its choruses and actors, and fostering the genius and inventiveness of its playwrights.

Aeschylus produced his masterpiece, the *Oresteia,* in 458 B.C.; and the years circa 440 to 406 saw both Sophocles and Euripides at the height of their powers. Visitors to Athens will have taken reports of tragedy back home, and the activities of traveling players may well have already begun before 400, though we have no direct evidence. We do know that the leading playwrights were in demand by rulers who wanted to lay on prestigious performances, such as the monarchs (tyrants) of Syracuse and Gela in Sicily, and the kings of Macedon in the north, eager to show that they were real Greeks. Such was the esteem in which tragedy was held that Athenians taken prisoner in Sicily in 414 are said to have won their freedom by remembering and teaching choruses of Euripides.

Figure 1.
Red-figured volute krater with Dionysos, the musician Pronomos, and the cast of a satyr play. Name vase of the Pronomos Painter. Attic, circa 410–390 B.C. Naples, National Archaeological Museum 3240. Photo courtesy National Archaeological Museum.

In view of all this, it is strange, and rather disappointing, that, so far as we can tell, tragedy left very little mark on the flourishing visual arts at Athens, or at least on pottery painting, which is, of course, the medium that has mainly survived. At least, if tragedy did influence the painters, for instance in their versions of mythological scenes, they have left no clear signals to indicate this. There are a few exceptions, above all the splendid Pronomos vase of circa 400—found at Ruvo in northern Apulia and now in Naples—which shows, and even names, a whole cast of chorusmen and actors with two musicians (one the central figure of Pronomos) and the playwright (fig. 1). But such exceptions highlight the situation by their very rarity.

The history of comedy does not advance in step with tragedy. Although it was made official in the 480s, it seems to have remained a relatively minor art form for some fifty years. But in the last thirty years of the century, it attracted increasing attention, even claiming, in its absurd way, a rivalry with tragedy. This is the great age of the so-called Old Comedy, which for us means mainly Aristophanes. It was fantastical, grotesque, rude, highly topical and political, exuberant, unpredictable, outrageous—an undignified, hilarious melange to set beside the moving and dignified beauties of tragedy.

The anti-prettiness of comedy was signaled above all by the mask, the physiognomies of which were distortions of all the most admired features of Classical beauty. The actors also wore quantities of padding, before and behind; and the male characters sported a large, dangling leather phallus, a typically comic aberration, in view of the Greek (male, that is) taste for small neat penises, at least when displayed by unaroused young men.[3]

Given the lack of representations of tragedy, it would not be surprising to find an absence of reflections of comedy in Athenian visual

Figure 2.
Red-figured oinochoe with a scene reflecting Old Comedy: Herakles with bow and club, Nike in a chariot drawn by centaurs, and an actor dancing with torches. Attributed to the Nikias Painter. Attic, circa 425–375 B.C. Paris, Musée du Louvre N3408. Photo courtesy Reunion des Musées Nationaux.

arts. It was a less prestigious occasion, and it was ugly, deliberately and provocatively so. Nonetheless, in the last decade or two of the century there are a few paintings, admittedly on cheap little jugs, which do seem to reflect comedy. It has always, until very recently, been assumed that this kind of "Old" comedy, was, unlike tragedy, never seen or even known outside Attica, because it is so very Athenian and topical. Yet the clearest reflection in Athenian art (a small jug now in the Louvre; fig. 2) was found at Cyrene in Libya.[4]

And, as will be discussed below, there is an increasing body of evidence from Greek South Italy, beginning in about 400 B.C., which shows that Athenian "Old" comedy was "exported".

The Post-Classical Era (300 B.C. and after)

During the fourth century B.C. theater spread from being predominantly Athenian to being a Panhellenic experience, although Athens retained a special status as the "mother" of drama. The most concrete evidence for this permeation is that all Greek cities with any cultural pretensions built theaters of their own. So did religious festival-centers—the magnificent auditorium at Epidaurus dates from around the 330s, for instance. We also know that cities and festival-centers paid a lot to lay on theatrical shows; and so did individuals such as Dionysios of Syracuse (who also wrote tragedies, badly) and Philip of Macedon (who was finally assassinated in his local theater). Furthermore some of the most celebrated actors—and this was an age of superstars—came from places far from Athens, and so even did some of the playwrights.

Not long after 300 the actors organized themselves into several guilds known as the Artists of Dionysos,[5] only one of which was based at Athens. These large troupes played a repertoire of "classics," especially but by no means only the tragedies of Euripides (and the comedies of Menander, see below). They were highly organized and commanded large fees to perform at all sorts of celebrations, whether religious, civic, or sponsored by individuals. This was the arrangement securely in place when bit by bit the Romans conquered mainland Greece and the rest of the Greek world; and it held good for several more centuries. Theater in the Hellenistic and Greco-Roman periods was everywhere, both in performance and in the visual arts.

Although there was a sense of the passing of a Golden Age with the deaths of Euripides and Sophocles in 407–406, the performance of the new tragedies continued to be the highpoint of the Great Dionysia at Athens. There are still not many tragedy-related vase paintings (or other artifacts) from Athens, but that may be, at least to some extent, because the production of Athenian fine painted pottery declined steeply. Meanwhile flourishing schools of vase painters grew up in

Sicily, Campania, and especially in the Greek cities of Lucania and Apulia (*Megale Hellas*, or Magna Graecia). Scenes that reflect the influence of tragedy are found on these vases from about 400 onward, at first especially on vases from the Metapontion area of Lucania, and then increasingly from Apulia, especially the large and prosperous city of Taras (Tarentum, modern Taranto).[6] The most spectacular are the huge, ornate funerary volute kraters produced in Apulia in the mid- to late fourth century. Their mythological scenes show their relation to tragedy most obviously in the highly decorated costumes with characteristic tight sleeves and fancy laced boots. Many, though far from all, of these scenes can be connected with particular tragedies, especially those of Euripides. It is important to note that these are not pictures of tragedies in performance, and they carry no conspicuous signals of their theatrical connections.

There are also vase paintings of actors holding tragic masks and an increasing number of terracotta masks. Figurines of characters from tragedy also seem to have become increasingly popular; many of these have been found in the cemetery at Lipara (Lipari in the Aeolian Islands).[7] No doubt if other art forms survived in larger quantity—metal tableware and ornaments, for example, or wall paintings—we would find more tragedy reflected in them. The first-century B.C. and first-century A.D. wall paintings that do survive from Pompeii and elsewhere reveal the wide-ranging presence of tragedy. Some are directly influenced by the theater in their treatment of mythological narratives; others in their background scenery, especially architectural perspectives; and some have tragic masks as motifs in their borders. The Fleischman lunette with the mask of Herakles (cat. no. 125) is a fine example.

While tragedy did not undergo any radical further developments in its form or content after the Golden Age of the fifth century, comedy went through fundamental transformations between the "Old" comedy of Aristophanes (died 380s) and the "New" comedy of the age of Menander (died circa 290), a century later. These did not come about steadily but in fits and starts. Nonetheless, at the beginning of the fourth century comedy had all those hallmarks that characterized it in 400 B.C., while by the end it had lost its polemical tone, topicality, obscenity, and fantasticality. Instead, New Comedy told stories about the tribulations and misunderstandings of familiar bourgeois life. The plot almost always revolves around a boy-meets-girl story, where they end up getting married despite various barriers and setbacks. But usually the real interest lies, rather, in the interactions of a varied cast of characters who are caught up in the various deceits, intrigues, errors, and revelations. The gallery of recurrent New Comedy characters includes angry old men, old men in love, ostentatious caterers, hangers-on, blustering soldiers, nasty tarts, tarts with a heart of gold, unscrupulous pimps, clever slaves, slaves in a hurry, uncouth rustics, upright young yeomen, and so forth—a resilient gallery of comic types.

So comedy became decent and domesticated and sophisticated. In step with this transition in tone and content, it also changed its appearance. Most of the padding went, the clothes became smarter, more like everyday wear, and the gross phallus was put away. The slave masks remained distorted and grotesque—or even became more so—while the old men and women were heavily wrinkled but not conspicuously ugly. But the masks of the younger characters were not ugly at all and even approached an idealized beauty.

Comedy as well as tragedy was played in repertoire by the Artists of Dionysos—though by different performers—from the early third century onward. New Comedy, and especially the plays of Menander, remained popular for another seven hundred years, and many papyrus fragments of Menander's works have been excavated. Comedy was especially popular in second-century B.C. Rome, and quite a few of the translation-adaptations into Latin by Plautus and Terence survived into the Renaissance and thence to the present.

And comedy, as well as tragedy, left its mark all over the place in the visual arts, though generally on humbler artifacts. Comic scenes and characters, many specifically from Menander, are found in wall paintings and mosaics, on painted glass, and as the subjects of innumerable figurines.[8] The scene in another Fleischman fresco fragment (cat. no. 127) might possibly reflect New Comedy to judge from the "scenery" and the rather masklike, pensive look on the woman's face—drinking a toast to a parted lover? (But it may be that *Romeo and Juliet* is misleadingly brought to mind!)

On smaller-scale artifacts it was the undignified figure of the slave, and especially his grinning mask, that was most popular. It seems that artists—and their customers—welcomed some less "respectable" antidotes to the generally idealized norms of high Classical art. A remarkably wide range of objects is represented in the Fleischman collection, in both terracotta and bronze, ranging in date from 300 B.C. to 300 A.D. The most familiar are the statuettes, representing a variety of roles and poses (cat. nos. 108, 117, 134, 148, 153). It is possible that some of these originally belonged to sets which could be assembled to "re-enact" a famous scene from a particular comedy. The amusement offered by the "ugly," without any such specificity, is illustrated by the adaptation of such other artifacts as lamps (cat. nos. 118, 154, 176), a handle (cat. no. 135), and a weight (cat. no. 99). Probably the earliest object, and possibly the finest in its artistry, is the bronze roundel (cat. no. 30), which is thought to have been made in Magna Graecia in the late fourth century. The individualizing of the mischievous mask suggests a craftsmanship that is enjoyed by maker and viewer alike.

The Comic Vases of South Italy

Although the material from Athens is so sparse, there is a fascinating corpus of painted pottery, and of other art forms, especially terracotta figurines, from the Greek cities of South Italy and Sicily. The vases date roughly from 400 to 320 B.C., and have been conventionally known as "phlyax vases," the title of A. D. Trendall's still-definitive catalogue of 1967.[9] On the broadest definition—any vase showing a comic mask—there are over 250 such vases; and there are over 100 which show comic scenes in performance. The single most important may also be the earliest: the krater in the Metropolitan Museum in New York, dated to circa 400 (fig. 3). In the scene on this well-known krater there are not only typical comic costumes and scenery, there are actually words issuing (rather like cartoon bubbles) from the characters' mouths. This is, then, a particular moment in a particular play. Furthermore the dialogue is in Attic Greek, not in the local Doric dialect of Taras, where it was most likely painted.

Since 1967 several of these vases have come to light which are of great importance for theater history. The assessment of their significance is still in progress. One point that is emerging is that the chronology and the original locality of each painting should be always borne in mind. Another is that it is essential to draw a distinction, even though the borderline is not always clear, between those vases which show a specific scene (like the New York krater), and those which simply include

Figure 3.
Red-figured calyx krater with a comic scene. Attributed to the Tarporley Painter. Apulian, circa 400 B.C. New York, Metropolitan Museum of Art 24.97.104.

a comic actor or mask as a decorative motif, usually because of the amusing connection with Dionysos and with wine. There is a tendency with time to move from the former kind to the latter.

This shift away from theatrical particularity can be seen most clearly in the later comedy-related (phlyax) vases, especially those from Paestum (the Greek city of Poseidonia), which by the mid-fourth century B.C. was firmly under the control of local Italic people. These date from about 360 and later, and a few of them show actual scenes from plays; but the great majority simply show a comic actor either by himself, or more often in the company of Dionysos, or in a Dionysiac group. Another nice example of this shift can be found in the Fleischmans' Apulian cup (cat. no. 60): the actor is there as an associate of Dionysos, not because he is in a play. The situation is similar in the two comedy-related Apulian vessels in the collection which are the latest in date, a situla (cat. no. 63) and an askos (cat. no. 59). The situla is painted in the polychrome Gnathia technique which grew in popularity in Apulia from the mid-fourth century onward. This lone actor, with his elaborate wreath and discreet phallus, might have been recognized by his painter or purchaser as playing a particular part in a particular play, but it is far more likely that he was perceived simply as a delightful decoration for a vessel that was part of the equipment for a drinking party. The two figures on the unusual and attractive askos are also unlikely to be taken from a particular play. As Professor Trendall observes in his discussion of the piece, scenes with a master pursuing or punishing a slave are common; they are typical, that is, rather than specific.

These observations supply some background for the interesting questions raised by the krater (cat. no. 64), which was painted, according to Professor Trendall, circa 360 and is either early Paestan, or was produced in Sicily by one of the school who moved to Paestum. Dionysos and the white-haired comic actor carrying torches are typical of the later Paestan Dionysiac scenes; the satyr is also typical, though the piggyback Eros is not common—and it is not very Dionysiac to be so dejected. But the feature that seems to point to a particular narrative is that Dionysos is carrying a lyre—more like Apollo—and the satyr is playing the *aulos* (double pipes). These bring to mind the story of Marsyas and his musical contest.[10] It seems to be an insolubly open question whether this does or does not allude to a specific comedy.

Three Comic Scenes

I turn now to the three clearly scene-specific comic vases in the collection, all of them Apulian kraters dating before 360 B.C. In the first (cat. no. 58; probably 360s) there is, as in many of this core type, an actual representation of the stage on which the scene is being played. Clearly the body-padding and the phalluses of the male characters are still gross, and the masks, including that of the woman, are extremely unbeautiful. The figure on the left, who is shown as "stage-naked," is remarkably similar to the left-hand figure on the New York scene of thirty-five years earlier (see fig. 3). Could he be a stock type? The "king" on the right seems to be Zeus, who does appear in a similar form on several other comic vases.[11] It is not obvious why he is showing such an interest in the woman, who is neither young nor beautiful, but presumably it is a matter of lust. And why is some of her clothing billowing in the wind? This is typical of the kind of conundrums that are posed by these entertaining vases. To anyone who (like us) does not know the comedy in question, the picture sets insoluble puzzles; to anyone who had seen the play it would have been an amusing reminder of the particular plot that makes sense of this bizarre combination of images.

All this is even more applicable to another bell krater (cat. no. 57; dating from circa 370s). This same painter, the Rainone Painter, was responsible for another conspicuously enigmatic comic scene, which may be connected with a burlesque of the Antigone story.[12] The two old men on this krater are typical, as is the stage, and the stage door, though it is unusually ornate (compare the New York krater, fig. 3). But the creature that they have found inside the large ornamental container is very weird. He is small but sturdy, either a boy or a dwarf; he has a ram's head and a formidable erection. Professor Trendall makes a strong case for a connection with the Athenian myth of Erichthonios; but there are still problems. The usual form of Erichthonios, when he is half-human, is that he is half-snake, not half-ram; and there is no evident explanation for his priapic state. It is indeed more than likely that this play involved some sort of mythological burlesque, but this may not be the myth in question. I am reminded, for example, that there is some resemblance to Cratinus' comedy *Dionysalexandros,* where the lustful Dionysos was transformed into a ram.[13] If only the painter had added name-labels, then we might have a little more idea of what on earth is going on.

Fortunately there are labels on the most important of all the images in this collection of comic vases. The *Choregoi* vase (cat. no. 56), was painted in Apulia probably in the 380s. First published in 1992, it has given a name to its notable artist—the Choregos Painter—and has already aroused warm discussion.[14] In many ways the picture is typical of the best of these vases. It presents an unexpected scene, impenetrable to those who do not know the play in question, while transparently amusing to those who do. Here we have the typical comic costumes and masks, the wooden stage with steps, the half-open stage door; furthermore the character on the upturned wool-basket has the common slave-name ΠΥΡΡΙΑ[Σ] (Pyrrias). But there are two further conspicuous features which are unique among the vases known to us so far.

First, two characters have the same label, ΧΟΡΗΓΟΣ (choregos), which is not a proper name, as is usual, but a function. This word might mean "chorus leader," but its most common application, at Athens and elsewhere, was to the rich citizens who were expected to cover the expenses of the training and equipment of a dramatic chorus (something like the modern patrons known as "angels," but less optional). In my book, *Comic Angels* (pp. 55–63), I argued that they are two representative members of a whole chorus of such "choregoi." There are problems with this solution, however—most notably that the two figures are differentiated by hair-color, and they are up on the stage behaving like individual characters.

The other even stranger feature is the handsome figure on the left, labeled ΑΙΓΙΣΘΟΣ (Aigisthos)—one of the least savoury characters of tragic myth. Furthermore, he has all the appurtenances of *tragedy*-related iconography. It is as if he has walked off a completely different kind of vase painting, yet his feet are firmly planted on this *comic* stage. He is likely to occupy much future scholarly attention, but there is at least one suggestion about him in my book which is, I think, likely to attract general agreement: that we have here some kind of meeting or contest between the two genres of tragedy and comedy, in which Aigisthos somehow "represents" tragedy and Pyrrias comedy. The two choregoi, whether members of a divided chorus or representative individuals, are then opposing partisans of the two kinds of theater. So this is an unprecendented juxtaposition of the iconographic images of the two genres, which is also a vivid reminder of their contrasting coexistence within the Greek theater.

It would, of course, make all the difference to interpretation if only we had the text of the comedy that this scene captures. The use of the Attic dialect (*choregos* as opposed to the Doric *choragos*) fits with the view that this is likely to be an Athenian comedy which was performed in South Italy. There may be some correspondences with a lost comedy by Aristophanes called *Proagon* (*Preview*), which probably juxtaposed tragedy and comedy at the proagon-ceremony, which was held at Athens three days before the dramatic performances. But this speculation is far from secure.

The discussion of these and other fascinating questions raised by the Choregoi vase are likely to keep interested scholars busy for some while. I would like to point to two ramifications that have already become clearer to me since I wrote *Comic Angels* in 1992. First, the image of Aigisthos as an emblematic figure of tragic horrors has been filled out by two new paintings by the Darius Painter, a central artist of the

Figure 4.
Red-figured amphora with the death of Atreus. Attributed to the Darius Painter. Apulian, circa 340–330 B.C. Boston, Museum of Fine Arts 1991.437, Collection of Shelby White and Leon Levy, and Gift of The Jerome Levy Foundation. Photo courtesy Museum of Fine Arts, Boston.

Figure 5.
Red-figured bell krater with comic actors and a monumental bust of Dionysos. Attributed to the Choregos Painter. Apulian, circa 390–380 B.C. The Cleveland Museum of Art 89.73, John L. Severance Fund.

high Apulian style in the third quarter of the fourth century.[15] Until 1987, when the first of these paintings came to light, the iconography of Aigisthos was limited entirely to the edifying episode of his death.[16] In one of the two new scenes the baby Aigisthos, offspring of the incestuous union of Thyestes and Pelopeia, is handed over to a slave by his father (all watched over by a Fury). In the other Thyestes and Aigisthos stand, weapons in hand (in the presence of Pelopeia), after killing Thyestes' brother Atreus on his throne (fig. 4). A Fury (*Poine*, Retribution) approaches the throne. This brings home that Aigisthos, though he looks so handsome, was in fact one of the most sinister figures of tragedy.

Second, with the help of Professor Trendall, the artistic characteristics of the Choregos Painter are beginning to emerge (see note 14). His skill can be seen, for instance, in the rendering of Aigisthos' costume, in the poses of the two choregoi, and in the wooden carpentry of the stage. The ancient repairs are also evidence that the vases's original owner regarded it as too good to throw away. Another of the finest of this whole category of vases, the "Cake-Eaters" in Milan, is by the same painter.[17] And the Cheiron scene in the British Museum and the "Birth of Helen" in Bari,[18] two more of the best, both for the quality of painting and interest of the subject matter, seem to be by closely associated artists.

We can now add to this group a marvelous and highly puzzling krater, recently acquired by the Cleveland Museum (fig. 5).[19] The comic actor picking grapes and the undignified old Silenos are strikingly juxtaposed with the mysterious calm and beauty of the bust of Dionysos.

The initial question posed is: are we looking at a symbolic scene of Dionysos with two of his genres of drama, comedy and satyr play? Or are we looking at a scene from a comedy in which Silenos is a character? In that case, the gigantic bust of Dionysos would be a stage property, something possibly paralleled by the giant form of Peace which is hauled out of a pit in Aristophanes' play of the same name. Either way, this painting confirms the Choregos Painter as an exceptional artist.

Memorial for a Comic Actor?

I have left till last a comedy-related vase of quite a different sort, the volute krater by the Underworld Painter, dating to about the 330s (cat. no. 62). The youth, who is sitting in a kind of shrine-monument that has become known as a *naiskos,* is holding up a comic mask. So many of this kind of Apulian large-scale ceramics have recently been published that their iconography, symbolism, and function are due for a complete reassessment.[20] It seems clear, however, that they are funerary; and it is highly likely that the figure in the naiskos somehow represents the deceased and the other figures his/her bereaved relatives. The dead person is usually portrayed as young and beautiful, which was presumably the convention, even when they had died old. Also, they are often surrounded by a variety of objects or creatures that seem to represent their favorite activities when alive and young.

Only four out of the dozens of such scenes include theatrical masks, two tragic and two comic.[21] These probably signal, then, men who were especially interested in the theater, and quite likely actors. Both those with tragic masks also have a lyre; in the other comic scene (Lecce 3544) the mask hangs above a youth whose primary attribute is a hound.[22] In the Fleischman krater, besides a ball and a staff (indicating the gymnasium?), there is a remarkable further object: a papyrus roll, with writing marked on it, lies on the floor. This surely indicates a close connection with dramatic literature (as is perhaps also suggested by the satyr bust on the neck).

Thus the dead man might be an actor who specialized in comedy and learned his roles from written copies. We do know of actors from this part of the world at this period—at least two of them rose to fame, Aristodemos of Metapontion and Archias of Thourioi.[23] Alternatively we might have here a comic playwright. One of the most famous of this era, Alexis, was born at Thourioi in the 370s. This piece cannot mark his funeral since he lived on well into the next century, but it might have been commissioned for someone who acted in Alexis' plays, or who was a fellow dramatist in Magna Graecia. In any case, the roll of papyrus is a reminder that comedy in the fourth century was far from primitive and improvised. It was, on the contrary, a literary form as sophisticated as the artistic representations that reflect it.

Notes

1. Notably by W. R. Connor in *Aspects of Athenian Democracy,* Classica et mediaevalia dissertationes xi (Copenhagen, 1990), pp. 7–32.
2. The vase paintings in question are well surveyed by J. R. Green, "A Representation of *The Birds* of Aristophanes," *Greek Vases in the J. Paul Getty Museum 2,* Occasional Papers on Antiquities 3 (Malibu, 1985), p. 95ff.
3. For discussions of this curious piece of comparative sexual ethnography, see K. Dover, *Greek Homosexuality* (London, 1978), pp. 124–134, and J. Winkler, "Phallos Politikos: Representing the Body Politic in Athens," *Differences* 2 (1990), pp. 29–45.
4. For a brief recent survey of these paintings, see O. Taplin, *Comic Angels* (Oxford, 1993), pp. 9–10, with refs. [hereafter *Comic Angels*]. A special case is the "Getty Birds," first published in 1985, see J. R. Green (note 2 above). For recent discussions of this highly anomalous piece, see *Comic Angels,* pp. 101–104; E. Csapo, "Deep Ambivalence: Notes on a Greek Cockfight," *Phoenix* 47 (1993), pp. 1–28, 115–124.
5. The standard account of the Artists of Dionysos is A. Pickard-Cambridge, *Dramatic Festivals of Athens,* 2nd ed., revised by J. Gould and D. Lewis (Oxford, 1968 [reissued with addenda, 1988]), p. 263ff; see also P. Ghiron-Bistagne, *Recherches sur les acteurs dans la Grèce antique* (Paris, 1976), esp. p. 163ff.; and I. E. Stephanis, Διονυσιακοὶ τεχνῖται(Heraklion, 1988).
6. The most useful collection of these scenes can be found in A. D. Trendall and T. B. L. Webster, *Illustrations of Greek Drama* (London, 1971); see also A. D. Trendall "Farce and Tragedy in South Italian Vase-Painting," in *Looking at Greek Vases,* T. Rasmussen and N. Spivey, eds. (Cambridge, 1991), pp. 169–182. I began to consider the issues for myself in *Comic Angels,* pp. 21–27; they have been further developed in the chapter on "The Visual Record" in *The Cambridge Companion to Greek Tragedy,* P. Easterling, ed. (Cambridge, forthcoming).
7. Tragic material is included along with primarily comic artifacts in L. Bernabò-Brea, *Menandro e il teatro greco nelle terracotte Liparesi* (Genoa, 1981).
8. For example, see T. B. L. Webster's *Monuments Illustrating New Comedy* (London, 1969), which is being updated by A. Seeberg and J. R. Green. Other collections of illustrations include L. Bernabò-Brea (note 7 above) and M. Bieber, *The History of the Greek and Roman Theater,* 2nd ed. (Princeton, 1961), pp. 87–107.
9. A. D. Trendall, *Phlyax Vases,* 2nd ed. (= *BICS* Suppl. 19 [1967]) (hereafter *Phlyax Vases*). Professor Trendall, in collaboration with J. R. Green, is working on a third edition.
10. The recently published (1992) *LIMC* entry (vol. 6 [A. Weis]) on Marsyas reveals him as quite often pictured as dejected (and sometimes as hairy).
11. Six are listed in *Comic Angels,* p. 60, n. 11.
12. See *Phlyax Vases,* no. 59; see also *Comic Angels,* pp. 84–88 (illustrated in both).
13. See Cratinus 1 (p. 140, lines 31–33) in *Poetae Comici Graeci* IV, R. Kassel and C. Austin, eds. (Berlin, 1983). I was reminded of this by Professor Dirk Obbink of Barnard College, New York.
14. See *RVAp,* Suppl. II, pt. 1, pp. 7–8; pt. 3, p. 495, and *Comic Angels,* chapter 6.
15. These are the "Baby Aigisthos" (Boston, Museum of Fine Arts, 1987.53), which is fully discussed by E. Vermeule in *Proceedings of the Cambridge Philological Society* 213 (1987), pp. 122–152, esp. 124–131, and catalogued as *RVAp,* Suppl. II, p. 151, 18/65c, pl. XXXVII.4, and "Aigisthos at the Death of Atreus" (Boston, Museum of Fine Arts, 1991.437), catalogued as *RVAp,* Suppl. II, p. 148, 18/47b, pl. XXXVI.1 (see fig. 4).
16. See *LIMC,* vol. 1, s.v. "Aigisthos" (R. M. Gais).
17. See *Phlyax Vases,* no. 45, pl. 2 and *Comic Angels,* pl. 12.5.
18. For "Cheiron," see *Phlyax Vases,* no. 37 and *Comic Angels,* pl. 12.6; for "Birth of Helen," see *Phlyax Vases,* no. 18 and *Comic Angels,* pl. 19.20.
19. For Cleveland Museum of Art 89.73, see A. D. Trendall, "A New Early Apulian Phlyax Vase," *The Bulletin of the Cleveland Museum of Art* 79, 1 (1992), pp. 2–15; *RVAp,* Suppl. II, p. 495.
20. Especially in *RVAp,* Suppl. II (note 14 above).
21. See A. D. Trendall "Masks on Apulian Red-figure Vases," in *Studies in Honour of T. B. L Webster,* vol. 2, J. Betts et al., eds. (Bristol, 1988), p. 137ff; cf. *Comic Angels,* pp. 92–93.
22. See *Phlyax Vases,* no. 168.
23. See *Comic Angels,* p. 92, with references to P. Ghiron-Bistagne (note 5 above).

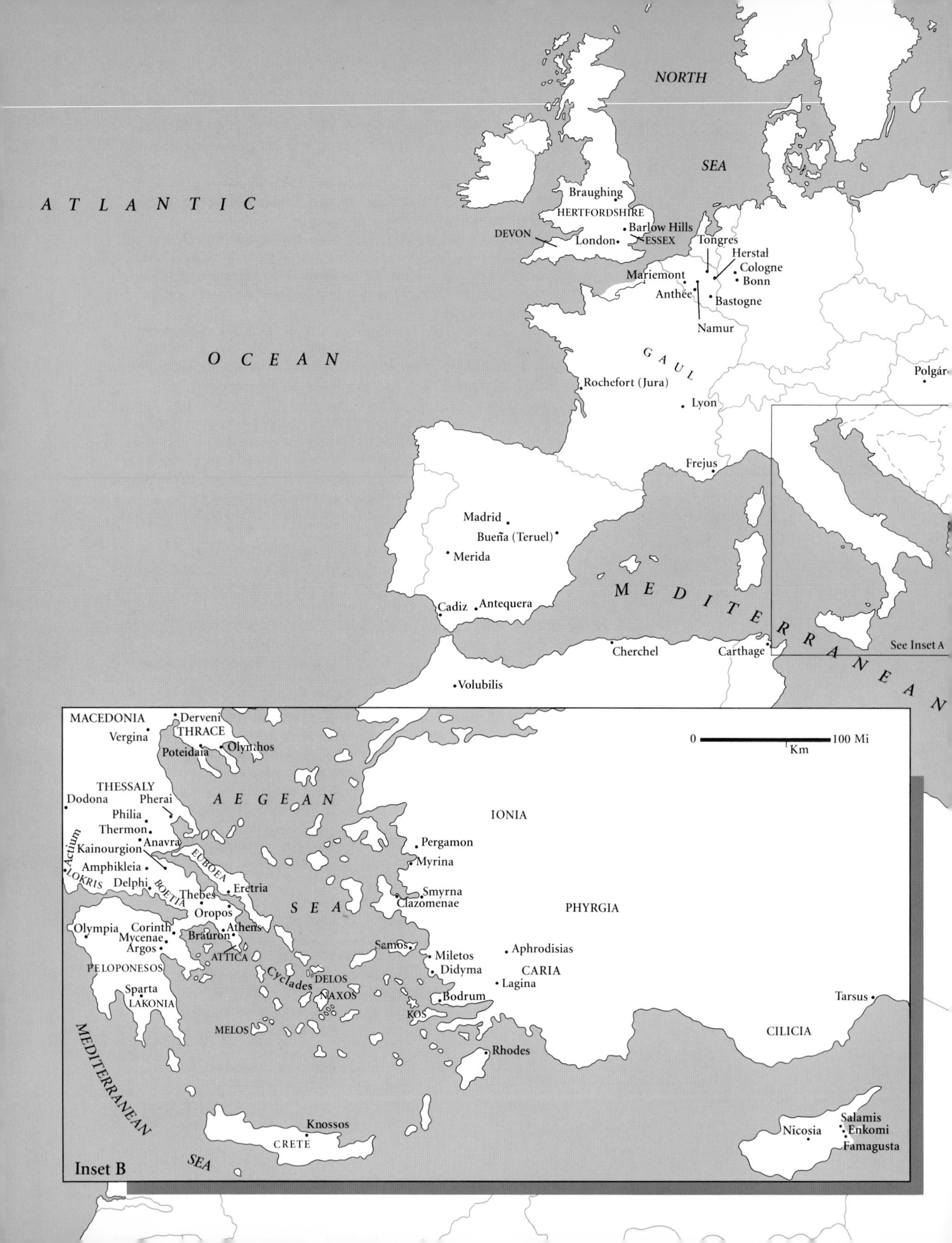

NORTH
SEA
ATLANTIC
OCEAN
Braughing
HERTFORDSHIRE
Barlow Hills
ESSEX
DEVON
London
Tongres
Herstal
Cologne
Bonn
Mariemont
Anthée
Bastogne
Namur
GAUL
Rochefort (Jura)
Lyon
Polgár
Frejus
Madrid
Bueña (Teruel)
Merida
Cadiz
Antequera
MEDITERRANEAN
Cherchel
Carthage
See Inset A
Volubilis
MACEDONIA
Derveni
THRACE
Vergina
Poteidaia
Olynthos
0
100 Mi
Km
THESSALY
Dodona
Pherai
AEGEAN
Philia
Thermon
Actium
Kainourgion
Anavra
Amphikleia
LOKRIS
Delphi
BOETIA
EUBOEA
Eretria
Thebes
Oropos
SEA
Olympia
Corinth
Mycenae
Argos
Brauron
Athens
ATTICA
PELOPONESOS
Sparta
LAKONIA
Cyclades
DELOS
NAXOS
MELOS
IONIA
Pergamon
Myrina
Smyrna
Clazomenae
PHYRGIA
Samos
Miletos
Didyma
Aphrodisias
CARIA
Lagina
Bodrum
KOS
Rhodes
Tarsus
CILICIA
MEDITERRANEAN
SEA
Knossos
CRETE
Nicosia
Salamis
Enkomi
Famagusta
Inset B

Verona
Aquileia
Florence
UMBRIA
Montepulciano
Bolsena
Volsini
Orvieto
Vulci
Viterbo
ETRURIA
Tarquinia
Civita Castellana
Pyrgi
Caere
Veii
LATIUM
Rome
Ostia
Praeneste
Trebenište
Skutari
Canosa
Ruvo
Bari
Gnathia
CAMPANIA
APULIA
Herculaneum
Pithekoussai
Baia
Pompeii
Naples
Metapontum
Boscoreale
Lecce
Paestum
LUCANIA
Taranto
Lipari
MAGNA
Palermo
Himera
GRAECIA
Assoros
0
20 Mi
Km
Inset A
CRIMEA
Anape
SCYTHIA
gozen
Nikolaevo
Panagjurište
Kirklareli
Istanbul
IONIA
PHYRGIA
CARIA
CILICIA
See Inset B
Adana
Seleucia ad Calycadnum
Seleucia Pieriae
Antioch
Palmyra
Tripolis
ASSYRIA
Susa
SEA
PERSIAN GULF
Alexandria
THE
FAYUM
RED
Aswan
Philae
SEA
Meroe
0
300 Mi
Km

Abbreviations

ABV
J. D. Beazley, *Attic Black-figure Vase-painters* (Oxford, 1956)

ABFV
J. Boardman, *Athenian Black Figure Vases* (New York, 1974)

AM
Mitteilungen des Deutschen Archäologischen Instituts, Athenische Abteilung

ARV[2]
J. D. Beazley, *Attic Red-figure Vase-painters,* 2nd ed. (Oxford, 1963)

BCH
Bulletin de correspondance hellénique

Beazley Addenda[2]
Beazley Addenda: Additional References to ABV, ARV[2]*, and Paralipomena,* 2nd ed., T. H. Carpenter with T. Mannack and M. Mendonça, comps. (Oxford, 1989)

BEFAR
Bibliothèque des Écoles Françaises d'Athènes et de Rome

BICS
Bulletin of the Institute of Classical Studies of the University of London

Boucher 1976
S. Boucher, *Recherches sur les bronzes figurés de Gaule pré-romaine et romaine,* BEFAR (Rome, 1976)

BWPR
Winckelmannsprogramm der Archäologischen Gesellschaft zu Berlin

Comstock/Vermeule 1971
M. B. Comstock and C. C. Vermeule, *Greek, Roman, and Etruscan Bronzes in the Museum of Fine Arts, Boston* (Greenwich, Conn., 1971)

CVA
Corpus Vasorum Antiquorum

Faider-Feytmans 1979
G. Faider-Feytmans, *Les bronzes romains de Belgique* (Mainz am Rhein, 1979)

GettyMusJ
The J. Paul Getty Museum Journal

Gods Delight
The Gods Delight: The Human Figure in Classical Bronze, exh. cat. (Cleveland Museum of Art, 1988)

Haynes 1985
S. Haynes, *Etruscan Bronzes* (London and New York, 1985)

LIMC
Lexicon Iconographicum Mythologiae Classicae

Master Bronzes
D. G. Mitten and S. F. Doeringer, eds., *Master Bronzes from the Classical World* (Mainz, 1967)

ÖJh
Jahreshefte der Österreichischen Archäologischen Instituts in Wien

Para
J. D. Beazley, *Paralipomena: Additions to Attic Black-figure Vase-painters and to Attic Red-figure Vase-painters* (Oxford, 1971)

RM
Mitteilungen des Deutschen Archäologischen Instituts, Römische Abteilung

Rolley 1986
C. Rolley, *Greek Bronzes* (Fribourg, 1986)

RVAp I/II
A. D. Trendall and A. Cambitoglou, *The Red-figured Vases of Apulia,* 2 vols. (Oxford, 1978 and 1982)

RVAp, Suppl. I/II
A. D. Trendall and A. Cambitoglou, First/Second Supplement to *The Red-figured Vases of Apulia* (= *BICS* Suppl. 42/60) (London, 1983 and 1991–92)

RVP
A. D. Trendall, *The Red-figured Vases of Paestum* (Rome, 1987)

Note to the Reader

The development of the Fleischman collection has not been determined by the chronology of the ancient Mediterranean world nor by the sequence of the styles we associate with Greek and Roman art but rather by the personal interests of its owners. The exhibition upon which this catalogue is based reflects these interests—objects having been grouped to highlight related subject matter or to emphasize continuity of visual form, regardless of medium or date.

The catalogue, however, as the first major reference book on the collection as a whole, has been organized chronologically and by medium to facilitate the finding of specific objects or groups of objects. There are five sequential sections and within each section objects are grouped by medium—fresco, metal (bronze, lead), precious metals and glass, terracotta, and stone.

Within each entry bibliographic sources have been grouped into two sections: *Bibliography,* i.e, publications on the specific work being discussed, and *Related References.* In the latter section, publications referred to within the entry text are listed first (in the order cited), followed by more general references. Some objects in the collection have been previously published or exhibited with their Fleischman inventory numbers. In these cases, the Fleischman number is provided in brackets following the bibliography heading, i.e. [F56].

Authors

A. D. T.	Arthur Dale Trendall
A. H.	Ariel Herrmann
A. K.	Arielle Kozloff
A. O.	Andrew Oliver, Jr.
A. v.d. H.	Annewies van den Hoek
E. B. T.	Elana B. Towne
J. B. G.	Janet B. Grossman
J. H.	John Herrmann
J. R. G.	J. Robert Guy
K. H.	Kenneth Hamma
K. W.	Karol Wight
L. M.	Leo Mildenberg
M. A.	Maxwell L. Anderson
M. J-N.	Marit Jentoft-Nilsen
M. R.	Michelle Roland
M. T.	Marion True
S. H.	Sybille Haynes

Catalogue of the Exhibition

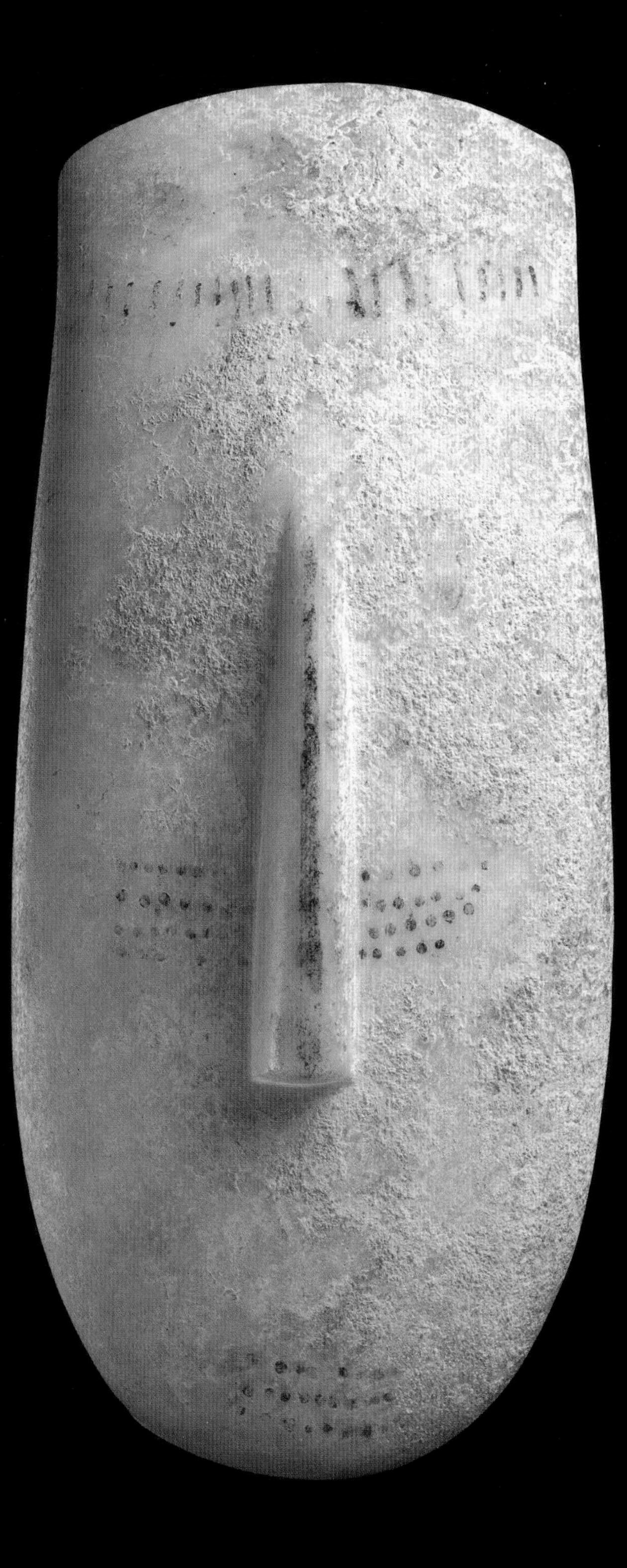

Greece in the Bronze Age

1

Beak-spouted Jug

Middle Cycladic III–Late Cycladic I, circa 1550 B.C.
Terracotta; H: 14.7 cm; MAX. DIAM: 11.3 cm
Condition: Small star-shaped crack repaired on proper left side; some minor abrasions.

Beak-spouted jugs, with their pulled-back necks, are a common type of Cycladic pouring vessel. Although painted bands around the body are common ornamental devices, the decoration, all in dark brown, on this jug is somewhat atypical. The edge of the mouth down to the handle has a wide border. Two bands, a broad one above a narrow one fringed with a zigzag, encircle the jug at the join of the neck to the shoulder. Another broad band framed by a narrow band above and below marks the belly, and a plain band encircles the base. Reaching from the belly to the neck are thin vertical lines with what appears to be a foliate motif, sets of long fingerlike leaves on wavy stems. That there seems to be no exact parallel for this motif, and that the vase does not have the breasts in relief which usually adorn vases of this type should not be too surprising. Even though they made vast numbers of ordinary vases, Cycladic vase makers were not unoriginal. They also created some novel shapes and decoration. Perhaps the motif was inspired by some foreign work such as the widely varying foliate patterns seen on Late Minoan pottery (see Popham). Whatever the inspiration for the motif, the vase is an example of the fine ware that was frequently imported to and imitated on mainland Greece.

BIBLIOGRAPHY: Unpublished.

RELATED REFERENCES: M. Popham, "Late Minoan Pottery, A Summary," *Annual of the British School at Athens* 62 (1967), pl. 76 h, i. For the development of these spouted jugs, see R. Higgins, *Minoan and Mycenaean Art*, rev. ed. (New York, 1981), p. 58, figs. 59–61. For two similarly shaped vases, found in Grave Circle B at Mycenae, see G. Mylonas and J. Papademetriou, "The New Grave Circle of Mycenae," *Archaeology* 8 (March 1955), p. 45, fig. 2.

—M. J-N.

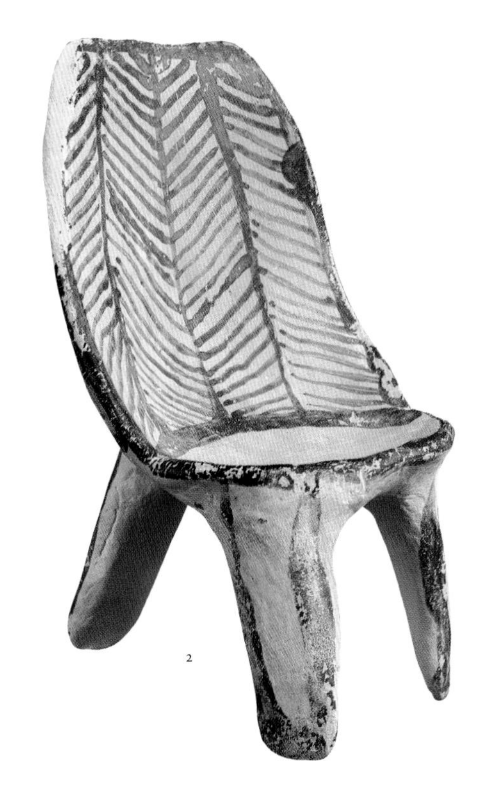
2

2

Miniature Throne

Mycenaean III, circa 1425–1100 B.C.
Terracotta; H: 10.4 cm; W: 7 cm; DEPTH: 7 cm
Condition: One front leg broken and reattached; some abrasion along top edge of back.
Provenance: Formerly in the collection of Sir Clifford and Lady Norton.

As suggested by the number of extant small seated figurines, miniature terracotta thrones that are unoccupied, and similar chairs with figures attached, this little throne may well have served as the seat for a separately made figure. Found at several Mycenaean sites, the unoccupied thrones have three legs, either latticed or solid backs, and a variety of decorative schemes. The ornament can sometimes be complex or clumsy, but here it is spare and neat. The legs and seat are outlined in brown glaze, and the inside as well as the outside of the high back is decorated with a vertical herringbone pattern.

Their meaning remains uncertain, but G. Mylonas contends that if the thrones were miniature copies of real chairs they would most probably have four legs. Noting that some prehistoric objects of a sacred nature have three legs, he proposes a similar significance for these thrones as seats for a divine figure, and thus it is quite likely that a figurine, perhaps that of a goddess, at one time sat in this throne.

BIBLIOGRAPHY: F. Nicholson, *Ancient Life in Miniature: An Exhibition of Classical Terracottas from Private Collections in England,* exh. cat. (Birmingham Museum Art Gallery, 1968), no. 16.

RELATED REFERENCES: For unoccupied thrones, see R. Higgins, *Greek Terracottas* (London, 1967), pl. 4C; J. Dörig, ed., *Art antique: Collections privées de Suisse romande* (Geneva, 1975), nos. 80A–C; G. Mylonas [who also discusses their significance], "Seated and Multiple Mycenaean Figurines in the National Museum of Athens, Greece," *The Aegean and Near East: Studies Presented to Hettie Goldman* (New York, 1956), pp. 110–121, esp. pls. XIV.5a–c, 6a–b; XV.7–8). For a study of thrones both with and without figures, see E. French, "The Development of Mycenaean Terracotta Figurines," *Annual of the British School at Athens* 66 (1971), pp. 167–172.

—M. J-N.

3
Askos in the Form of a Hedgehog

Mycenaean III, circa 1425–1100 B.C.
Terracotta; H: 7.5 cm; L: 12.3 cm
Condition: End of animal's snout, handle, and top of bottle neck missing.

This rotund little animal standing on four short, stubby legs has reddish brown decorations over a cream-colored slip. Although he has lost his pointed nose, he retains around his small head the ruff characteristic of a hedgehog. The ruff's edge is emphasized by a broad band of paint, and the modeled pop-eyes stare out from large painted circles. The mouth is a small hole from which liquid could have been dispensed. On the hedgehog's right shoulder are two vertical slashes. On the tail section of the body, the left side is marked by irregular diagonal lines, and the right by horizontal lines. Each midsection is covered by the elongated figure of an animal with four dangling legs that terminate in backward curls. Each has a long-eared head atop a long, ladder-patterned neck and a solid body. Above and below the body is a pattern of concentric dot-edged semicircles, perhaps an unconventional plant motif, and rising from the rear end is a ladder pattern. The animal on the hedgehog's left side has a drooping tail and faces toward the back of the vase while that on the right faces the front.

The provenance of this hedgehog is unknown, and since the decoration appears to be without specific parallels, it is difficult to determine its precise origin. Cyprus may be a good possibility given the existence of later Bronze Age Cypriote Mycenaean amphorae on which pairs of chariot-pulling horses with elongated bodies, legs, ears, and large bug-eyes seem to be more evolved descendants of the creatures painted on this vase (see, for example, Vermeule/Karageorghis). Askoi dated to circa 1425–1225 B.C. and found on both mainland Greece and on Cyprus have more systematic patterns as decoration but are very comparable in shape (see Misch). The term *askos* itself, the ancient Greek word for a leather bag or wineskin, is used to designate vases used to hold liquid. Such vases, while not necessarily in the shape of an animal skin, are generally asymmetric with a handle arching across the rounded top of the vessel to its mouth, which can be off center.

BIBLIOGRAPHY: Unpublished.

RELATED REFERENCES: E. Vermeule and V. Karageorghis, *Mycenaean Pictorial Vase Painting* (Cambridge, Mass., and London, 1982), no. V.14 (ills). P. Misch, *Die Askoi in der Bronzezeit* (Jonsered, 1992), p. 146, fig. 122; p. 169, fig. 143.

—M. J-N.

3

4

Aryballos in the Form of a Bull's Head

Mycenaean III B, thirteenth century B.C.
Terracotta; H: 4.2 cm; L: 8.4 cm; W (ACROSS HORNS): 8.2 cm
Condition: Part of left ear and horn, all of handle on underside missing.
Provenance: Formerly in the Erlenmeyer collection.

The pottery produced by ancient vase makers includes not only ordinary vessels with shapes appropriate to their function but also large numbers of zoomorphic vases that are more decorative, sometimes even whimsical, in appearance. This vase takes the shape of the head of a bull. From the bulge at the jaw-line, it tapers in toward the muzzle and then flares out at the nose where the nostrils are indicated by two large depressions. The mouth and large almond-shaped eyes are carefully modeled. On the underside of the vessel are the root of a handle and a hole from which a vase-neck has broken off; the missing handle must have been joined to this neck.

The whole of this *aryballos* (a shape of vase intended to hold oils or ointments) was first covered in a creamy white slip and the physical details then added in black and dark reddish brown paint. The muzzle and horns are covered in the dark reddish brown, which is also the color of the line that runs down the middle of the face. On each side of that line is another that runs alongside it and then curves around under the eye and terminates in a spiral on the cheek. The

4

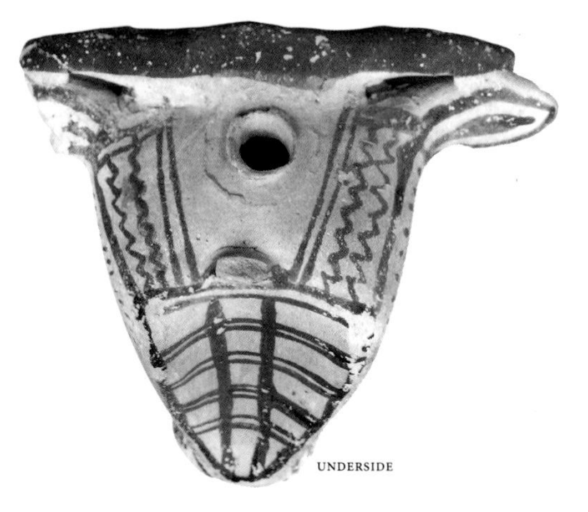

UNDERSIDE

wrinkles on the face are drawn in long rippling lines and the short, stiff hair on the rest of the head is suggested by stippling. The eyes, heavily outlined and with their whites enhanced by added white, are indicated by large dots in circles. Painted under the ears are leaflike shapes. The underside of the head is decorated at the front with a grid of double lines and at the sides with framed panels of zigzag lines.

Although the structure, modeling, and shape may be suggestive of archaic orientalizing art, the decorative scheme seems to this author Mycenaean in nature (compare especially the curlicues on the cheeks with the ends of the tentacles of the cuttlefish on the Mycenaean hydria [cat. no. 5]; see also the Mycenaean bull's head rhyton from Enkomi, dated to the thirteenth century B.C.: Karageorghis). H.-G. Bucholz and V. Karageorghis illustrate several Late Bronze Age rhyta in the shapes of heads of various animals, of which that of a fox, or dog, seems most similar to this vessel in ornamentation; noting that vases in the shape of animals and animal heads are indicative of the ancient desire to give lifelike form to inanimate objects, they also point out that most animal-head vases served as cult vessels for ritual purposes.

BIBLIOGRAPHY: Unpublished.

RELATED REFERENCES: V. Karageorghis, *The Civilization of Prehistoric Cyprus* (New York, 1983), p. 194, no. 159. H.-G. Bucholz and V. Karageorghis, *Prehistoric Greece and Cyprus* (London, 1973), p. 377.

—M. J-N.

5
Hydria

Mycenaean III B, circa 1300–1200 B.C.
Terracotta; H: 14 cm; DIAM (OF MOUTH): 7.1 cm; (OF FOOT): 5.4 cm; W (INCLUDING HANDLES): 15.7 cm
Condition: Intact; some surface wear.

The sides of this wide-mouthed jar for carrying water look almost as if they are already sagging with the weight of the vessel's intended burden. This type of vase made its first appearance around this time and became a standard shape throughout the history of Greek vase making. In a color scheme characteristic of much Bronze Age pottery, this hydria was covered first with a cream slip before its brownish black decoration was added. Painted bands encircle

5

the lip, the inside of the mouth, the join of the neck to the body, and the base. The outside of the back handle is marked along its length by a broad line, and the two side handles are accented at both their tops and roots by short broad strokes.

Many of the pictorial motifs in the art of the Mycenaeans, a seafaring people, were inspired by their familiarity with dolphins, fish, octopuses, shells, and even seaweed. The entire body of the hydria is covered with a stylized image of a cuttlefish, which has eight tentacles like an octopus, as well as two longer ones for catching prey. Here those two long, sinuous arms coil around the vase and terminate in curlicues under the back handle. The eyes are indicated by pairs of concentric circles painted outside the body; the details on both body and tentacles are highlighted in added white. This cuttlefish has evolved from the more elaborate type of the period Mycenaean III A into a simpler, stylized form.

BIBLIOGRAPHY: Unpublished.

RELATED REFERENCES: For a comparable cuttlefish, see the kylix from Rhodes: F. Stubbings, *Mycenaean Pottery from the Levant* (Cambridge, 1951), pl. XIII.3.

—M. J-N.

MARBLE

6

Head of an Idol

Cycladic, circa 2600–2500 B.C.
Greek island marble; H: 22.8 cm; W: 8.9 cm; DEPTH: 6.4 cm
Condition: Broken off at the neck; some encrustation, particularly heavy on proper left side of face.
Provenance: Formerly in the collections of Mr. and Mrs. A. Leuthold and Asher Edelman.

The marble idols crafted in the Cyclades during the Early Bronze Age evolved into forms that became traditional and varied surprisingly little over half a millennium. Although their abstract simplicity is very modern in appearance, these Cycladic sculptures actually represent the earliest art created by the Greeks. Most of the fewer than two thousand known examples are of small-scale reclining female figures with folded arms, and those that even begin to approach the almost life-size monumentality of the figure from which this head must have come are exceedingly rare. What exact purpose they served has yet to be understood with absolute surety, but it is generally agreed that they were used for religious and funerary rituals.

This head has been classified as an Early Spedos type, so-called after an important Cycladic grave site on the island of Naxos. Long and elegantly proportioned, it preserves strong evidence of the colorful detail that was originally applied to many, if not all, Cycladic sculptures, but which seldom survives. On the face, purplish red pigment defines a rarely seen band of short vertical strokes across the forehead, a stripe down the nose, and rows of dots, which may indicate tattooing, on the cheeks and chin. Because of its more fugitive nature, the blue paint that indicated the eyes, brows, and hair has faded, leaving behind only dark ghostlike impressions of those details.

BIBLIOGRAPHY: P. Getz-Preziosi, *Sculptors of the Cyclades: Individual and Tradition in the Third Millennium B.C.* (Ann Arbor, 1987), pl. XI; idem, *Early Cycladic Art in North American Collections* (Richmond, Va., 1987), pl. III, no. 43; idem, "A Head of an Early Cycladic Marble Figure," *Sotheby's: Art at Auction, 1988–1989*, p. 300ff.; Merrin Gallery, *Masterpieces of Cycladic Art* (New York, n.d. [1989?]), no. 17.

—M. J-N.

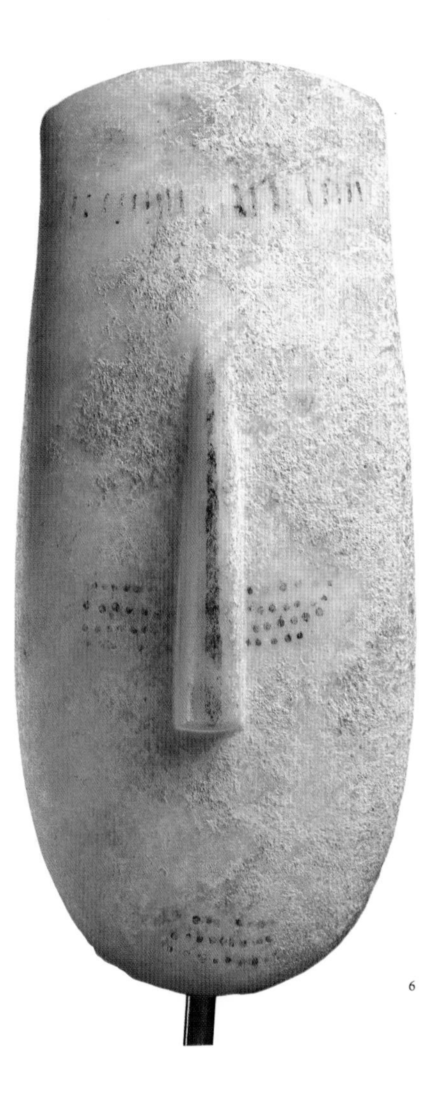

6

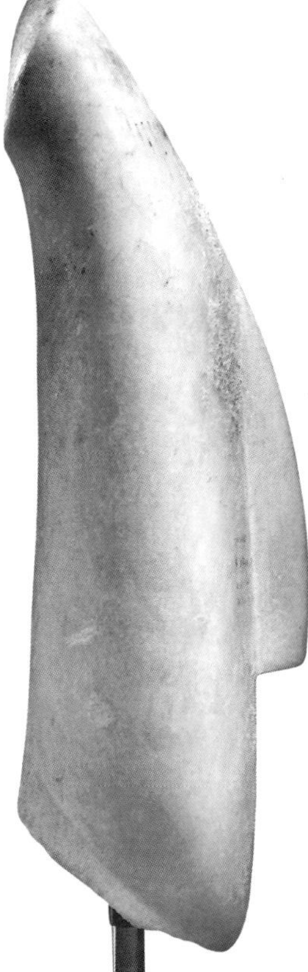

Greece from the Geometric Period to the Late Classical

7

BRONZE AND LEAD

7

Pendant in the Form of a Bird Atop a Disk

Greek, Geometric, circa 750–700 B.C.
Bronze; H: 6.9 cm; DIAM (OF DISK): 3.8 CM; DEPTH: 1.3 cm
Condition: Some losses along edge of disk; some surface encrustation on both sides of disk.

A long-billed bird with a flattened tail sits atop a disk incised with concentric circles. The body is pierced from the front of the breast to behind the neck for suspension. Incised lines decorate the base of the neck and tail.

Bird-disk pendants are known from a variety of sites in the Peloponnesos, at sanctuaries such as Olympia and the Argive Heraion, and in Thessaly. Many of the Thessalian examples have been found at Pherai, in the excavations of the temple of Artemis Ennodia that lies atop an early Geometric necropolis, and in Lokris, at Amphikleia, in Geometric cist graves. Thus the pendants seemed to have served as both funereal and votive offerings.

I. Kilian-Dirlmeier divides the corpus of bird-disk pendants into four categories based on the number of elements separating the body of the bird from the disk. The Fleischman piece belongs to her Type 2, which is characterized by a single ring or rib between the bird and the spherical knob. It is the only example of this group to have incised circles on the disk; on the others, the patterns take the form of a cross. Three other pendants have disks with incised circles similar to the Fleischman piece; their find spots are unknown. All belong to Kilian-Dirlmeier's Type 3. One is in the Ashmolean Museum (1938.738: *Select Exhibition;* Kilian-Dirlmeier), the second is in the Ny Carlsberg Glyptotek (see Poulsen; Kilian-Dirlmeier), and the third example is in the Antikensammlung, Munich (3450: Kilian-Dirlmeier).

BIBLIOGRAPHY: Unpublished.

RELATED REFERENCES: Ashmolean Museum, *Select Exhibition of Sir John and Lady Beazley's Gifts to the Ashmolean Museum 1912–1966* (London, 1967), p. 154, no. 584, pl. LXXVII. V. Poulsen, "Geometrisk Kunst Paa Glyptoteket," *Meddelelser Fra Ny Carlsberg Glyptotek* 19 (Copenhagen, 1962), pp. 13 (fig. 6), 16. For the corpus of bird-disk pendants, see I. Kilian-Dirlmeier, *Anhänger in Griechenland von der mykenischen bis zur spätgeometrischen Zeit,* Prähistorische Bronzefunde 11, vol. 2 (Munich, 1979), p. 154ff., nos. 887–937, pls. 49–51 (Vögel aus Scheibe type); for the Ashmolean, Copenhagen, and Munich pendants, respectively, pl. 49, nos. 919, 920, 922.

—K. W.

8

Pendant in the Form of a Bird

Greek, Geometric, late eighth century B.C.
Bronze; H: 5.3 cm; W: 6.05 cm; DEPTH: 2.3 cm
Condition: Top of face and beak and exterior of left leg corroded; area of deep corrosion, or perhaps a miscast section, on proper left side of back; crack along proper left side of bird's body.

The body of the bird is highly decorated: the front of the breast and edge of the tail are notched, and there are incised lines at the outer edges of the legs, at the join of the tail and body, on the front of the face (although the area is corroded and hard to see), on the neck behind the head, and at the join of the neck to the body. There is an incised double-lined zigzag pattern across the back.

The bird belongs to a group I. Kilian-Dirlmeier calls "Thessalian hens with [suspension] loops on their backs," of which there are numerous examples. The bird closest in overall body shape was formerly in the de Kolb collection (*Master Bronzes,* p. 40, no. 23; Kilian-Dirlmeier). Like the Fleischman bird, many of the hens from this group have elaborate body markings. Almost all of the pieces in the group with a known provenance come from the excavations of the temple of Artemis Ennodia in Pherai, which is located above a Geometric necropolis. One other excavated example comes from Delphi, indicating that the birds were probably used as both funereal and votive offerings. Because so many of this type of pendant have been found in Pherai, Kilian-

8

9

Dirlmeier believes them to be the products of a Thessalian workshop.

BIBLIOGRAPHY: Unpublished.

RELATED REFERENCES: For the corpus as a whole, see I. Kilian-Dirlmeier, *Anhänger in Griechenland von der mykenischen bis zur spätgeometrischen Zeit,* Prähistorische Bronzefunde 11, vol. 2 (Munich, 1979), p. 139ff., nos. 760–788, pls. 41–44; for the de Kolb bird, p. 140, no. 772, pl. 42.

—K. W.

9
Pendant in the Form of a Siren

Greek, Geometric, late eighth century B.C.
Bronze; H: 6.9 cm ; W: 6 cm; DEPTH: 1.4 cm
Condition: Consistent light green patina; corroded in some places, most notably on front of siren's head and top of suspension loop.

The siren is shown with her head thrown back as if in song. Her long neck is articulated by three spaced rings, perhaps meant to represent necklaces. Lines are incised onto her head to indicate hair, and the underside of her tail is cross-hatched to suggest overlapping feathers. The tips of her upswept wings are separated from the body by a grooved undercut section between the wings and tail. Below her body is a finial composed of a shaft with six rings and a flattened end. A similar, but less stylized bronze siren is in the Canellopoulos Museum (Bronze 235) in Athens. Other pendants with finials of this type have been found at Olympia, Pherai, Philia, and Thermon.

This pendant, from a period when representations of figures from myth and epic are rare, is one of the earliest Greek sculptural renditions of a siren with a female head atop a bird's body. As recorded in Homer's *Odyssey,* sirens lured sailors to their deaths with the hypnotic melody of their sweet songs. Sirens also played a role in the afterlife, conducting the souls of the deceased to the underworld. In this role, they have been related to the Egyptian Ba-bird, which had a similar function and is frequently depicted in Egyptian art with a bird's body and a female head.

The figural type is also thought to derive from imported Near Eastern cauldron attachments of winged birds with bearded male heads. When duplicated by Greek artisans, however, the figures were generally depicted as beardless. As a type, the female-headed siren was popular among Greek artists and was frequently depicted on ceramics of the seventh and sixth centuries B.C.

BIBLIOGRAPHY: Unpublished.

RELATED REFERENCES: For pendants with similar finials, see I. Kilian-Dirlmeier, *Anhänger in Griechenland von der mykenischen bis zur spätgeometrischen Zeit,* Prähistorische Bronzefunde 11, vol. 2 (Munich, 1979). For

sirens in Greek art, see B. Cohen, "The Sirens," in B. Cohen and D. Buitron, eds., *The Odyssey and Ancient Art: An Epic in Word and Image,* exh. cat. (Annandale-on-Hudson, N.Y., 1992), p. 108ff.; E. Hofstetter, "Sirenen im archaischen und klassischen Griechenland," *Archäologie* 19 (Würzburg, 1990). For the relationship between sirens and Egyptian Ba-birds, see J. D. Cooney, "Siren and Ba, Birds of a Feather," *The Cleveland Museum of Art Bulletin* 55 (1968), pp. 262–271, figs. 1–3.

—K. W.

10

Statuette of a Horse

Greek, Geometric, late eighth century B.C.
Bronze; H: 7 CM; W: 6.49 cm; DEPTH: 1.6 cm
Condition: Light green to black patina; surface fairly uniform, one area of corrosion on exterior of left rear leg.

Incision is used to enhance parts of this mannered and angular horse. The mane is indicated by short strokes along the length of the back of the neck. Two parallel lines run on both sides of the head from the underside of the neck to behind the ears and may indicate a halter. In contrast to the other Fleischman Geometric horse (cat. no. 11), this one has no base and balances on the tips of its four spatula-like hooves. It can be placed in J.-L. Zimmermann's Lokrian group, the style of which was influenced by the Thessalian, Attic, and northeastern Peloponnesian groups. Some of the stylistic characteristics of the Lokrian horses are their arched backs, slender curved feet and legs, and rectangular shoulders.

Closest to this statuette are two horses in the Lamia Museum (230 and 235), both of which are in Zimmermann's Lokrian group. They are from Anavra, in Thessaly. In common among all three are the incised mane, the straight, forward-leaning neck, the small squared-off head, and the flattened hooves. The majority of Geometric horse statuettes are votive gifts, found primarily in sanctuaries, although some have been excavated from burials.

BIBLIOGRAPHY: Unpublished.

RELATED REFERENCES: For Geometric bronze horses in general, see J.-L. Zimmermann, *Les Chevaux de bronze dans l'art géométrique Grec* (Mainz, 1989); for the Lokrian group, p. 218ff., pls. 49–51.

—K. W.

11

Statuette of a Horse

Greek, Geometric, circa 700 B.C.
Bronze; H: 5.9 cm; W: 5 cm; DEPTH: 1 cm
Condition: Green patina, somewhat crusty, although smoother on proper right side; what appears to be a casting flaw along lower exterior end of proper left rear leg.

This diminutive alert horse epitomizes the large corpus of statuettes manufactured throughout Greece.

The strong shoulders and haunches are offset by the thinness of the tubular body, and the thick arching neck is emphasized by the shortness of the muzzle and forward-leaning ears. A short line across the front of the muzzle may indicate the mouth. The horse is probably a stallion. The small unshaped mass of bronze between the hind legs seems intentional, as the artist would have been able to remove any residual wax while preparing his model for casting. The figurine stands atop a flat base the underside of which is less thick toward the front legs. It was perhaps manufactured using an unevenly folded sheet of wax.

This horse relates stylistically to groups manufactured in northern Greece, primarily Macedonia, but also Thessaly, a region renowned for its horse breeding. The closest parallel is a horse in the de Menil collection (x2 126-6: Hoffman), that is placed by J.-L. Zimmermann in his Macedonian group. The raised hindquarters, and the asymmetry between the shoulders and hind legs of the Fleischman horse indicate that it was made toward the end of the Geometric period. Although some horses have been found in funereal contexts, the majority of the statuettes seem to be votive gifts at religious sanctuaries.

BIBLIOGRAPHY: Unpublished.

RELATED REFERENCES: H. Hoffmann, *Ten Centuries That Shaped the West* (Houston [?], 1971), p. 124, no. 42. For Geometric bronze horses in general, see J.-L. Zimmermann, *Les Chevaux de bronze dans l'art géométrique Grec* (Mainz, 1989), for Macedonian horses, p. 260ff.; for the de Menil horse, p. 261, no. 3, pl. 61.3.

—K. W.

12

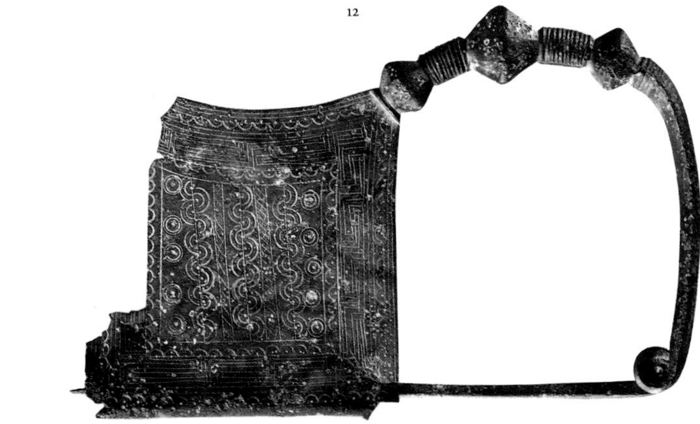

12

Catchplate Fibula

Greek, Geometric, late eighth–early seventh century B.C.
Bronze; H: 7.7 cm; W: 12.1 cm; DEPTH: 1 cm
Condition: Good condition with some areas of corrosion; brownish encrustation along the interior of the catchplate; losses along the outer edge of the catchplate, and along both lower corners near pin.

The ancient equivalent of the modern safety pin, *fibulae* were used in antiquity to hold the various elements of the draped fabric together, especially at the shoulders. The designs of ancient fibulae range from the simple to the ornate, as the craftsmen sought to embellish this necessary element of the wardrobe. One means of ornamenting the pins was to enlarge and engrave the catchplate, which held the pin in place after it had been fed through the various layers of cloth.

The Fleischman fibula is elaborately fashioned and decorated. The faceted pin ends with three biconical beads separated by ribs. Both sides of the catchplate are incised. The exterior panel has alternating bands of double circles with central dots, semicircles offset to form an undulating double line, and parallel lines with crosshatching between. The panel is framed by two bands of dotted semicircles separated by a double-line meander. The frame for the panel on the interior side is the same, but the panel decoration is figural, with three fish whose bodies, including the four fins, are formed with double lines. A diamond with a line extending to the tip of the mouth marks the head; the bodies are decorated with four lines to represent the gills, and four pairs of short hatched lines to indicate scales. In the corner closest to the pin attachment is a stylized sun.

The fibula belongs to K. Kilian's Thessalian Type E V c, which is characterized by three biconical beads along the pin. He lists four examples, all of which are from Pherai. One of these, now in the National Archaeological Museum in Athens, also has representations of swimming fish. The closest stylistic parallel for the type of fish found on the Fleischman fibula, however, is a fragmentary fibula in the de Menil collection (see Hoffmann), where similar fish decorate the reverse side of the catchplate. On the front of the de Menil catchplate, a lion devours his prey, represented by a decapitated animal head and a disembodied leg. The de Menil fibula is said to be northern Greek, perhaps Thessalian. A fragmentary catchplate from the site of Kainourgion (Lamia Museum 75: Kilian) has four fish with double lines of decorative patterns and a sun in one corner. In the complexity of their incision, these fish are similar to those found on the Fleischman fibula, although the latter are more elaborate, having more fins, plus the diamond element on their heads. A corner fragment of a catchplate from Pherai (Volos Museum 285: Kilian), preserves the head of a fish with a lozenge-shaped element similar to the diamond, but not enough survives of the body to indicate the number of fins.

Many of the fibulae from Thessaly have come from votive deposits, but like other small bronzes of this period, they also served as grave goods. Fibulae could be seen as appropriate offerings at both Pherae and Philia, as the sanctuaries at both of these Thessalian sites were dedicated to goddesses, Artemis Ennodia and Athena Itonias, respectively. A gift of fine jewelry from both male and female dedicants undoubtedly pleased the feminine deities.

BIBLIOGRAPHY: Unpublished.

RELATED REFERENCES: H. Hoffmann, *Ten Centuries That Shaped the West* (Houston[?], 1971), pp. 150–152, no. 71, figs. 71a-c. For the corpus of Thessalian catchplate fibulae of Type E V c, see K. Kilian, *Fibeln in Thessalien von der mykenischen bis zur archaischen Zeit*, Prähistorische Bronzefunde 14, vol. 2 (Munich, 1975), p. 105ff., esp. pp. 115–133, nos. 1522–1524, pl. 54; for the fibulae in Athens and Lamia, respectively, p. 132, no. 1525, and p. 161, no. 1884, pl. 62; for the Volos Museum fragment, p. 114, no. 1337, pl. 47.

—K. W.

13
Griffin Protome

Greek; circa seventh century B.C.
Bronze, possibly some lead filling; H: 28.5 cm; W: 10.7 cm; DEPTH: 9.3 cm
Condition: Tip of tongue missing; right ear possibly reattached; ears bent inward; thick, glossy light green patina.

"The Samians . . . made therewith a bronze vessel, like an Argolic cauldron, with griffins' heads projecting from the rim all round; this they set up in their temple of Hera" (Herodotus IV.152). The Fleischman griffin head, along with two or three others, would once have adorned the shoulder of a cauldron such as Herodotus describes, attached by means of the three holes in the flanged base. The earliest of the cauldron protomes were hammered, with stocky proportions, and date to the end of the eighth century B.C. By the middle of the seventh century B.C., such protomes, like this one, were produced by casting. The piece's slender form and sinuous lines, which are not as exaggerated as those of later protomes, also conform to this date.

Although it possesses the typical attributes of the griffin protome, this piece does not fit stylistically into one of H. V. Herrmann's groups. Like other griffin protomes, it has a serpentlike neck, which is decorated with a spiral coil on either side, a birdlike beak, and a baggy pouch behind the mouth that recalls a lion's mane. The griffin's unfaltering vigilance is expressed through the upright ears, the open mouth, and the wide open eyes, which were once inlaid with some other material. The shape of the Fleischman protome's head is similar to several found on the island of Samos (see Jantzen 1955). Above the eyes is a knoblike protuberance, with a smaller one closer to the beak. The top knob might be a reference to the crest of the peacock, an animal associated with immortality in antiquity. Multiple protrusions are most often seen in the group of protomes designated Type II, Combined Technique protomes by Herrmann and are especially clear in representations on vases.

Widespread colonization and the reorganization of the Olympic games in 776 B.C. facilitated the transference of ideas between Greece and the East. The griffin, a creature which combines the characteristics of a lion, a bird, and a snake, was originally a widely used form in the Aegean culture during the late Bronze Age. The motif was adopted by the peoples of the Near East and then introduced to the Greeks of the Archaic period. Although neither the cauldron nor the griffin were new forms, the Greeks were the first to unite the two.

The griffin cauldron, which is found extensively at Olympia and Samos, most likely served as a votive offering. The griffin would have been an appropriate motif for such a valuable gift, as these creatures were the ones responsible for guarding Apollo's gold, which he had entrusted to the Hyperboreans.

13

BIBLIOGRAPHY: Unpublished.

RELATED REFERENCES: U. Jantzen, *Griechische Greifenkessel* (Berlin, 1955), nos. 72–75. U. Jantzen, "Greifenprotomen von Samon: Ein Nachtrag," *AM* 73 (1958), pp. 26–49, Beil. 28–52. J. L. Benson, "Unpublished Griffin Protomes in American Collections," *Antike Kunst* 2 (1960), pp. 58–70. H. V. Herrmann, *Die Kessel der Orientalisierender Zeit* (Berlin, 1979), pp. 122–130, pls. 55–60. A. Dierichs, *Das Bild des Greifen in der frühgriechischen Flachenkunst* (Münster, 1981).

—E. B. T.

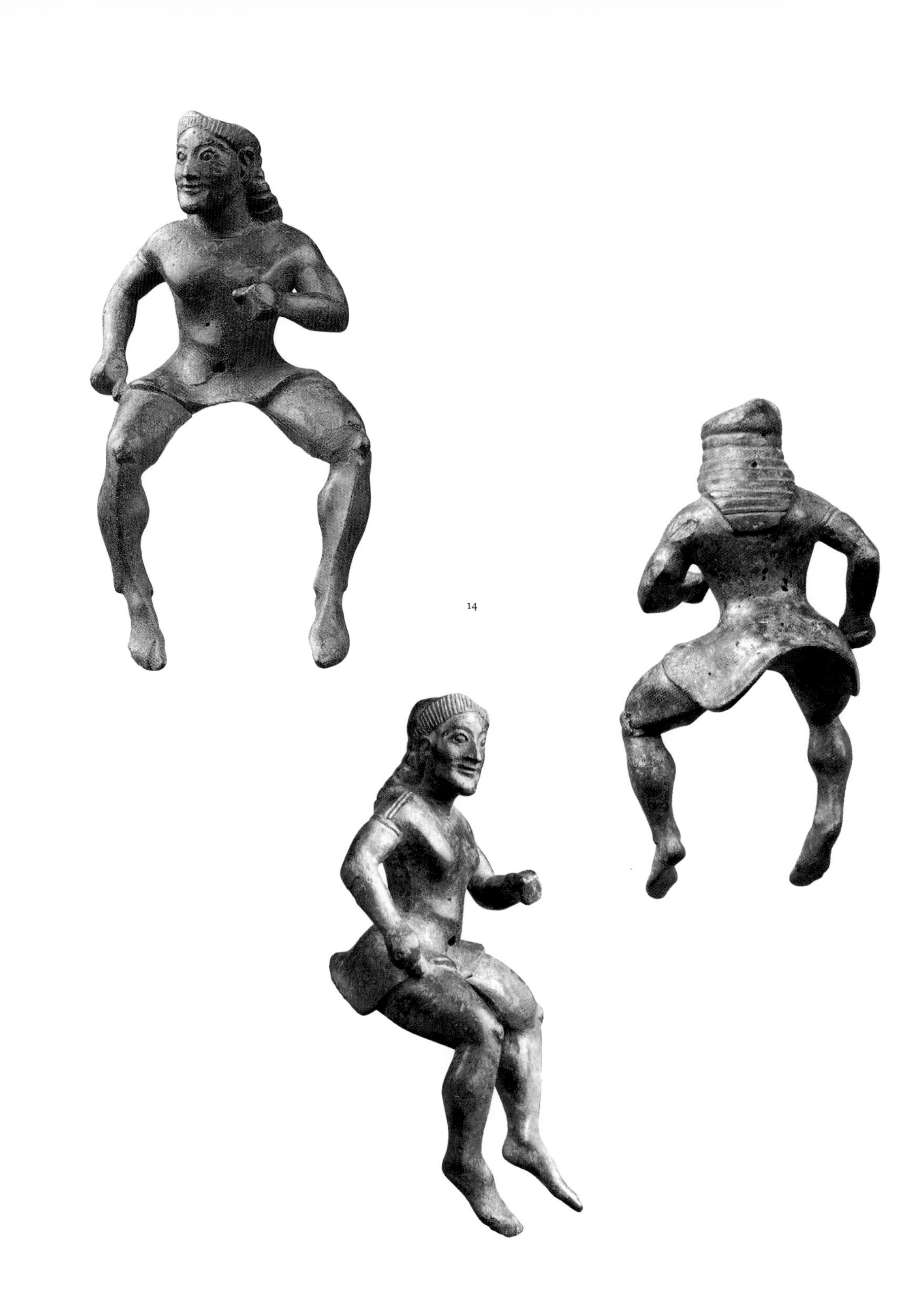

14

14
Statuette of a Rider

Corinthian, circa 550 B.C.
Bronze; H: 8.5 cm
Condition: Intact, with a fine greenish gray patina and numerous flaws in solid-cast bronze; figure probably originally wore a hat.

Filled with youthful exuberance, this small figure originally sat astride a large horse, his muscular legs tensely clinging to the sides of the mount and his head turned to the right. He is dressed in a short tunic decorated at the shoulders with a kind of zigzag pattern. His long hair is carefully dressed in horizontal waves down the back, ending in a long, broad point, with a short fringe around the forehead and over the ears. His proper left arm, bent at the elbow, is raised before him and the closed left hand originally held the reins. The proper right arm extends out at the side, also bent at the elbow, and may also have held a rein, or a goad. The top of the head is flattened, suggesting that the rider may originally have worn a hat (*petasos*), like the rider in Boston (MFA 98.659: Comstock/Vermeule 1971, p. 35, no. 33).

The figure is remarkably close in costume, pose, and style to the rider found in excavations at the sanctuary of Zeus in Dodona and now in the National Archaeological Museum, Athens (Karapanos Collection 27: Karouzou). The two figures also show the same porous surface filled with casting flaws. The Fleischman bronze is somewhat livelier in execution, however. He turns his head alertly to his right, his expression filled with intensity, and the posture of his arms is more spontaneous. As if to echo the enthusiasm expressed by his posture, the skirt of his tunic seems to ripple around his body. By contrast, the Athens example appears calm, sedate, and more mature.

The Athens rider has long been associated with a horse in the Louvre (BR 148: Karouzou; Rolley, *Monumenta*), to which it is thought it was originally attached. In fact, a second horse from Dodona exists in the collection of the National Archaeological Museum, Athens (16547: Rolley 1986, p. 102; Calligas), and scholars have long speculated that another rider might also be found.

S. Karouzou and C. Rolley have attributed the manufacture of the Athens bronze to Corinth, though it was dedicated in Dodona. Given its similarity, the Fleischman rider must be assigned to the same Corinthian source, though its patina would suggest that it, like the Athens example, may have been found in Dodona, or at least in northern Greece, and was also a votive offering dedicated at a sanctuary.

BIBLIOGRAPHY: Unpublished.

RELATED REFERENCES: S. Karouzou, "ΟΙ ΙΠΠΕΙΣ ΤΗΣ ΔΩΔΩΝΗΣ," *Theoria: Festschrift für W. H. Schuchhardt* (Baden-Baden, 1960), pp. 231–250, pls. 1–10. C. Rolley, *Monumenta Graeca et Romana V*, fasc. 1, p. 5, no. 49. P. G. Calligas in *Mind and Body: Athletic Contests in Ancient Greece*, exh. cat. (National Archaeological Museum, Athens, 1989), pp. 141–142, no. 32.

—M. T.

15
Statuette of a Seated Lion

Lakonian, mid-sixth century B.C.
Bronze; H: 9.3 cm; L: 13.3 cm; W: 5 cm
Condition: Tip of tail missing; glossy dark brown patina.

The lion is shown sitting, with its head facing to the right and its tail curling up onto the back in a graceful curve. The face, full and round, tilts up slightly. The animal's large mouth is closed, giving it an intent look, which is emphasized by the arched eyebrows and bulging eyes. The mane, a thin ruff with incised radiate lines denoting hair, frames the lion's face. The fur under its chin and on the back of its head is indicated by means of an incised wavy pattern. The statuette is hollow cast, with most of the core intact.

The Fleischman lion is of the Lakonian type, best exemplified by a bronze lion that was dedicated at the Heraion on Samos by a Spartan (Vathy Museum B 5). It displays certain stylistic features common to both Lakonian reliefs and sculpted depictions of lions—the incised "collar-mane," the S-curve of the tail, the tullelike design on the body, and the round face and ears. With its "collar-mane" and mild demeanor, our lion appears to have been influenced more by Hittite representations of lions rather than by Assyrian models.

Several parallels for the Fleischman lion exist. The closest examples are two bronze lions from Olympia, one now in Berlin (Ol. 4790) and one in the National

15

Archaeological Museum, Athens (2150: Gabelmann). There is also a series of Lakonian lion fibulae (see Blinkenberg). Several lions decorate the handle of a bronze hydria found in Grachwil, Switzerland, which is thought to have come from the Dorian West Greek colonies or from Sparta (Historisches Museum Bern 11620: Jucker).

BIBLIOGRAPHY: Unpublished.

RELATED REFERENCES: H. Gabelmann, *Studien zum frühgriechischen Löwenbild* (Berlin, 1965), p. 69ff., no. 68a, pl. 12. C. Blinkenberg, *Fibules grecques et orientales*, vol. 13, pt. 1 of *Det Kgl. Danske Videnskabernes Selskab* (Copenhagen, 1926), pp. 280–282, fig. 319. H. Jucker, "Altes und Neues zur Grächwiler Hydria," *Zur Griechischen Kunst: Festschrift Bloesch = Antike Kunst* Beiheft 9 (1973), p. 42ff.

—M. R.

16

Fragmentary Appliqué in the Form of a Bearded Male

Lakonian, circa 550–525 B.C.
Bronze; H: 12.2 cm
Condition: Figure complete except for loss of left ankle and foot; part of a larger composition that included one or perhaps two other figures, of which only the right foot and shin of one preserved; right end of base missing; light gray-green patina, slightly crusty in places.

The tall, slender nude male strides purposefully to the right, drawing his sword from its scabbard as he moves. In characteristic Archaic fashion, his head and legs are in profile, while his torso is turned toward the viewer. His hair, combed behind the ear, hangs over his right breast in four elaborately dressed coils, and, though bearded, he has no mustache. At his feet is preserved the right foot and shin of another figure, who was also facing right, either falling or collapsed on the ground.

This openwork relief, slightly hollowed out on the back, was made to be fastened to a flat surface by means of rivets, the holes for which are clearly visible in the thigh of the standing figure and the preserved left end of the base. In technique, it recalls two other fragmentary plaques, one in Delphi (2560: *Guide de Delphes*) and one in a private collection in New York (see Buitron/Cohen), that depict a figure tied beneath the body of a ram, probably Odysseus escaping from the cave of Polyphemos. These plaques must have been intended to decorate a chest or piece of furniture made of either wood or bronze. The beauty of these figural compositions recalls Pausanias' famous description (v.17.5–19.10) of the chest elaborately decorated in ivory and dedicated by Kypselos at Olympia. Rather similar openwork reliefs in ivory have been found at Delphi and attributed in the most recent study to Lakonia. The style of this figure is most comparable in its details to bronzes also attributed to the Lakonian workshop, especially those dated to the third quarter of the sixth century B.C.

Although it is impossible to be certain of the subject of the original composition, it is unlikely that

16

the preserved figure's opponent is at his feet, since he looks straight ahead and not down. As he is not dressed in armor, it is improbable that the scene is one of battle over a fallen comrade. Rather, the plaque would seem to be a mythological composition. The pose of the figure with sword drawn stepping over the leg of another recalls scenes of Theseus in combat with the Minotaur. However, Theseus in this period is more conventionally a beardless youth, and he generally uses one hand to hold the monster, who is often in a kneeling position, as he prepares to stab it. Another possibility might be Orestes preparing to kill Klytaimnestra, having already slain Aigisthos, who lies at his feet. A third possible identification might be Odysseus drawing his sword against Circe, the sorceress who has turned his comrades to animals.

BIBLIOGRAPHY: Unpublished.

RELATED REFERENCES: École Française d'Athènes, *Guide de Delphes: Le Musée* (Paris, 1991), p. 168, fig. 35. D. Buitron and B. Cohen, eds., *The Odyssey and Ancient Art: An Epic in Word and Image,* exh. cat. (Annandale-on-Hudson, N.Y., 1992), pp. 48, 66–67, no. 13. For the suggestion that the Delphi ivories may have been Lakonian in style, see E.-L. Marangou, *Lakonische Elfenbein- und Beinschnitzereien* (Berlin, 1969), pp. 191–192. For the most recent consideration of the Delphi plaques and their association with Lakonia, see J. Carter, *Greek Ivory Carving in the Orientalizing and Archaic Periods* (Ph.D. Diss., Garland Series, 1985), pp. 163–173; idem, "The Chests of Periander," *American Journal of Archaeology* 93 (1989), pp. 355–378; esp. figs. 2, 11 (the treatment of the hair of the bearded male on the right end of plaque 9918 is particularly close to that of the Fleischman figure). For the most recent consideration of Lakonian bronzes, see M. Herfort-Koch, *Archaische Bronzeplastik Lakoniens,* Boreas Beiheft 4 (1986), p. 71 (on appliqués); p. 115, nos. K129–130, pl. 17.1–6 (two standing figures from the rim of a vessel found at Olympia [B 5000, B 25] are especially comparable to the Fleischman figure). For Theseus in combat with the Minotaur, see F. Brommer, *Theseus* (Darmstadt, 1982), pp. 35–64, ills. For Orestes attacking Klytaimnestra with a sword, see the gilded silver repoussé band in the Getty Museum (83.AM.343): F. Brommer, "Ein Silberstreifen," *GettyMusJ* 12 (1984), pp. 135–138. For Odysseus and Circe on vases, see the merrythought cup in the Museum of Fine Arts, Boston (99.518), and the stamnos in Parma: F. Brommer, *Odysseus* (Darmstadt, 1983), pp. 76–77, pl. 27 and fig. 35 respectively (on both pieces the hero holds a drawn sword and scabbard). For a three-dimensional representation of a transformed comrade of Odysseus in a position similar to the fallen figure on the Fleischman plaque, see the small bronze in the Walters Art Gallery (54.1483): D. K. Hill, *Catalogue of Classical Bronzes in the Walters Art Gallery* (Baltimore, 1949), p. 122, no. 282, pl. 54.

—M. T.

17
Statuettes of Three Banqueters

Northern Greek, circa 550–525 B.C.
Bronze
Figure A. H: 3.8 cm; L: 7.7 cm
Figure B. H: 3.7 cm; L: 7.2 cm
Figure C. H: 3.5 cm; L: 7.2 cm
Condition: Figure A, intact; Figure B, missing attribute once attached to left hand; Figure C, missing fingers and perhaps attached attribute from left hand; all have a fine, gray-green patina.
Provenance: Said to be from northern Greece.

This charming little trio of symposiasts probably once decorated the rim of a vessel used for mixing and serving water and wine, either a cauldron or a krater. All are bearded males who recline on soft cushions irregularly compressed beneath their bodies. They support themselves on their left elbows with left legs resting on the *klinai,* right legs bent at the knee and raised. All wear himatia draped around their lower bodies, the folds of which are indicated with wavy incised lines. Each figure, however, is distinguished in a variety of subtle ways. Banqueter A rests his right hand on his raised knee and looks to his left, toward the *keras* (drinking horn) that he holds in his left hand. His long wavy hair is carefully dressed, and he sports a raised fillet decorated with diagonal incisions. Resting his elbow on his raised knee, Banqueter B gazes straight out before him. His right hand is extended forward, as if gesticulating in conversation, while his missing left fingers may also have held a drinking cup of some kind. Around his head is a beaded wreath, and his long tresses are dressed in the same fashion as those of Banqueter A. The most distinctive of the three, Banqueter C looks to his right, holding his right hand to his abdomen, as if to pat his belly. His extended left palm is pierced with a hole for a pin that must have secured a flat, open vessel in his fingers, perhaps a phiale. He wears no wreath around his head, but his hair, crimped and longer at the back than that of his companions, is arranged in a thin roll around his forehead with a single long spiraling lock hanging down over the front of each shoulder. The abdomen of Figure C is treated differently from those of his comrades: while their abdominal arches are indicated by semicircles with short strokes radiating from the navel to define the muscles, his muscles are described with a herringbone pattern within a pointed arch.

Hollow cast and open at the bottoms, with much of the core material preserved, the figures were intended for attachment to the rim of a large bronze vessel or tripod,

17

as the rivet holes at either end of their bases indicate. Figures of banqueters were popular subjects for the decoration of the rims and handles of vessels and tripods associated with the symposium, just as their painted representations are common on the walls of terracotta vases used for mixing and serving wine. Given a series of such figures, the artist would do his best to vary their poses and attributes, both to avoid monotony in the composition and to endow each participant with individuality. In his extensive study of the ancient banquet, J. M. Dentzer has pointed out that the motif of the banqueter with raised and flexed right leg, seen in profile, probably began in graphic representations of symposia and was adopted for small bronze figures of banqueters around the middle of the sixth century B.C. This change in representation also enhanced the three-dimensional quality of the figures, as the knees project in space in different directions.

Dentzer provides a list of twenty-six examples of small bronze reclining banqueters. The Fleischman figures find their closest parallel in a small bronze once in the Antiquarium in Berlin (10586: Neugebauer). The Berlin figure's overall dimensions, as well as the treatment of the drapery, the irregular cushion, and the stylized abdomen are quite close to the Fleischman trio, especially Figures A and B, though the Berlin example is beardless. Found in the sanctuary of Zeus at Dodona, the Berlin bronze was assigned by K. A. Neugebauer to a northern Greek workshop on the basis of the style. As the Fleischman trio shows the same combination of lively invention and somewhat unrefined simplification of details, they may be similarly classified.

BIBLIOGRAPHY: Unpublished.

RELATED REFERENCES: K. A. Neugebauer, *Katalog der statuarischen Bronzen im Antiquarium,* vol. 1 (Berlin, 1931), p. 111, no. 217, pl. 3. *The George Ortiz Collection: Antiquities from Ur to Byzantium,* exh. cat. (Berne, 1993), no. 116 (a larger bronze banqueter with more refined details). On the ancient banquet and banqueters in general, see J. M. Dentzer, *Le Motif du banquet couché dans le Proche-Orient et le monde grec du VIIième au IVième siècle avant J.-C.,* BEFAR 246 (Rome, 1982), esp. pp. 216–221 (for the Greek bronzes). For banqueters on the rim of a tripod in Belgrade, see L. Popović et al., *Anticka Bronza u Jugoslaviji 1844–1969* (Belgrade, 1969), p. 71, no. 32 (ill). For banqueters decorating a handle, see Dentzer, op. cit., fig. 179, pl. 29.

—M. T.

18

18

Appliqué in the Form of a Horse and Rider

Lakonian, circa 500 B.C.
Bronze; H: 7.3 cm; L: 9.5 cm
Condition: Intact except for missing horse's tail and separately attached rein; detail worn from surface; irregular green and brown patina.

In this small relief image, a slim boy sits bareback astride a large mount which is in full gallop to the right. Leaning back slightly to accommodate the horse's motion, he presses his legs and feet tight against the horse's sides to secure his seat; his right hand, loosely closed, rests on his right thigh. Separately added reins, now lost, originally ran from the hole at the bit ring of the bridle to the rider's left hand above the animal's withers. Both the long hair combed back over the figure's shoulders and his nude body betray his youth. The mane and forelock of the spirited horse are blown back, suggesting the speed of the gallop, and the animal has its ears flattened.

Open at the back and slightly curved to fit a rounded surface, with holes for rivets in the front and rear quarters of the horse, this lively image was originally made for attachment, most likely to the neck of a large bronze krater, such as the magnificent example in Belgrade from Trebenište (Nationalmuseum BR 174/I: Vulic). Three other similar single rider appliqués, in Mainz (Römisch-Germanisches Zentralmuseum 0.21296), Athens (National Archaeological Museum, Karapanos 36), and Paris (Louvre BR 4445) are known (see Hitzl), but the Fleischman example stands somewhat apart. In all of the other pieces, the rider turns his head out to face the viewer. The Fleischman rider, in contrast, does not turn to face the viewer, but sits in a comfortable three-quarter posture—anatomically more correctly modeled and more convincing than any of the other riders. His mount is also more persuasively naturalistic, its body not so extended in the gallop as those of the other three horses. Like the Fleischman example, the Mainz appliqué had a separately added rein, a feature that reflects the refined workmanship, while the horse of the Athens example also has its ears flattened.

In fact, the Fleischman rider is closer to a more fully modeled rider appliqué from Dodona, now in the National Archaeological Museum, Athens (Karapanos 26: Karouzou; Rolley). This slim nude figure with his hair combed back over his shoulders turns his head slightly to the right and has much of the refinement visible in the simpler Fleischman attachment. As S. Karouzou points out, this new position of the head in relation to the viewer is an advance over the earlier relief images. Since the Trebenište krater and related rider appliqués have been dated around 520 B.C., the Fleischman appliqué may be dated to around 500 B.C.

The place of manufacture of the large bronze kraters decorated with such appliqués has been much discussed. A. Rumpf argued rather convincingly that the Vix krater should be assigned to Sparta and placed the other large kraters there as well; other scholars have identified Corinth as the place of manufacture for the Trebenište krater and the other known rider appliqués. If the freestanding bronze horses and riders found at Dodona discussed above (cat. no. 14) should be attributed to a Corinthian workshop, however, as S. Karouzou and C. Rolley have suggested, the kraters and appliqués should not, since the treatment of both animal and human figures is markedly different, even allowing for the difference in date. Rather, the krater and appliqués may be better associated with the Lakonian workshop.

BIBLIOGRAPHY: Unpublished.

RELATED REFERENCES: N. Vulic, *ÖJh* 27 (1932), p. 1ff. K. Hitzl, "Bronzene Applik vom Hals eines Volutenkraters in Mainz" *Archäologischer Anzeiger* (1983), pp. 5–11. S. Karusu [Karouzou], "ΤΕΧΝΟΥΡΓΟΙ ΚΡΑΤΗΡΩΝ, Fragmente Bronzener Volutenkratere," *AM* 94 (1979), pp. 80–81, pl. 17.1–4; Rolley 1986, p. 102, fig. 71 (as Corinthian). For the Vix krater's attribution to the Lakonian workshop, see A. Rumpf, "Krater Lakonikos," *Charites: Festschrift für Ernst Langlotz* (Bonn, 1957), pp. 127–135. For the attribution of the Trebenište krater and related appliqués to Corinth, see K. Hitzl, op. cit., pp. 10–11, with additional bibl.

—M. T.

19
Handle of a Vessel

Greek, circa 500 B.C.
Bronze; H: 25 cm
Condition: Tail of left satyr missing; minor restoration on edge of upper volutes; volutes that attach basal palmette to satyrs reinforced; crusty pale green patina.
Provenance: Perhaps from Ionia.

This substantial vertical handle was probably one of a pair cast for attachment to a large amphora or storage vessel. The openwork relief decoration at the handle base depicts two satyrs holding up between them an enormous kantharos; this suggests that the contents of the vase would have been wine since these half-horse, half-human followers of the god of wine, Dionysos, were infamous for their love of drink. Perched rather precariously on a pair of scrolls above a pendant palmette, the satyrs each have one hoof balanced on a central spine and a bent knee on an outer volute. Their complementary poses are basically the same, each extending his left hoof and arm to the center with the right knee bent back and the right arm raised to support one upright handle of the kantharos. This composition allows the left figure to expose his muscular frontal torso and erect penis to the viewer, while the right figure displays his back and his bushy horse's tail.

In its repetition of floral motifs, the handle as a whole evokes a sense of organic growth. The broad expanse of the central grip seems to grow out of the spiraling tendrils and small central palmette that spring from the bowl of the kantharos carried by the satyrs. These small scrolls, with relief oculi, are

19

echoed at the top of the handle by two much larger volutes, finished on both sides, that originally extended horizontally along the rim of the vessel to form a secure join. Small half-palmettes border the central rib at the point where it separates into the two stems, and a large palmette crowns the top of the handle, its fronds overlapping the elegant console that originally curled over the rim of the vessel. Pointed relief oculi extend from the centers of the volutes on either end. The plant motifs on the underside of the top of the handle are also finished with remarkable care. There, two scrolls develop from the thick stems of the freestanding volutes that extended along the rim, spiraling outward to frame a delicate five-frond palmette that grows between them.

The ornament of this handle, in particular the palmettes and fleshy volutes, has been associated with the workshops of Ionia on the coast of Asia Minor. In its original publication, the handle was compared with the simpler, non-figural handle said to be from Miletos, now in the Museum of Fine Arts in Boston (62.1105 a: Comstock/Vermeule 1971, p. 294, no. 420), that shows a similar treatment of the floral elements.

BIBLIOGRAPHY: Michael Ward, Inc., *Form and Ornament: The Arts of Gold, Silver and Bronze in Ancient Greece*, exh. cat. (New York, 1989), no. 6.

—M. T.

20

Olpe

Greek, circa 500–480 B.C.
Bronze; H (OVERALL): 19.9 cm; (TO THE RIM): 18 cm; DIAM (AT LIP): 4.5 cm
Condition: Intact, abrasion above the *nu* of ΗΙΠΠΑΡΧΟΝ in the inscription and on the base below this area; smooth blackish green patina on body, areas of crusty light green patina on handle and rim.

This elegant biconical serving vessel belongs to the class of oinochoai Shape 5B, as described by J. D. Beazley, and to T. Weber's Shape III B in his classification of bronze jugs. Both the base and the shoulder are articulated with fine ridges that give precise definition to the structure of the vase shape, but only the rim and handle are ornamented. A fine beaded molding surrounds the rim and the rounded lip is decorated with an ovolo pattern. The high-arched handle, U-shaped in cross section, is attached at the top to the underside of the lip by means of a leaf-shaped plate secured by two rivets. At its lower end, the handle broadens into the head and fore-paws of a pendant lion's skin for attachment just below the offset shoulder. Weber dates the production of the bronze jugs of Shape III B to the later sixth century and the first half of the fifth, mostly on the basis of comparisons with ceramic versions, considering 525–475 B.C. to be the primary period. The majority of his list of thirty-one examples of the type are from sites in Greece, but several have been found in South Italy, with singletons in North Africa, Serbia, and Bulgaria.

Among Weber's examples of the III B shape, the closest parallel for the profile of the swelling lower body of the Fleischman piece appears to be the elaborately incised olpe in Mainz (Römisch-Germanisches Zentralmuseum O.28 840). Although it has lost its handle, this piece does show a similarly ornamented rim and lip. Similar to the Mainz vase in both profile and decoration is an olpe in Karlsruhe (Badisches Landesmuseum F 596), which does preserve its handle, high-arching and U-shaped in cross section with a lion's skin at the base. On the basis of the development of the shape, following Beazley's tradition, "from stout to slender," these vases may be dated around 500 B.C. or in the early years of the fifth century.

Inscriptions on the handle and just below the shoulder of the vessel add another dimension to its importance. On the handle is incised: ΤΟΙΝ ΔΙΟϟΚΟΡΟΙΙ ([dedicated] to the Dioskouri [with the dual case used]). As the excavations in the Altis of Olympia have proved, olpai were important vases in ritual offerings, and the dedication to the Dioskouroi on the handle indicates that this vase was also intended as a votive gift. Just beneath the shoulder the dedicator is named: ΑΛΚΙΜΑΧΟϟ ΗΙΠΠΑΡΧΟΝ (Alkimachos, holding the office of hipparch). The only non-Attic letter form in the inscription is the X, being + in Attic script and ↓ here. This may indicate an Eretrian origin.

The connection among the knights (ἱππεῖς), the *hipparch* (the elected general of the cavalry [ἵππαρχος]), of which there were two in Athens, and the immortal Dioskouroi (the sons of Zeus, Kastor and Polydeukes, who were often represented as horsemen) is evidenced by a mid-fifth century B.C. dedication by the knights of two slightly under life-size sculptures of men leading horses outside the entrance to the Athenian acropolis. Two Roman Imperial inscriptions for the rededication of these sculptures survive as well as one original inscription from the fifth century B.C., which provided the model for the other two. The inscriptions on the bases name the knights as dedicators, list three hipparchs, and name Lykios of Eleutherai, son of Myron, as the sculptor. Pausanias, who saw the two Roman rededications (XXII.4), exhibits a certain confusion in describing these statues. He is unable to decide between identifying them as the sons of Xenophon (one of the hipparchs named) or as horsemen of merely decorative value. Very likely they represented the Dioskouroi. Although only one inscription survives from the fifth century, there is no reason to suppose that there were not originally two sculptures as in the Roman rededication. Given the close connections between Athens and Eretria, and the widespread belief in the helpfulness of the Dioskouroi in battle and at sea, a dedication to the Dioskouroi by a hipparch in Eretria, if that is indeed the origin of this olpe, at the turn of the sixth century is not entirely unlikely, although no evidence has yet appeared from the excavations at Eretria for similar dedications to the Dioskouroi. But the presence in Eretria at that date of a leading citizen named Alkimachos is attested. Herodotus (VI.101: Εὔφορβός τε ὁ Ἀλκιμάχου) remembered his son, Euphorbos, as one of the two men who betrayed the city to the Persians not long after this olpe was dedicated. (The authors thank Professor A. E. Raubitschek for his observations on the inscriptions and their importance.)

BIBLIOGRAPHY: Unpublished.

RELATED REFERENCES: For the olpe Shape 5B, see *ARV*^{2}, p. l; T. Weber, *Bronzekannen,* Archäologisches Studien 5 (Frankfurt, 1983), pp. 148–174, 372–389, pl. xiv; idem, "Eine spätarchaisch-korinthische Bronzeolpe in Mainz," *Archäologischer Anzeiger* 2 (1983), pp. 187–198, esp. fig. 1. On the significance of the olpai found at Olympia, see T. Weber, *Archäologischer Anzeiger,* op. cit., esp. p. 188, n. 10, and p. 196. On the identification of the statues as Dioskouroi, see *Pauly's Realencyclopädie der classischen Altertumswissenschaft,* vol. 13, pt. 2, p. 2293, s.v. "Lykios" (G. Lippold); U. Thieme and F. Becker, *Allgemeines Lexikon der bildenden Künstler,* vol. 23, p. 492, s.v. "Lykios" (M. Bieber); W. Judeich, *Topographie von Athen* (Munich, 1931), p. 229, n. 1. For a discussion of the inscriptions on the statues, see A. E. Raubitschek, *Dedications from the Athenian Akropolis* (Cambridge, Mass., 1949), nos. 135, 135 a, b. For a summary and bibliography of excavations at Eretria, see P. Auberson and K. Schefold, *Führer durch Eretria* (Bern, 1972), passim; for the closeness of Athens and Eretria, p. 29ff. For a recent discussion of the Dioskouroi, see T. Lorenz, "Die Epiphanie der Dioskuren," *Kotinos* (Mainz, 1992), pp. 114–122.

—M. T./K. H.

20

21

21

Statuette of a Goat Rising to Its Feet

Northern Greek, circa 500–480 B.C.
Bronze; H: 14.2 cm; L: 19 cm
Condition: Left ear, point of left horn, and right back hoof missing; left horn appears to have been twisted out to the side and left rear hoof is bent upward against the body; irregular light green and brown patina.
Provenance: Formerly in the collection of George Ortiz, Geneva; said to have been found in Thessaly (together with cat. no. 22).

Half-sitting, half-reclining, this finely cast animal seems to be in the process of rising from the ground. It turns its head sharply back to the right, the deep-set eyes wide open, as if alert to some noise or approaching danger, and its whole posture betrays a nervous tension. The accurate depiction of the momentary pose is contradicted somewhat by the detailed decorative treatment of the outward-curling horns, the fringe of fur around the beast's forehead, and the neat, well-combed beard, all of which reflect the lingering conventions of the Archaic style, as does the regular pattern of long fringed fur down the spine. However, the refinement of the execution and the expressive, sensitive treatment of the head suggest that this figure must be dated toward the end of the Archaic period.

Figures of reclining goats with their heads turned back were popular subjects for bronze casters in the Archaic period, and many show similar, though less precise treatment of the horns, fringed fur, and beard. None of the known examples is as large as the Fleischman animal, however, or appears to be shown in the act of standing. Neither does any known example show the same attention to surface details.

The small tab attached to the beast's left haunch and the short metal plate that connects the front hooves are both perforated and were used to attach the figure to another surface, most likely the rim of a tripod or a large, open vessel. The rim of a handleless krater from Trebenište (see Filow) preserves a smaller figure of a reclining goat, demonstrating the widespread popularity of the motif, but this animal conveys nothing of the impressive dignity of the Fleischman goat. Although made as an ornamental attachment, together with the lion described in the following entry (cat. no. 22), with which it is said to have been found, this figure surely ranks among the finest examples of late Archaic animal sculpture that have survived from antiquity.

BIBLIOGRAPHY: Unpublished.

RELATED REFERENCES: B. Filow, *Die archaische Nekropole von Trebenischte am Ochrida-See* (Berlin, 1927), p. 53, no. 69, figs. 52–54. *Master Bronzes*, no. 63 (from Vassar College, no acc. no.), with references to other examples.

—M. T.

22

Statuette of a Seated Lion

Northern Greek, circa 500–480 B.C.
Bronze; H: 13.4 cm; L: 18.5 cm
Condition: Intact; irregular, greenish patina.
Provenance: Formerly in the collection of George Ortiz, Geneva; said to have been found in Thessaly (together with cat. no. 21).

Like the goat in the preceding entry (cat. no. 21), with which it is said to have been found, this imposing, beautifully executed image of a seated lion could easily have functioned as an independent work but was intended to serve a decorative purpose. Holes through both front paws and the left rear paw indicate that it was once fastened to the rim of a large tripod or open vessel. The head turned slightly to the right and the coiled tail together contribute an aspect of tension to the pose, while the ears, laid back against the head, the slightly open mouth, and the deep, wide-set eyes give the head a menacing expression entirely appropriate to a beast of prey. Only the stylized, flamelike locks of the mane and the symmetrical whiskers (or creases) rendered like palmette fronds on the snarling muzzle reflect the love of ornamental patterns typical of the Archaic style, while the elongated, muscular body and threatening visage reflect the artist's developing interest in a more naturalistic depiction of the creature, most appropriate to a date in the early fifth century B.C.

This solid-cast figure belongs among lions of the "Late Archaic unified type," as defined by H. Gabelmann in his study of early Greek lion images, and incorporates oriental features such as the

22

stylized whiskers (or creases) on the muzzle and the flame-shaped locks particularly common among the Ionian examples. A fine fragmentary marble lion's head from Ephesos, now in the British Museum (B 140: Pryce; Brown; Gabelmann), shows similar sensitivity to the structure of the head as well as a comparable treatment of the folded ears, though they are less flattened, and of the radiating locks around the face, which do not form a truly separate ruff that stands away from the rest of the mane. This last feature was considered by Gabelmann to be a point of distinction between Ionian lions and the lions of Greece and Etruria, as the manes of Ionian lions tend to be more unified. In his study of Etruscan lions, W. Brown noted that the folded ear flattened in anger and the petal-shaped folds on the muzzle ultimately derived from Assyrian images that were often reflected in Ionian art. The reported northern provenance for the Fleischman lion would make sense as its place of manufacture also, as the northeastern workshops were often influenced by Ionian style.

BIBLIOGRAPHY: Unpublished.

RELATED REFERENCES: H. Gabelmann, *Studien zum frühgriechischen Löwenbild* (Berlin, 1965), pp. 91–105; for the lion's head from Ephesos, pp. 92, 121, no. 132, pl. 27.3. F. N. Pryce, *Catalogue of Sculpture in the Department of Greek and Roman Antiquities of the British Museum,* vol. 1, pt. 1 (London, 1928), p. 63, fig. 71. For the Assyrian sources of conventions in the depiction of lions, see W. Brown, *The Etruscan Lion* (Oxford, 1960), p. 15; for the Ionian lion's mane, pp. 91–92; for the marble lion's head from Ephesos, p. 76, pl. LXIII a.

—M. T.

23
Statuette of a Maiden Dressed in a Peplos

Argive, circa 460–450 B.C.
Bronze; H: 16.1 cm
Condition: Intact except for loss of proper left foot and forepart of right foot; also missing is the element the figure originally supported; dull, rather uniform blackish patina.

The calm austerity of this peplos-clad female is well suited to her purpose as a support, either for a thymiaterion or a candelabrum. She stands in perfect repose, her weight on her engaged right leg and her right hand resting on her hip. Her more relaxed left leg steps forward, breaking the columnar folds of the peplos, and her arm, bent sharply at the elbow and raised before her, is held beneath its overfold, with the muffled hand raising the fabric near the neckline. The soft fabric of the sakkos that covers her head is patterned with circles and zigzags and forms a series of soft ridges down the back of the head. The top of the head supports a segmented stem, which is hollow at the top for the attachment of some functional element.

The type of attachment on the head suggests that this figure supported a thymiaterion or a candelabrum rather than a mirror. A similarly clad peplophoros in Delphi (7723: Rolley) supports a large open bowl on her head, though there is no stem between the head and bowl of the thymiaterion. Closer in terms of the attachment are a kore from Skutari in Albania (now Louvre MNB 2854: Weber) and a kore from Metapontum that preserves a small thymiaterion on its tall segmented stem (Rolley 1986, pp. 128–129, figs. 107, 108; *Ortiz Collection*). Another peplophoros in the National Archaeological Museum, Athens (6490: Tölle-Kastenbein) supports the tall, undecorated stem of a candelabrum.

The pose of the figure, with her left hand hidden beneath the overfall of the peplos, is known in three examples of bronze mirror caryatids, the finest one in the National Archaeological Museum, Athens (7576: Karouzou); one in Copenhagen (Danish National Museum 102: Congdon; Tölle-Kastenbein); and one in Mariemont (Musée de Mariemont G. 95: Congdon; Tölle-Kastenbein). As S. Karouzou indicated in her discussion of the Athens example, J. Six was the first to point out that this pose was found in the figure of Eriphyle in the painting of the *Nekyia* (Descent into the Underworld) by Polygnotos as described by Pausanias in his guide to the sanctuary of Delphi (x.29.7–8), and Karouzou herself suggested that it probably did not originate in bronze sculpture. The heavy features of the Fleischman figure and her confident pose, similar to those of the other three examples, must reflect a monumental Classical original. It has been suggested that the pose with the left hand on the hip was introduced by the sculptors of Argos. On the basis of the treatment of her peplos' drapery, her heavy facial features, lantern jaw, and broad neck, this figure seems to fit quite well in the Argive workshop. Furthermore, the edge of the peplos overfall on a caryatid figure in the British Mu-

23

23

seum (188), which Tölle-Kastenbein connects with the Argive workshop, presents a pattern similar to that found on the sakkos of the Fleischman piece (though the figure is dated earlier).

BIBLIOGRAPHY: Unpublished.

RELATED REFERENCES: C. Rolley, *Les Statuettes de bronze: Fouilles de Delphes V* (1969), pp. 155–160, no. 199, pls. XLIV–XLVII. *The George Ortiz Collection: Antiquities from Ur to Byzantium*, exh. cat. (Berne, 1993), no. 124. T. Weber, "Eine spätarchaisch-korinthische Bronzeolpe in Mainz," *Archäologischer Anzeiger* 2 (1983), p. 195, fig. 9. R. Tölle-Kastenbein, *Frühklassische Peplosfiguren Originale* (Mainz, 1980), pp. 106–107, no. 14b, pl. 74 (Athens candelabrum); pp. 43–46 (Copenhagen mirror); pp. 107–108, pl. 66 (Mariemont mirror); p. 235, no. 42a, pl. 164 (British Museum caryatid). S. Karouzou, "Attic Bronze Mirrors," *Festschrift for David M. Robinson* (St. Louis, 1951), pp. 565–587. L. Congdon, *Caryatid Mirrors of Ancient Greece* (Mainz, 1981), pp. 201–202, no. 93, pls. 90, 91 (Copenhagen mirror); p. 174, no. 63, pl. 59 (Mariemont caryatid). For the pose of Eriphyle in the *Nekyia* of Polygnotos, see J. Six, "Die Eriphyle des Polygnot," *AM* 19 (1894), pp. 335–339. For the Argive style, see E. Langlotz, *Früh Griechische Bildhauerschulen* (Nuremberg, 1927), pp. 54–67; S. Karouzou, *National Museum Illustrated Guide* (Athens, 1984), pp. 114–118.

—M. T.

24
Rear Handle of a Kalpis

Greek, circa 450–425 B.C.
Bronze; MAX. H: 17.5 cm
Condition: Intact, but surface much worn; glossy greenish brown patina.

This elegant handle was originally attached upright to the back of a kalpis. The convex circular attachment plate that originally held the top of the handle to the neck of the kalpis is decorated with a pattern of

24

radiating tongues and separated by a beaded collar from the fluted shaft of the handle grip. At its base, the plain circular root of the handle is enhanced by the floral elements and the frontal figure of a siren executed in openwork relief. This mythological hybrid creature, a human-headed bird, stands with her wings spread over a symmetrical complex of five-frond palmettes and spiraling tendrils that terminates at the center of the base in a pendant nine-frond palmette. The spaces among the scrolling tendrils are open, and the entire composition would have provided a lacy effect when applied to the polished surface of the vessel's body.

Approximately sixty examples of kalpis handles with attachments decorated with frontal sirens over openwork floral patterns are now known, indicating that this motif must have been a very popular one in the mid-fifth century B.C. Similar siren-scroll compositions are found at the bases of oinochoe handles (see Comstock/Vermeule 1971, pp. 296–298, no. 423 [acc. no. 99.481]), and sirens with spread wings also serve as mirror supports and finials (see *Master Bronzes,* pp. 92–93, no. 88 [Walters Art Gallery 54.769]). No doubt the mythological figure with wings spread gracefully suited the need for a broad area of attachment, especially at the base of a handle or beneath a mirror disk. E. Diehl divided her list of thirty-seven bronze kalpides decorated with sirens into sub-categories based on the hairstyles of the female-headed sirens and the treatment of the wings. The Fleischman example falls into Diehl's Group C—sirens with hair parted in the center of the forehead and rolled around a fillet with long locks hanging down over the shoulders. These pieces are generally dated to the third quarter of the fifth century B.C.

BIBLIOGRAPHY: Unpublished.

RELATED REFERENCES: For siren handles on bronze hydriae, see E. Diehl, *Die Hydria* (Mainz, 1964), pp. 34–39, 219–220; D. von Bothmer, Review of *Die Hydria* by Diehl, *Gnomon* 37 (1965), pp. 599–608, esp. p. 603 (20 additional examples); E. Hofstetter, *Sirenen im archaischen und klassischen Griechenland* (Würzburg, 1991), pp. 144–151, pl. 32; see also a handle in the White Levy Collection, New York: D. von Bothmer, ed., *Glories of the Past*, exh. cat. (New York, 1990), p. 108, no. 89, and Michael Ward, Inc., *Form and Ornament: The Arts of Gold, Silver and Bronze in Ancient Greece*, exh. cat. (New York, 1989), no. 10. For siren attachments on a mirror, see O. Waldhauer, "Ein Askos aus der Sammlung Chanenko in Kiew," *Jahrbuch des Deutschen Archäologischen Instituts* 44 (1929), esp. pp. 262–263; fig. 23 (Hermitage 12959); Hofstetter, op. cit. For a discussion of the relationship between siren mirror supports and hydria attachments, see P. Oberländer, *Griechische Handspiegel* (Hamburg, 1967), pp. 150–178.

—M. T.

25

25
Tritemorion Weight with Amphora

Attic, second half of fifth century B.C.
Bronze filled with lead; H: 5.2 cm; W: 5.3 cm; DEPTH (OF BACKGROUND): 1.5 cm; WEIGHT: 277 g (ATTIC TRITEMORION)
Condition: Intact; glossy dark green patina.

The weight, a thick, square plaque, bears the representation of an amphora in high relief. The underside is plain except for an off-center rectangular cutting into which lead has been poured to adjust the weight. The Greek inscription T ᑭ I (tritemorion, "a third") in large incised letters runs from top to bottom beside the vase at the left. A stamp of a standing owl within a vaguely triangular incuse area has been applied twice, in the lower right corner of the top and in the middle of the right edge, where it is flanked by two large dotted letters, a *delta* and an *epsilon.*

That this weight is bronze, rather than the cheaper and more easily altered lead, indicates that it had official status and was kept in a public office (presumably in the Agora, given the two owl countermarks).

The Panathenaic Type I amphora recalls those on sixth-century *Wappenmünze* (coins with weapons [shields] as devices) and is earlier than the fourth-century vessels depicted on the tritemoria found in the Agora. The script with the triangular ᑭ and the prereform E instead of the later H (ΔE standing for ΔEMOΣION) is also early. Furthermore, the image of the *glaux* (the perched owl of the Athenian tetradrachms) used here (as countermarks) indicates that the weight cannot be later in date than 400 B.C. The weight of 277 grams is not a tritemorion of the fourth-century Attic mina of 450 grams but rather an exact third of the earlier mina, or stater, of 840 grams. Our outstanding and extremely rare bronze weight thus belongs to the fifth century B.C.

BIBLIOGRAPHY: J. Neils, *Goddess and Polis* (Hanover, N.H., 1992), pp. 51, 191, no. 70.

RELATED REFERENCES: *Dictionnaire des antiquités grecques et romaines d'après les textes . . .*, vol. 4, pp. 548–549, s.v. "Pondus" (E. Michon). E. Babelon and A. Blanchet, *Catalogue des bronzes antiques de la Bibliothèque Nationale* (Paris, 1895), no. 2234. E. Pernice, *Griechische Gewichte* (Berlin, 1894), nos. 44, 45 (lead weights in London and Berlin). M. Lang and M. Crosby, *Weights, Measures and Tokens*, vol. 10 of *The Athenian Agora* (Princeton, 1964), nos. LW 17–21 (for third-stater lead weights with amphoras of later shape). On countermarked weights and the status of bronze weights, see J. H. Kroll, "Three Inscribed Greek Bronze Weights," *Studies Presented to George M. F. Hanfmann* (Mainz, 1971), pp. 87–93. For an amphora *Wappenmünze*, see C. T. Seltman, *Athens* (Cambridge, England, 1924), no. 9; K. Schefold, *Meisterwerke griechischer Kunst* (Basel, 1964), no. 421. For Athenian bronze weights, see Y.-E. Empereur, *Bulletin de Correspondance Hellénique* 105 (1981), pp. 538–542.

—L. M.

26

27

26
Weight with Volute Krater

Greek, last quarter of fifth century B.C.
Bronze; H: 2.8 cm; W: 2.4 cm; DEPTH: 0.5 cm; WEIGHT: 272 g
Condition: Intact, somewhat worn.

The weight has a volute krater in high relief. The underside has been hollowed out to adjust the weight. The intended weight of 272 grams represents one thirtieth of the early Attic mina, or stater, and thus corresponds to five Persian silver sigloi of the period.

The device of a volute krater is seen on silver coins from Thebes, staters struck around 400 B.C. (see Kraay). The known Theban weights, however, depict Boeotian shields.

BIBLIOGRAPHY: Unpublished.

RELATED REFERENCES: C. M. Kraay, *Archaic and Classical Greek Coins* (Berkeley and Los Angeles, 1976), no. 359. For several Theban square-shaped quarter minas with shields, see J. F. de Rochesnard, *Album des poids antiques,* vol. 2 (n.d.), p. 35.

—L. M.

27
Statuette of a Reclining Satyr

Greek (or possibly South Italian), circa 450–425 B.C.
Bronze; L: 8.4 cm
Condition: Solid-cast; intact except for missing tail(?); irregular green and brown patina.

Half-seated, half-reclining, this nude satyr rests back with his ankles crossed, apparently supporting himself on his left elbow. He turns to his left, his right arm drawn across his chest and his right hand rather coyly tucked beneath his chin, and extends his open left palm, as if to a missing second party. Around his head he wears a fillet. His sinewy body, completely human in form, is modeled with extraordinary attention to detail, both front and back. Only the balding domed forehead, the low-placed, goatlike ears, and flattened nose betray the creature's bestial nature. The large, expressive features are typical of later fifth-century satyrs.

The Fleischman figure is the twin of a figure formerly in the Pomerance Collection and now in a private collection in London (see Merrin Gallery), and it is likely that both were part of the same original composition. Although the figures are carefully modeled on all sides and show no obvious signs of attachment, they may have once rested on the handles of a large vessel. The small X's incised on the proper right back of both figures may have been marks for placement, as these are otherwise difficult to explain.

Reclining satyrs were popular figures for the decoration of vessel rims and handles. An Archaic trio from Olympia seem to be a similar combination of sitting and reclining (Olympia Museum B 4232, B 4200: Herfort-Koch). The conceit of the crossed legs is found in a small Greek satyr now in St. Louis (City Art Museum 95:65: *Master Bronzes,* no. 95). However, the features of the Fleischman figure and its duplicate seem to find their closest parallel in a small bronze satyr found at Himera in Sicily and now in Palermo (see Jantzen). Although this dancing figure does not show the same attenuated proportions, its anatomy is carefully executed and the treatment of the head, with furrowed brow, large expressive features and snub nose, as well as the low ears and flattened left hand, in particular recalls the Fleischman and ex-Pomerance examples. These figures have consistently been associated with Greece, however, and it is possible that the Palermo bronze was an import; it is also possible that the Fleischman and Pomerance pieces were exported for donation at some Greek sanctuary.

BIBLIOGRAPHY: Unpublished.

RELATED REFERENCES: Merrin Gallery, *The Majesty of Ancient Egypt and the Classical World,* exh. cat. (New York, 1986), p. 25. M. Herfort-Koch, *Archaische Bronzeplastik Lakoniens,* Boreas Beiheft 4 (1986), pp. 119–120, pl. 21.1–3 (Olympia B 4232), 4 (Olympia B 4200), nos. K 147, K 148. U. Jantzen, *Bronzewerkstätten in Grossgriechenland und Sizilien,* Jahrbuch des Deutschen Archäologischen Instituts Ergänzungsheft 13 (Berlin, 1937), p. 55, pl. 24. For reclining satyrs in banquet settings, see J.-M. Dentzer, *Le Motif du banquet couché dans le Proche-Orient et le monde grec du VII ième au IV ième siècle avant J.-C.,* BEFAR 246 (Rome, 1982), pp. 217–218.

—M. T.

28

28
Plaque with Head of Pan

Greek, second half of fourth century B.C.
Bronze; H: 6.7 cm; W: 7.05 cm
Condition: Pieces missing from edge at upper right corner and under right edge of Pan's neck; tip of nose dented.

The god Pan appears as a beautiful, wild-looking youth with pointed animal ears. Small horns emerge from his long, windblown hair. His eyes have the engraved irises often seen before the Roman period in metalwork and in small objects made of materials that could not be painted. The plaque is armor-weight, robust but not too encumbering. It is punched all around its edge with little holes for sewing on a lining, probably of leather, to strengthen and cushion it. The shape of the piece suggests that it may have functioned as a *pteryx* (one of the flaps along the lower edge of a cuirass).

In typology and stylistic feeling the head recalls the young Pan on a superb box mirror in the Metropolitan Museum of Art (25.78.44 a–d: Mertens). As instiller of panic in the enemy, Pan is an appropriate subject for the decoration of armor. A cuirass from Laos in the Reggio Calabria Museum (see Lattanzi), datable in the second half of the fourth century, is adorned in front by a head of Pan and in back by a youthful, animal-eared head not unlike ours in style but without horns. If the identification of our plaque as a pteryx is correct, it would be interesting as an unprecedentedly early relief-decorated example (D. Cahn, verbal opinion).

BIBLIOGRAPHY: Unpublished.

RELATED REFERENCES: J. R. Mertens, "Greek Bronzes in the Metropolitan Museum of Art," *Metropolitan Museum of Art Bulletin* 43 (1985–86), no. 29. E. Lattanzi, *Il Museo Nazionale di Reggio Calabria* (Rome and Reggio Calabria, 1987), p. 130f.

—A. H.

29

Box with Relief-decorated Lid

Greek, mid- to late fourth century B.C.
Bronze; H: 4.7 cm; W: 8.95 cm; DEPTH: 6.4 cm
Condition: Essentially intact.

The box is a miniaturized version of the large gold ossuary chests from the royal tombs at Vergina (see Andronikos, figs. 135, 155), which imitate the construction of wooden models (see Vaulina/Waşowicz). Two pegs on the front side of the box correspond to two on the edge of the hinged lid and must have been used for ties to hold the little container shut. The top is decorated with a repoussé relief, separately made and soldered in place. In a rocky landscape, a nude, bearded silenos sits on the ground, holding a large-topped thyrsos. In front of him is a gnarled tree trunk, recalling those conspicuous in the painted hunt scene on the façade of the "Tomb of Philip" at Vergina (see Andronikos, figs. 57–63). The figural relief, in a style familiar from mirror covers, armor, and relief vessels of the fourth century B.C., seems to have been attached to the box in antiquity. However, like the relief made for armor but applied to a mirror cover now in Boston (see Comstock/Vermeule 1971, no. 367), it might originally have been intended for some other use.

29

BIBLIOGRAPHY: Unpublished.

RELATED REFERENCES: M. Andronikos, *Vergina: The Royal Tombs and the Ancient City* (Athens, 1984), figs. 135, 155. M. Vaulina and A. Waşowicz, *Bois grecs et romains de l'Ermitage* (Wroclaw, 1974), nos. 3, 12.

—A. H.

dictionary, is well attested on such pottery (see Webster/Green). These parallels suggest that the roundel was made in Apulia and is datable at the beginning of Hellenistic times.

BIBLIOGRAPHY: Unpublished.

RELATED REFERENCES: Association Hellas et Roma, *Art grec insolite: Terres cuites hellénistique de Grande Grèce dans les collections privées genevoises*, exh. cat. (Geneva, 1988), no. 25. E. Berger, ed., *Kunstwerke der Antike*, exh. cat. (Lucerne, 1963), no. E4. Münzen und Medaillen A.G., Basel, *Classical Antiquities*, auct. cat. 14, June 14, 1954, no. 16. T. B. L. Webster and J. R. Green, *Monuments Illustrating Old and Middle Comedy*, BICS Suppl. 39 (1978), p. 14, type A, pl. VIII, fig. a (Bari, Museo Archeologico 3279).

—A. H.

30

30
Roundel with Comic Mask

West Greek (?), end of fourth–beginning of third century B.C.
Bronze; DIAM: 6.9–7 cm
Condition: Losses from border below tip of beard; holes in front of hair and beard on proper right side; two horizontal bronze bands applied to the back, each fastened with two rivets and shaped to allow insertion of a rather thick, narrow (1.8 cm) strap running vertically.

Within a delicate double-cable border, a comic mask, worked in fairly high repoussé relief, is shown in three-quarter view against a plain background. The mask has full, straight hair brushed back from a receding hairline, a thick mustache, and a small pointed beard. The quill-like locks are finely chased. The eyebrows are raised, causing the forehead to pucker, and the eyes, with engraved irises and pierced pupils, roll sideways to stare out at the beholder. Their impish expression is enhanced by crow's feet. The full lips are slightly parted and are shown naturalistically, rather than as the mouthpiece of a mask with a "real" mouth inside.

Thick enough to be robust but not too heavy, the roundel probably decorated a bridle like those worn by Canosan terracotta horses (see *Art grec* and checklist no. 242 below). A set of such bridle ornaments, but in silver and with ideal heads in the roundels, was found at Taranto and is now divided between Basel, Princeton, and Boston (see Berger). A bearded comic mask seen in three-quarter view, quite similar to the one on our roundel, appears on a type of fourth-century Tarantine antefix (see *Classical Antiquities;* other examples are in the Taranto Museum). Our piece's air of elfin guile recalls the most charming of the theatrical masks on Gnathia vases. The mask type, perhaps the "sphenopogon" cited by Pollux in his

SILVER

31
Group of Four Drinking Vessels

Greek, second half of fourth century B.C.
Silver

A. Bowl

H: 6 cm; DIAM: 9.6 cm; WEIGHT: 138.5 g
Condition: Intact.

The bowl was probably raised (although casting or a combination of techniques cannot be ruled out) and then finished with chasing and lathe turning. Forty-six flutes radiate from a composite floral on the underside consisting of three overlapping rosettes of eight petals each. The floral was certainly hammered because its contours show on the inside. On the shoulder is a band of Lesbian kymation ornament set off with beading; a third line of beading runs below a plain band at the widest point of the body of the bowl. The designs are especially crisp and well preserved, and the edge of the rim is sharp.

This form, often called an Achaemenid bowl, is one of the more common shapes from fourth- and third-century B.C. tombs in Macedonia (where they have been excavated at Stavroupolis, Nikisiana, Sedes, Derveni, and Vergina), in Bulgaria, and even in south Russia, at Karagodeuashk. But this geographical distribution of find spots speaks more for local burial customs than for sites of manufacture. Bowls of this design were imitated in black-glazed pottery in Athens and other localities. Since some of the Attic versions have relief heads centered on the interiors as emblemata in exactly the same manner as in some of the silver bowls found in Macedonia, one can argue that Athenian potters were copying locally produced Achaemenid-style bowls.

BIBLIOGRAPHY: Michael Ward, Inc., *Form and Ornament: The Arts of Gold, Silver and Bronze* (New York, 1989), no. 14.

RELATED REFERENCES: On silver Achaemenid bowls, see M. Pfrommer, *Studien zu alexandrinischer und grossgriechischer Toreutik frühhellenistischer Zeit* (Berlin, 1987), pp. 234–236, pls. 43–44. On ceramic black-glazed Achaemenid bowls, see B. A. Sparkes and L. Talcott, *Black and Plain Pottery*, vol. 12 of *The Athenian Agora* (Princeton, 1970), pp. 121–122, 285, nos. 691–695, pl. 28; M. Pfrommer, op. cit., pp. 214–215, 218–219, pl. 46.

B. Oinochoe

H: 14.8 cm; WEIGHT: 230 g
Condition: Handle missing; part of rim bent; circular dent on shoulder with traces of solder where handle would have been attached.

The oinochoe was shaped and decorated by hammering. On the lower body are two sets of overlapping nymphaea nelumbo petals recognizable by their down-turned tips. The petals are set off by convex moldings, two below, one above. On the offset between the shoulder and neck is an egg-and-dart molding. On the underside of the foot, within a convex molding, is an eight-pointed floral rosette. The edge of the foot is rounded, and the edge of the rim is sharp.

Two oinochoai with nymphaea nelumbo petals on the body and palmettes or scrolls on the shoulder are known from burials in Macedonia: one in Salonica from Tomb B at Derveni (see *Treasures of Ancient Macedonia*), and another, fragmentary one in Kavalla from Nikisiana (see Lazaridis et al.). Closely related is a silver jar in Istanbul (see Pfrommer 1985) found at Kirklareli in Thrace in 1892, which features a triple overlapping rendition of nymphaea nelumbo petals.

The discovery at Rogozen, Bulgaria, in 1985, of a hoard of 165 silver vessels of which 54 were jugs of this shape (see Katintscharov et al.) has caused one to ask where jugs like these were made and for what principal customers. On close inspection, most of the jugs from Bulgaria, particularly those with figural decoration, have a provincial look, as if made locally or at least with an inland Thracian market in mind. Among the non-figured jugs only ten are decorated with nymphaea nelumbo petals, and of these only four are done in a style that makes them a match for the Fleischman oinochoe.

This oinochoe shape is derived from Achaemenid metalwork, specifically jars or amphorae that were remodeled by Greek craftsmen and outfitted with a single handle. The pattern of nymphaea nelumbo petals, though widely used and associated with Hellenistic metalwork from Egypt, is likewise of Achaemenid origin and is here too incorporated in Greek work. Despite the Macedonian and Thracian provenances of the silver oinochoai with nymphaea nelumbo petals, they need not have been made in Macedonia and could equally well have been made in Athens or another Greek city.

BIBLIOGRAPHY: Michael Ward, Inc., *Form and Ornament: The Arts of Gold, Silver and Bronze* (New York, 1989), no. 14.

RELATED REFERENCES: Archaeological Museum of Thessaloníki, *Treasures of Ancient Macedonia* (Athens, [1978]), p. 64, no. 197, pl. 26. D. Lazaridis et al., *The Tumulus of Nikisiani* (Athens, 1992), p. 26, pl. 10 top. M. Pfrommer, "Ein Bronzebecken in Malibu," *GettyMusJ* 13 (1985), pp. 14–15, fig. 6. R. Katintscharov et al., *Der thrakische Silberschatz aus Rogozin Bulgarien* ([Sofia],[ca. 1990]), esp. nos. 143, 145–147.

C. Ladle

L: 25.5 cm; DIAM (OF BOWL): 6.5 cm; WEIGHT: 202 g
Condition: Junction of bowl to handle cracked.

The ladle was probably cast and then hammered to its finished form. The broad rim of the bowl curves inward. There are minimal moldings at the junction of the bowl to the handle. The handle is flat, its inner face plain, the outer

31A

31B

31C
31D

face grooved at the edges. The top of the handle is looped and ends in a duck's head with incised features. Ladles of this design were standard items in tableware from as early as the fourth century B.C. to the early Roman Imperial period.

BIBLIOGRAPHY: Michael Ward, Inc., *Form and Ornament: The Arts of Gold, Silver and Bronze* (New York, 1989), no. 14.

RELATED REFERENCES: On Hellenistic ladles in general, see M. Castoldi and M. Feugère, "Les simpulums," in M. Feugère and C. Rolley, eds., *La Vaisselle Tardo-Républicaine en Bronze,* Université de Bourgogne, Centre de Recherches sur les techniques gréco-romaines, no. 13 (Dijon, 1991), pp. 61–88.

D. Strainer

MAX. W: 21.6 cm; DIAM (OF BOWL): 10.5 cm; WEIGHT: 156 g
Condition: Intact; one edge of bowl bent.

The strainer was probably cast and then hammered, cut out, and turned on a lathe to achieve the final desired form. The edge of the bowl has a perfectly formed set of concentric moldings surely made by turning the object on a lathe. A whirligig pattern of holes is punched in the bowl. The handles are in the form of duck's heads with sinuous necks springing from trapezoidal plaques, which are made in one piece with the bowl and are edged with curlicues and engraved with palmettes on the upper surface.

Several varieties of silver strainers are known. Most have a broad flange around the perforated bowl and a pair of matching duck's-head handles. Others have a loop in place of one handle (see *Ancient Macedonia*), or merely one handle (see Pfrommer 1983), while still others have bowls of different design, one even in the shape of a grape leaf (see Oliver). Two-handled strainers are also known in bronze (see Mertens).

Silver strainers like the Fleischman piece have been found over a wider geographical area than is generally thought to be the case, although it is true that the majority have a Macedonian provenance. In addition to five in the Salonica Museum, from Poteidaia, Derveni (Tomb B), Stavroupolis (see Thessaloníki), and Vergina (from two tombs, see Andronikos), and one in the Walters Art Gallery reported to be from Kavalla, there are two found further afield. One in Istanbul, found with a silver kylix and pitcher, came from Kastamonu in Asia Minor in 1900, and one in Boston came from Meroe, Sudan, from the tomb of King Arakakamani, who reigned from 315 to 297 B.C.

BIBLIOGRAPHY: Michael Ward, Inc., *Form and Ornament: The Arts of Gold, Silver and Bronze* (New York, 1989), no. 14.

RELATED REFERENCES: Greek Ministry of Culture and International Cultural Corporation of Australia, *Ancient Macedonia,* exh. cat. (Athens, 1988), p. 297, no. 248 (from Dion). M. Pfrommer, "Ein Silberstreifen," *GettyMusJ* 11 (1983), pp. 135–139, figs. 1, 6. A. Oliver, Jr., *Silver for the Gods* (Toledo, 1977), p. 47, no. 16 (from Minneapolis). J. R. Mertens, "A Hellenistique Find in New York," *Metropolitan Museum Journal* 11 (1976), pp. 71–73, 78–80, figs. 4–5, 15–17. Archaeological Museum of Thessaloníki, *Treasures of Ancient Macedonia* (Athens, [1978]), p. 74, no. 281. M. Andronikos, *Vergina: The Royal Tombs and the Ancient City* (Athens, 1984), pp. 148, 211, figs. 108, 178. For silver strainers like the Fleischman piece, see A. Oliver, Jr., op. cit., p. 45, with bibl.

—A. O.

TERRACOTTA VASES

32

Plastic Vase in the Form of a Female Bust

Rhodian, circa 580 B.C.
Terracotta; H: 10.3 cm; W: 6.25 cm; DEPTH: 5.9 cm
Condition: Rim broken, otherwise intact.

The bust of a woman is depicted on this plastic vase. She is topped by a small vase-neck with a disk mouth. Typical of Rhodian plastic vases, the mouth is decorated with black dots, placed in between the four rays. The artist has also incorporated the standard Archaic facial features seen on vases produced in Rhodes at this time—a long face, broad forehead, long nose, and large eyes. The hair falls in three long locks over the front of each shoulder and hangs loose on the shoulders in back, with each lock ending in a point, reminiscent of Archaic korai. Attention to detail is seen in the rendering of the mantle's texture by modeled vertical undulations. As on other vases of this type, the woman is adorned with earrings, the paint on which is now lost, a necklace, and a headband.

Plastic vases, vessels in the form of humans, animals, and plants, were mass produced and exported at many centers from the end of the seventh century to the middle of the sixth century B.C. The Fleischman vase, like others that were mass produced, was molded in two pieces. It belongs to the main type of Rhodian female bust vases, known as "C: Series normale" (see Ducat). Similar vases can be found

32

in the British Museum (1608, 1609, 1610: Higgins), and in the Antikenmuseum Berlin (30732: Heilmeyer). These objects would, most likely, have held scented oils, for use by women.

BIBLIOGRAPHY: Unpublished.

RELATED REFERENCES: J. Ducat, *Les Vases plastiques rhodiens* (Paris, 1966), pp. 33–35, nos. 5, 6, pl. IV. R. A. Higgins, *Catalogue of the Terracottas in the Department of Greek and Roman Antiquities, British Museum* (London, 1975), pp. 7–15, pls. 5, 6. W. D. Heilmeyer, *Antikenmuseum Berlin: Die Ausgestellten Werke* (Berlin, 1988), p. 51, fig. 10. M. I. Maximova, *Les Vases plastiques dans l'antiquité* (Paris, 1927).

—E. B. T.

33
Black-figured Lip Cup

Attributed to the Workshop of the Phrynos Painter [J. Haldenstein]
Attic, circa 550 B.C.
Terracotta; H: 15 cm; W: 29 cm; DIAM (OF BOWL): 20.9 cm; (OF FOOT): 9 cm
Condition: Reconstructed from large well-preserved fragments.

The shape of this lip cup, decorated by a Little-Master painter, is typical of the class, a cup with a straight lip set at a slight angle to the bowl, which is supported by a tall stem with a broad flat base. The decorative scheme also closely follows what was usually painted on a lip cup: figural decoration consisting of one or two figures on an otherwise reserved lip, a line at the top of the handle zone, an inscription between the handles, and a reserved band lower on the cup wall. Rather than the expected palmettes at either side of the handles, here there are panthers heraldically arranged, facing back toward each other across the broad expanse of the cup. Added red was used for the manes of the panthers, the manes and some interior markings of the lions, and the hair and beard of Herakles. The miniaturist style of the figures, which are confined to a single zone, exemplifies the work of a group of sixth-century vase painters known as the Little Masters.

On either side of the cup Herakles wrestles the Nemean lion. Crouching, he holds the lion around its upper body and by one forepaw. The lion has its other forepaw over Herakles' shoulder. By turning the lion's head away from Herakles, the painter has with minimal overlapping tightened the composition and provided a pleasing parallel treatment of the heads of the two figures. The two sides of the cup are distinguished from one another only by slight variations in the positions of the two figures: on one side the lion stands on its left rear leg with its right paw clawing at Herakles' shoulder, on the other the lion stands on its right leg and no paw is visible. Directly below, in the handle zone, is a nonsensical inscription applied solely for its decorative value. It, like the figural composition, is repeated with only slight variation. Such inscriptions—often injunctions to the drinker to enjoy the contents of the cup—are

33

not infrequently found next to the makers' signatures. On the lip cup in London from which the Phrynos Painter takes his name, we find both the potter's signature, "Phrynos made it," and such a note to the drinker. As on the Fleischman piece, next to the handles on either side are panthers, three of which raise their forepaws toward the handles.

This fine Little-Master cup has been attributed to the workshop of the Phrynos Painter, who decorated the exquisite cup in London with Herakles' entry to Olympos and the Birth of Athena, which was signed by Phrynos as potter. The Phrynos Painter also decorated a Gordion cup, a Type B amphora, and a neck amphora of the Botkin Class. This painter shares points of contact with the Painter of the Boston Polyphemos and the BMN and Oakeshott Painters, and it is in this stylistic sphere that the Fleischman cup belongs.

BIBLIOGRAPHY: Unpublished.

RELATED REFERENCES: On the Phrynos painter, see *ABV*, pp. 168–169, 688; *Para*, pp. 70–71; *Beazley Addenda*², p. 48; H. Brijder, "A Band-cup by the Phrynos Painter in Amsterdam," *Stips Votiva: Papers Presented to C. M. Stibbe* (Amsterdam, 1991), pp. 21–30; J. T. Haldenstein, "Four Attic Black-figured Cups at the Elvehjem," *Elvehjem Museum of Art, University of Wisconsin-Madison Bulletin/Annual Report* (1989–91), pp. 8–9, 12, nn. 14–22. On Little-Master cups in general, see K. Vierneisel and B. Kaeser, eds., *Kunst der Schale: Kultur des Trinkens* (Munich, 1990), pp. 18–181.

—K. H./J. R. G.

34
Black-figured Amphora (Type B)

Attributed to the Painter of Berlin 1686
[D. von Bothmer]
Attic, circa 540 B.C.
Terracotta; H: 42 cm; W (AT HANDLES): 27 cm; DIAM (OF BODY): 27.2 cm; (OF MOUTH): 18 cm; (OF FOOT): 13 cm
Condition: Reconstructed from fragments but essentially complete; breaks retouched, some chipping along edge of rim.

Depicted in both panels is a scene from one of Herakles' twelve labors—the theft of the cattle of Geryon. Herakles was ordered by Eurystheus to steal Geryon's cattle from the island of Erytheia on the extreme western edge of the world. In order to do so, Herakles had to slay the herdsman Eurytion, the two-headed dog Orthros, and finally the triple-bodied warrior, Geryon himself. In the scenes painted here, Herakles faces this last line of defense. Nonsense inscriptions are placed in the field on both sides.

Painted with nearly identical images on both sides, this vase is attributed to an artist who has paired scenes in a similar manner on at least six other vases. J. D. Beazley (1951) categorizes the Painter of Berlin 1686 as an "able painter of the second rank," whose work is equal to or better than most of the vases in Group E. The work of the Painter of Berlin 1686 is closest stylistically to that of the Princeton Painter, and the Type B amphora seems to have been his favorite vase shape; at least twenty are recorded.

Among the scenes compiled by P. Brize of Herakles battling Geryon with his club, this vase is clearest in composition; in most of the others, subsidiary figures are present. The figure shown most frequently filling the space between Herakles and Geryon is Eurytion, who crouches or sits. On the Fleischman vase, this space is filled by an eagle flying toward the eventual victor.

It is difficult to assign a primary and secondary side to the amphora as the scenes are nearly identical. The tripod device on Geryon's shield might be one determining factor, and the vase has been published with this side labeled as A, but for use of incision and added color, the opposite side is the more ornamental. An object of interest on the more ornamental side is the quiver that replaces the plain scabbard which Herakles carries on the opposite side.

BIBLIOGRAPHY: Atlantis Antiquities, *Greek and Etruscan Art of the Archaic Period* (New York, 1988), pp. 38–40, fig. 35 (side A illustrated in reverse), p. 41 (detail, panel of side B).

RELATED REFERENCES: For the Painter of Berlin 1686, see *ABV*, pp. 296–297, 692; *Para*, pp. 128–129; *Beazley Addenda*², pp. 77–78; J. D. Beazley, *The Development of Attic Black-figure* (Berkeley and Los Angeles, 1951), p. 73; J. Maxmin, "A New Amphora by the Painter of Berlin 1686," in E. Böhr and W. Martini, eds., *Studien zur Mythologie und Vasenmalerei* (Mainz, 1986), pp. 35–40. For representations of Herakles battling Geryon, see P. Brize, *Die Geryoneis des Stesichoros und die Frühe Griechische Kunst* (Würzburg, 1980); T. H. Carpenter, *Art and Myth in Ancient Greece* (London, 1991), pp. 126–127; *LIMC*, vol. 5, pt. 1, p. 73ff., esp. p. 77, nos. 2492–2497, s.v. "Herakles and Geryon" (P. Brize). For vases with repetitive scenes, see K. Schauenburg, "Zu Repliken in der Vasenmalerei," *Archäologischer Anzeiger* (1977), pp. 194–204. For a detailed discussion of amphorae, see M. B. Moore and M. Pease Philippides, *Attic Black-figure Pottery*, vol. 23 of *The Athenian Agora* (Princeton, 1986), pp. 4–18.

—K. W./J. R. G.

34 SIDE A

34 SIDE B

35
Black-figured Amphora (Panathenaic Shape)

Attributed to the Three-line Group [D. von Bothmer]
Attic, circa 530 B.C.
Terracotta; H: 29 cm; W (AT HANDLES): 17.3 cm; DIAM (OF MOUTH): 11.9 cm; (OF BODY): 18.2 cm; (OF FOOT): 10.4 cm
Condition: Intact except for a few small surface chips; small amount of surface encrustation.

The wedding procession of Alcestis (AΛKEϟTEϟ [retrograde]) and Admetos (AϟMETOϟ) adorns the principal side of this amphora. Admetos, king of Thessaly, was fated to die young, but Apollo won an extended period of life for him if he could find someone to die for him. He approached his aged parents, but they declined; only his wife Alcestis was brave enough to give her life for his. In the play of the same name by Euripides, Alcestis is lauded: "And by her bravery in death, she has been a credit—no, a glory—to her sex" (lines 623–624).

On this vase, the couple stands in a four-horse chariot, Admetos in a chiton and mantle holding the reins, Alcestis veiled in a dress decorated with incised squares enclosing circles. Behind them is a female figure in a patterned dress similar to Alcestis' with a mantle over her left shoulder. Perhaps she is Peitho (Persuasion), often present in wedding scenes to instruct the bride in her new role. Behind the horses, but in front of the chariot, stands Apollo (AΓOΛON), holding a *kithara* (lyre). Facing him is Artemis (APTEMIϟ), wearing a patterned dress and a polos and holding her hand up in front of her face in a gesture of greeting. Standing in front of the horses, facing the bride and groom, are two female figures, one wearing a dress decorated with squares enclosing alternate circles and dotted saltires, the other wrapped in a mantle decorated with a pattern of crosses and swastikas and holding flaming torches aloft in both hands. Their pairing and the torches suggest that these two may be Demeter and her daughter Persephone, or, by analogy with the splendid and markedly similar hydria in Florence (*ABV*, p. 260, no. 30), they could be Aphrodite and Semele. The Florence vase depicts the wedded pair Peleus and Thetis in the chariot, while Dionysos stands behind, accompanied by Thyone (Semele). An inscription on the Florence vase identifies a female figure hidden by the horses as Aphrodite. We might expect Aphrodite, as the goddess of sexual love and marriage, to be present here also. Behind the two female figures, also facing the bridal couple, stands Dionysos, identified by his ivy crown and the long branch of ivy he carries in his left hand. Dionysos is accompanied by a boy attendant wearing only a short mantle, who serves as a transition figure to the scene on the other side of the vase.

The scene on the reverse of the vessel, while unrelated iconographically to the obverse, is associated by type, featuring a four-horse quadriga centered and frontal. The heads of the driver and his passenger, both looking left, are just visible above the heads of the

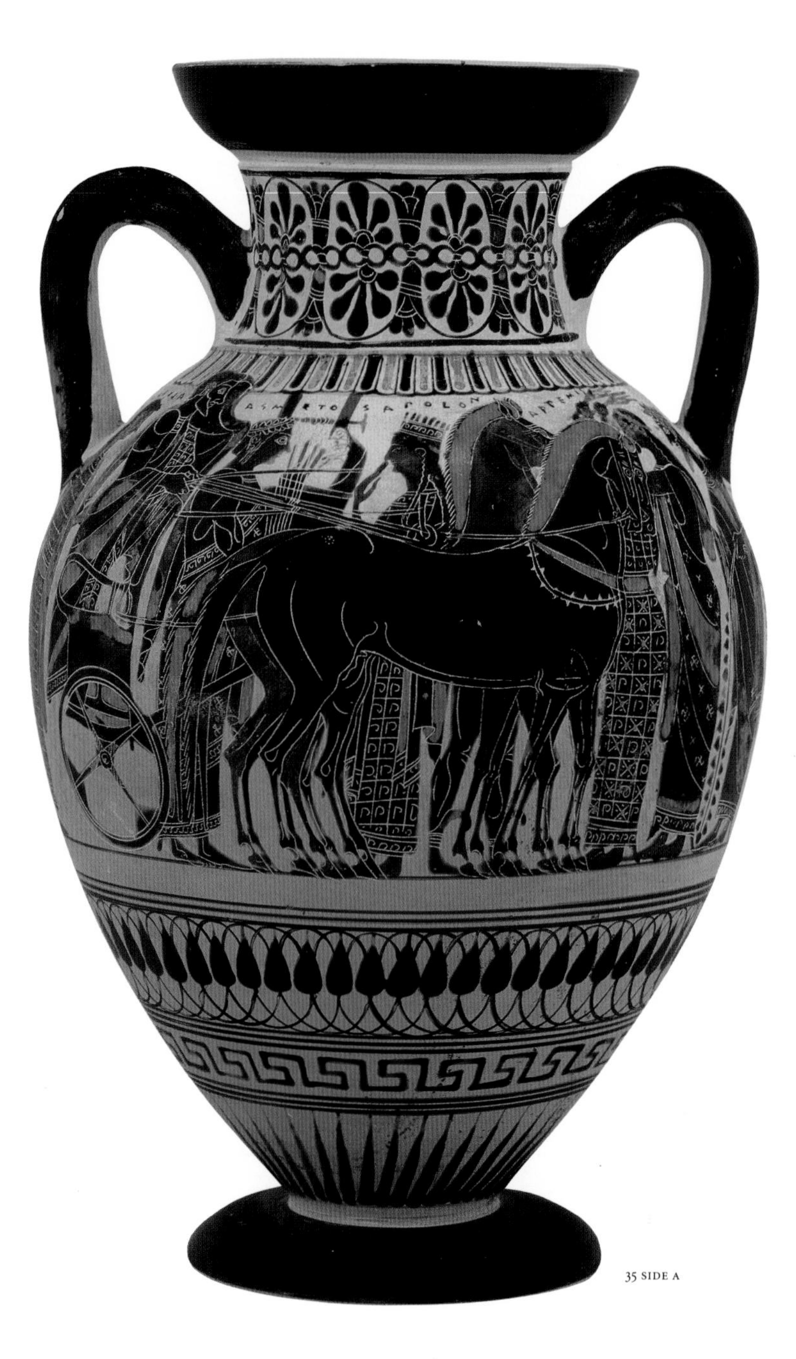

35 SIDE A

SIDE B

35

SIDE A–B

SIDE B–A

horses. Two pairs of figures facing inward flank the quadriga. At the left of the quadriga stand a female figure in a plain-bordered dress and a rosette-decorated mantle and a youthful male figure, beardless with long hair rolled up, wearing a mantle wrapped about his hips and carrying a spear in his right hand. To the right of the quadriga stands a female figure wearing a plain-bordered dress under a voluminous striped himation. Her hair is worn long at the back, short in the front with long tendrils in front of her ears. Behind her stands a boy attendant wearing a short chiton and mantle and holding a spear. His long, rolled-up hair is held in place by a fillet.

Most depictions of Alcestis show some aspect of her special deed of sacrifice, either dying or being led back from the underworld by her rescuer in the myth, Herakles. The marriage of Admetos and Alcestis is an uncommon theme in the iconography of vase painting. Other than this vase, the wedding procession is presented only once on fragments of an Attic red-figured loutrophoros by the Methyse Painter in the Acropolis Museum in Athens (*ARV*², p. 632, no. 1). A good parallel for the quality of the Fleischman piece is a work in Boston (MFA 01.8059: *CVA* Boston) attributed by D. von Bothmer to a painter of the Three-line Group, circa 530 B.C.

This amphora has an echinus (convex curved) mouth, round handles, and an echinus foot. There is liberal use of added red on alternate folds of the mantles, wreaths, horses' manes, details of the chariots, and alternate tongues on the shoulder. Added white, mostly not preserved, was used for the flesh of all female figures, some of the decoration of the clothing, the coat of the right pole horse, and a dotted circle with a central dot brand on the rump of the right trace horse. There is a reserved groove at the junction of the mouth and neck and a raised red fillet at the junction of the neck and shoulder. The top of the mouth, the insides of the handles and the junction of the body and foot are reserved. The glaze does not extend to the base of the foot, which is reserved on the underside.

The neck of the amphora is decorated with a palmette and lotus chain of four and one-half pairs on each side. On the shoulder are tongues in alternating red and black. Below the figured zone are three lines, a chain of upright lotus buds, three lines, a leftward-facing key, three lines, and at the base of the body, rays. The triple lines between these patterned friezes are a distinguishing characteristic of the Three-line Group to which this vase belongs.

BIBLIOGRAPHY: Unpublished.

RELATED REFERENCES: *CVA* Boston 1, pp. 26–27, pls. 36, 38.1,2. For the Three-line Group, see *ABV*, pp. 320–321, 693–694; *Para*, pp. 140–141; *Beazley Addenda*², p. 86; *ABFV*, p. 112. For other depictions of Alcestis, see *LIMC*, vol. 1, pp. 533–544, s.v. "Alkestis" (M. Schmidt). For mythical weddings, see B. Schweitzer, *Mythische Hochzeiten: Eine Interpretation des Bilderkreises an dem Epinetron des Eretriameisters* (Heidelberg, 1961). For a recent bibliography on brands on horses, see A. Clark, *CVA* Malibu 1, p. 28.

—J. B. G./J. R. G.

36

Black-figured Psykter

Attributed to the Lysippides Painter [J. R. Guy]
Attic, circa 530s B.C.
Terracotta; H: 33 cm; DIAM (OF MOUTH): 12 .8 cm; (OF BODY): 26 cm; (OF FOOT): 13.4 cm
Condition: Reconstructed from fragments; some losses in lower part of body and top of stem; profile complete from mouth to foot; top of mouth and reserved resting surface under foot show considerable wear.

Arrayed around the body of this wine cooler are two frontal quadrigae, each holding a charioteer and a warrior. One of the chariots is flanked by Scythian archers with tall caps, bows, and *gorytoi* (combination quiver and bow-case); the other by three warriors. The exterior of the rim is decorated with an addorsed ivy tendril; descending tongues, alternating black and red, encircle the top of the shoulder. Above the base of the stem is a hanging lotus-bud chain in which the odd-man-out is clearly visible below the charioteer on side B. Much of the detail is incised. Added red is used for the hair of the Scythians and for details of the weapons, chariots, and horses. White is used on the gorytoi, helmet crests, and shield devices.

This is the earliest complete psykter in Attic pottery, dating before the psykter in the de Menil collection (see Hoffmann). The lip of the Fleischman psykter is like that of the de Menil piece, but the base is not. There are no parallels for the band of hanging lotus buds on the foot, although a lidded psykter in Brussels has rays (A1321: Drougou).

36 SIDE A

36

SIDE A–B

SIDE B–A

SIDE B

Although attributed to the Antimenes painter by D. von Bothmer (personal communication of July 1988), this vase may be better compared to a fragmentary psykter by the Lysippides Painter in Frankfurt (144: *Para,* p. 116, no. 44 quater; Schauenberg).

Images of Scythians on Attic black-figured vases have been much discussed. G. Pinney summarizes the various theories and, along with K. W. Welwei, dismisses M. F. Vos's interpretation of their appearance on Attic black-figured vases as corresponding to the arrival of Scythian archers in Athens between 540 and 530 B.C., where they allegedly functioned as a specialized army corps in the service of the Athenian state until about 500 B.C. Pinney believes that when depicted on Attic black-figured vases, Scythians should be viewed as minor characters. On vases painted during the 530s, the role of the Scythian archer as squire becomes established, and indeed this is the role they play on the Fleischman psykter as they attend to the trace horses on one of the chariots. Pinney also discusses the relationship between Achilles and Scythia, and following her argument it is tempting to identify the chariot the Scythians flank as that of Achilles, but there are no inscriptions to assist in the identification of any of the figures on the psykter.

Painted representations of psykters in use are ambiguous as to function, but it is generally accepted that the psykter held wine and floated inside a larger vessel filled with water or snow. Lugs are present on the shoulders of some, an indication that the vessels may have been lowered into wells to chill their contents. A long-handled ladle was used to remove the chilled wine from the psykter to a drinking vessel; such an action is depicted on the exterior of a red-figured cup in Vienna (Kunsthistorisches Museum 137: *CVA* Austria).

BIBLIOGRAPHY: Unpublished.

RELATED REFERENCES: H. Hoffmann, *Ten Centuries That Shaped the West* (Houston [?], 1971), p. 378ff., no. 175. S. Drougou, *Der Attische Psykter* (Würzburg, 1975), p. 19, no. B3, pl. 13. 1. K. Schauenburg, "Ein Psykter aus dem Umkreis des Andokidesmalers," *Jahrbuch des Deutschen Archäologischen Instituts* 80 (1965), pp. 76–104. *CVA* Austria 1, Kunsthistorisches Museum 1, pl. 1:2. For the treatment of the mouth and in part for the martial subject, see a dinos in Salerno: F. Lisserrague, "Figures of Women," in *From Ancient Goddesses to Christian Saints,* vol. 1 of *A History of Women in the West* (Cambridge, Mass., 1992), p. 151, fig. 6. On psykters, see S. Drougou, op. cit. For Scythians, see K. Wernicke, "Die Polizeiwache auf der Burg von Athen," *Hermes* 26 (1891), pp. 51–75; M. F. Vos, *Scythian Archers in Archaic Attic Vase-painting* (Groningen, 1963); K. W. Welwei, *Unfreie im antiken Kriegsdienst,* vol. 1 (Wiesbaden, 1974); G. Ferrari Pinney, "Achilles, Lord of Scythia," in W. G. Moon, ed., *Ancient Greek Art and Iconography* (Madison, 1983), pp. 127–146.

—K. W./J. R. G.

37
Black-figured Volute Krater

Attributed to the Leagros Group [J. R. Guy]
Attic, circa 510–500 B.C.
Terracotta; H: 58.5 cm; (WITHOUT VOLUTES): 51 cm; DIAM (OF LIP): 45 cm; (OF FOOT): 24.5 cm
Condition: Reconstructed from fragments; a few losses in figural decoration especially on side B; two bronze rivets, remains of an ancient repair, visible on side B; some losses due to wear and surface abrasion on rim and neck of side B.

The top frieze on side A of this krater shows the battle between the gods and the giants. The two running protagonists near the center are nearly hidden in the dense composition with its overlapping figures, a condition made worse by the missing fragments. On our right is Herakles, wearing the Nemean lion's skin and a fiercely determined countenance as he races with shield and long spear toward his opponent. Opposite him, the giant in full armor seems about to hurl his lance. The complexity of the composition seems to have overwhelmed the painter: the giant is seen mainly behind but partly in front of the quadriga that shares his place in the frieze. Among the figures around these two, we can identify for certain only Hermes, the running figure to our far left, and Athena, in the chariot near Herakles. Beneath the quadrigae are large white boulders.

The lower frieze on side A shows a series of athletic contests or training scenes: from our left are two runners in armor, a runner, a diskos thrower, a javelin thrower, wrestlers, pankratists, and a nude figure holding a fillet or a weight. Scattered among these figures are numerous clothed trainers, and at our far right is a pillar with a fillet tied around it, perhaps the turning-post for the foot races. The center, more or less, of this frieze is punctuated by a standing figure playing the double flutes, more prominent for the strong white meander that decorates his clothing. The most serious of these events, toward the right end of the frieze, seems to be the *pankration* (a combination of wrestling and boxing in which vir-

37

37 SIDE A, DETAILS

tually nothing was disallowed and the fight continued until one contestant gave up). Both contestants bleed from their chests, and the one on the ground profusely from his nose; perhaps the finger he holds up is his sign of capitulation.

On side B, the top frieze is a chain of horizontal palmettes. The lower frieze shows the departure of warriors. In the center are two quadrigae, their drivers carrying shields on their backs and being bid farewell by women standing behind them. At either side are fully armed warriors, female figures, and finally horsemen, each carrying two lances and wearing a white petasos. At our far right is a lone figure wearing a pointed Scythian cap. Departure scenes are frequently encountered in Attic vase painting of the sixth century, though they are usually smaller and less complex than this one.

The body of the vase is entirely black except for rays above the red line marking the junction of the body and the foot of the vase. A reserved vertical edge sets off the heavy molding of the foot, and black and reserved tongues surround the bases of the handles. The edges of the volutes that spring from the handles to the rim are decorated with ivy, and a meander pattern runs the entire way around the rim. The neck is modeled in two bands offset from one another.

Added white is used throughout for female flesh, the large boulders, a column, and some details on clothing and implements. Added red is used for some hair and beards, interior details, and blood. Added red is also used along the outer edge of the rim, the edges of the volutes, and for wavy stalks of ivy chains on the outer faces of the volutes. The inside of the neck is glazed. A graffito, KE||||+V, on the reserved underside of the foot may be a price inscription. A. W. Johnston suggests that the first part of the inscription could be a monogram, KE or EK. As KE it might refer to the shape name *kelebe*, known from ancient sources as a large vessel with a wide mouth and two handles.

Although the Fleischman krater has been attributed by D. von Bothmer to the Acheloos Painter, it recalls in particular works by the Edinburgh Painter who, in J. D. Beazley's view, "stems from the Leagros Group and may even be counted as belonging to it" (*ABV*, p. 476). The neck fragment of a volute krater, once on the Basel market and now in Princeton (Art Museum, Princeton University 1987–5: Hitzl, p. 426, no. 149), bears comparison for style and date with the Fleischman vase, and it too is to be attributed to the Leagros Group. Also close to these is the impressive vase in the Faina collection, Orvieto (1955: Wójcik).

BIBLIOGRAPHY: Unpublished.

RELATED REFERENCES: K. Hitzl, *Die Entstehung und Entwicklung des Volutenkraters von den frühesten Anfängen bis zur Ausprägung*

37 SIDE B, DETAILS

des kanonischen Stils in der attisch schwarzfigurigen Vasenmalerei (Frankfurt, 1982). M. R. Wójcik, *Museo Claudio Faina di Orvieto: Ceramica attica a figure nere* (Perugia, 1989), pp. 171–174, no. 89. For the graffito, see A. W. Johnston, *Trademarks on Greek Vases* (Warminster, 1979), pp. 109, 201, figs. 5m, 5n, 14f (Type 9c). For the Leagros Group, see *ABV*, pp. 354–391, 665, 695–696, 715–716; *Para*, pp. 161–172, 519; *Beazley Addenda*², pp. 95–103; E. Moignard, "The Acheloos Painter and Relations," *Annual of the British School at Athens* 77 (1982), pp. 201–211. On the development of volute kraters, see K. Hitzl, op. cit.; D. von Bothmer, "Observations on Proto-Volute Kraters," in M. del Chiaro, ed., *Corinthiaca: Studies in Honor of Darrel A. Amyx* (Columbia, 1986), pp. 107–116; idem, Review of *Die Entstehung und Entwicklung des Volutenkraters* by K. Hitzl in *Gnomon* 57 (1985), pp. 66–71. On gigantomachies in Archaic black-figure, see M. B. Moore, "Lydos and the Gigantomachy," *American Journal of Archaeology* 83 (1979), pp. 79–99; idem, "Poseidon in the Gigantomachy," in M. B. Moore and G. Kopcke, eds., *Studies in Classical Art and Archaeology* (Locust Valley, N.Y., 1979), pp. 23–27; H. A. Shapiro, *Art and Cult under the Tyrants in Athens* (Mainz, 1989). For recent discussions of ancient athletics, see D. G. Kyle, "The Panathenaic Games: Sacred and Civic Athletics," in J. Neils, *Goddess and Polis* (Hanover, N.H., 1992), pp. 77–101; D. G. Kyle in *Mind and Body: Athletic Contests in Ancient Greece*, exh. cat. (National Archaeological Museum, Athens, 1989).

—K. H./J. R. G.

38

Black-figured Kylix (Type C)

Attributed to an artist close to the Theseus Painter [J. R. Guy]
Attic, late sixth century B.C.
Terracotta; H: 8.5 cm; W (AT HANDLES): 26.7 cm; DIAM (OF CUP): 19.7 cm; (OF FOOT): 8.35 cm
Condition: Reconstructed from fragments with areas of fill and some retouching of figures on both sides, especially in face of male figure on side B; severe secondary burning has discolored the glaze.

Simply decorated, this cup has sinuous tendrils under each handle with small upright pomegranates set between the handle roots. From each of these hangs a line on which another pomegranate is suspended. On one side, a man raises a *machaira* (broad-bladed knife) to cut up a tuna placed on an hourglass-shaped chopping block. Behind him is a *trapeza* (a table with thin curving legs), upon which are set already sliced steaks. The conical object on the ground might be a pilos, or perhaps the head of a fish. On the other side, a man strides to the right carrying a tuna. The rim of the bowl is reserved, and the inside is black except for a small reserved circle. Added red has been used for the fillets of both figures, for the slices of fish, and for the folds of the fish-chopper's garment.

Scenes of the butchering of fish or animals are rare in vase painting. Seven examples are: a black-figured oinochoe in Boston (MFA 99.257: *ABV*, p. 30, no. 25) showing a man with an assistant slaughtering a quadruped (only the rear legs are present); a black-figured olpe in Berlin (1915: *ABV*, p. 377, no. 247;

38 SIDE A, DETAIL

SIDE B–A

Durand) with two wreathed men preparing to cut up a tuna; a black-figured pelike in the Fondation Custodia in Parigi (3650: Durand) showing two men butchering a boar; a red-figured pelike by the Syleus Painter in Erlangen (486: *ARV*[2], pp. 250 [no. 21], 1639; Grünhagen) depicting a man chopping meat lying on a table assisted by a boy; a Boeotian red-figured pelike in Munich (2347: Sparkes), very close to the Syleus Painter pelike, showing a man chopping meat on a table; a South Italian bell krater recently on the London market (see Bielefeld; Christie's) with a large fish placed upon a table, a satyr holding a knife at left, and a woman pouring a liquid over the fish's body at right; and a Campanian krater in the Museo Mandralisco in Palermo (2: Bielefeld; Tullio) showing a butcher slicing a large fish atop a table for a customer, with another fish on the ground.

D. von Bothmer attributed the cup to the Painter of Nicosia C975, whose body of work consists of two other cups and four fragments of a third. However, despite direct correspondences in potting and ornament to the cups assigned to the Painter of Nicosia C975, the Fleisch-

SIDE B

man cup is a more polished product, in a finer style, and thus is closer to the unusual shallow cup skyphos of special type by the Theseus Painter (Agora P1383: *ABV*, p. 520, no. 31; *Beazley Addenda*², p. 129). The tendril ornament is ultimately derived from Kamiros palmettes, which provides a link with the origin of the shape in the Amasis Painter's workshop.

BIBLIOGRAPHY: Unpublished.

RELATED REFERENCES: J.-L. Durand, "Figurativo e processo rituale," *Dialoghi di Archeologia* 1 (1979), fig. 9 (Berlin); fig. 4 (Parigi). W. Grünhagen, *Antike Originalarbeiten der Kunstsammlung des Instituts* (Nuremberg, 1948), pl. 16. B. Sparkes, "Illustrating Aristophanes," *Journal of Hellenic Studies* 95 (1975), p. 132, pl. XVI(a). E. Bielefeld, "Ein Unteritalisches Vasenbild," *Pantheon* 24 (July/August 1966), p. 253, fig. 1 (London art market); p. 254, fig. 3 (Palermo). Christie's, London, sale cat., December 12, 1990, p. 39, lot 79. A. Tullio in *Museo Mandralisco* (Palermo, 1991), pp. 68–69, fig. 55. For the Painter of Nicosia C975, see *Para*, pp. 99–100; *Beazley Addenda*², p. 56; M. Moore and M. Pease Philippides, *Attic Black-figure Pottery*, vol. 23 of *The Athenian Agora* (Princeton, 1986), p. 314, no. 1825, pl. 117:1825. For the Amasis Painter, see *The Amasis Painter and His World* (Malibu, 1985); for the Oxford and Vatican cups, see D. von Bothmer, pp. 223–228, nos. 62, 63. For the palmette ornament, see B. Freyer-Schauenburg, "Kamiros-Palmetten," in J. Christensen and T. Melander, eds., *Proceedings of the Third Symposium on Ancient Greek and Related Pottery* (Copenhagen, 1988), p. 152ff. For the Syleus Painter pelike (Geneva, private collection), see *Hommes et dieux de la Grèce antique* (Brussels, 1982), pp. 243–244, no. 156; *La Cité des images* (Lausanne, 1984), p. 50.

—K. W./J. R. G.

39
Red-figured Cup (Type B)

Attributed to the Nikosthenes Painter and to the potter Pamphaios [J. R. Guy]
Attic, circa 510–500 B.C.
H: 12.4–13 cm; W: 43 cm; DIAM (AT RIM): 34 cm; (OF TONDO): 13.6 cm; (OF FOOT PLATE): 12 cm
Condition: Reconstructed from a few large fragments; complete except for small triangular loss through right lower leg of warrior and adjacent cloak end on far left of side A; surface well preserved with overall greenish glaze.

Within the interior tondo a youth wearing a cloak about his lower body and seated on a simple stool, binds the long thongs of his sandals. Hanging in the field are an aryballos and a sponge, symbolizing the gymnasium and exercise. On side A of the exterior, warriors assemble for maneuvers or for battle—as is so often the case in such generalized scenes, the purpose remains uncertain. Two figures, armed with helmet, spear, and shield, look on as their spear-bearing companions rein in spirited mounts. On side B satyrs frolic with maenads. Dionysos incites the revel with his potent gift of wine, signified by the oinochoe, drinking horn, and wineskin. Movement is rhythmic and graceful, the three pairs of figures finely balanced and drawn with a degree of care unusual for the Nikosthenes Painter.

The Nikosthenes Painter, prin-

39 SIDE A, DETAIL

39 SIDE B, DETAIL

39

cipally a decorator of cups, the majority of which are of mediocre quality, takes his name from three vessels other than cups, all of which are signed by the potter Nikosthenes. On cup surfaces, the Nikosthenes Painter collaborated regularly with the potter Pamphaios whom J. D. Beazley suggests may have been the younger partner of the potter Nikosthenes. Pamphaios, who signs often as potter, also shaped vases for the vase painters Epiktetos, Oltos, and Paseas. The Fleischman cup, which alone among the artist's abundant works bears fair comparison with a cup in London also signed by Pamphaios (E12: *ARV*[2], p. 120, no. 24), secures the attribution of both to the Nikosthenes Painter: one encounters the same sureness of line and composition, a refreshing attention to detail (notably in the use of dilute glaze), identical exterior ground lines of silhouetted palmettes, and tightness of potting.

BIBLIOGRAPHY: Unpublished.

RELATED REFERENCES: On the Nikosthenes Painter, see *ARV*[2], pp. 123–127, 1627; *Para*, p. 333; *Beazley Addenda*[2], pp. 175–176; B. Cohen, *Attic Bilingual Vases and Their Painters* (New York, 1978), pp. 489–493 (esp. A92 for a possible link to the potter Hischylos); M. Robertson, *The Art of Vase-painting in Classical Athens* (Cambridge, 1992), pp. 38–39. On Pamphaios, see H. Bloesch, *Formen attischer Schalen* (Bern, 1940), pp. 62–69; *ARV*[2], pp. 127–131; *Beazley Addenda*[2], pp. 176–177; B. Cohen, op. cit., passim. For a recent stimulating and controversial study of satyrs, see G. M. Hedreen, *Silens in Attic Black-figure Vase-painting* (Ann Arbor, 1992); also F. Lisserrague, "On the Wildness of Satyrs," in T. H. Carpenter and C. A. Faraone, eds., *The Masks of Dionysos* (Ithaca and London, 1993), pp. 207–220, with additional bibl.

—J. R. G.

40

40

Red-figured Amphora

Attributed to the Berlin Painter [J. R. Guy]
Attic, circa 480–470 B.C.
Terracotta; H: 47 cm; DIAM (OF MOUTH): 18.9 cm; (OF FOOT): 13.5 cm
Condition: Restored from fragments with numerous losses; greater part of neck on side A reconstructed; breaks in places retouched, but figure work largely free of in-painting; gap through middle of warrior on side A clearly visible.

The body of this vase is entirely black except for two figures that "float" without ground lines. A lunging hoplite is depicted on side A, in greaves, cuirass, and a pushed-back Corinthian helmet. On his left arm he holds out a large round shield, over which is draped a shield apron. Side B presents a Scythian fleeing to our right, glancing anxiously back over his shoulder. He is bedecked in typical Scythian costume—a tall cap that extends onto his shoulders with flaps at the back and sides of the neck, and tight, patterned garments: a long-sleeved jacket worn over a shirt, long leggings, and a short cloth with a separate front panel worn like a loincloth. On his feet are soft pointed shoes fas-

40 SIDE A

SIDE B

tened with laces. In his left hand he holds out a bow. A *machaira* (broad-bladed sword) is in his right hand, and attached to his waist is a *gorytos* (the combined quiver and bow-case used by peoples to the north and east of Greece). The long red beard and snub-nosed physiognomy characterize this figure as un-Greek in contrast to the hoplite on side A. For an interesting link between the Greek view of the half-human and the barbarian, one can compare here the elderly *komast* (reveler) on side B of the Berlin painter's neck amphora in the Getty Museum (86.AE.187: *ARV*², p. 200, no. 51; *Para*, p. 342; *Beazley Addenda*², p. 192) and the bearded komast on side A of the neck amphora in London (E266: *ARV*², p. 198, no. 21); both works are early in the painter's career, as is the Fleischman vase. Also similar are the facial features of the satyrs on the artist's name piece in Berlin (*ARV*², p. 196, no. 1) and on his Munich panathenaic amphora (*ARV*², p. 197, no. 9).

In contrast to the scenes on black-figured vases (see cat. no. 36), this illustration may be a reference to the Greek triumph over the Persians in 479 B.C., a recent event when this vase was painted. The Scythians, a more northern people, were allied to the Persians during that invasion.

Erratic relief contours are more extensive on side A of this vase. A lightly incised preliminary sketch is clearly visible. Dilute glaze was used for many details throughout, while added red was used sparingly: for the ties securing the ends of the warrior's shoulder flaps, for the quiver harness encircling the barbarian's waist, and for the forward edge of the his cap. Under the foot there is a graffito in Attic script ON, possibly a price inscription.

BIBLIOGRAPHY: Unpublished.

RELATED REFERENCES: On the Berlin Painter, see *ARV*², pp. 196–214; D. C. Kurtz, *The Berlin Painter* (Oxford, 1983); idem, "Gorgos' Cup: An Essay in Connoisseurship," *Journal of Hellenic Studies* 103 (1983), pp. 68–86; G. Ferrari Pinney, "The Nonage of the Berlin Painter," *American Journal of Archaeology* 85 (1981), pp. 145–158 (a controversial study of his earliest work); S. Matheson Burke and J. J. Pollitt, *Greek Vases at Yale* (Yale University Art Gallery, New Haven, 1975), pp. 50–51, no. 45; C. M. Robertson, "The Berlin Painter at the Getty Museum and Some Others," *Greek Vases in the J. Paul Getty Museum 1*, Occasional Papers on Antiquities 1 (Malibu, 1983), pp. 55–72; idem, *The Art of Vase-painting in Classical Athens* (Cambridge, 1992), pp. 66–83. For the representation of "barbarians" on Greek vases, see L. Schnitzler, "Darstellungen von Vorderasiaten auf griechischen Vasen," *Zeitschrift der deutschen morgenländischen Gesellschaft* 108 (1958), pp. 54–65; V. Zinserling, "Physiognomische Studien in der spätarchaischen und klassischen Vasenmalerei," in *Die Griechische Vase*, Wissenschaftliche Zeitschrift der Universität Rostock 16 (1967), pp. 571–575, pls. 128–130; D. Metzler, *Porträt und Gesellschaft: Über die Entstehung des griechischen Porträts in der Klassik* (Münster, 1971); G. Ferrari Pinney, "Achilles, Lord of Scythia," in W. G. Moon, ed., *Ancient Greek Art and Iconography* (Madison, 1983), pp. 127–146. For bibliographical references on the Scythians, see cat. no. 36 above. For information on the historical interaction of Greeks and Persians, see T. Hoelscher, *Griechische Historienbilder des 5. und 4. Jahrhunderts v. Chr.* (Würzburg, 1973), pp. 38–49; W. Raeck, *Zum Barbarenbild in der Kunst Athens im 6. und 5. Jahrhunderts v. Chr.* (Bonn, 1981). A. A. Barrett and M. Vickers, "The Oxford Brygos Cup Reconsidered," *Journal of Hellenic Studies* 98 (1978), pp. 17–24. On shield aprons, see E. R. Knauer, *Ein Skyphos des Triptolemosmalers*, BWPr 125 (1973), p. 9. For the graffito, see A. W. Johnston, *Trademarks on Greek Vases* (Warminster, 1979), pp. 226–227, fig. 12 (Type 10F).

—E. B. T./J. R. G.

41
White-ground Lekythos

Attributed to the Painter of Athens 1826 [D. Kurtz]
Attic; circa 460–450 B.C.
Terracotta; H: 29.1 cm; W: 9.3 cm; DIAM (OF MOUTH): 5.8 cm; (OF FOOT): 6.3 cm
Condition: Intact; some minor surface spalls and encrustation below handle; slip worn down to clay ground in places; forelocks of female flaked away where glaze was superimposed on second white.

On the front of this oil vessel, a young woman holding a sakkos and a draped youth with a walking stick face each other over a low stool. The youth offers the woman an egg, a love gift also symbolic of rebirth, which she accepts. A small bird floats above the stool, and a black lekythos hangs between the figures from the border pattern. A silhouetted palmette frieze decorates the shoulder, and the figural scene has a continuous meander border. The top of the mouth is reserved.

The vessel has been attributed by D. Kurtz to the Painter of Athens 1826, who was thought to have been a specialist who painted only lekythoi until quite recently when a white-ground oinochoe attributed to him appeared on the New York art market (see Hesperia Arts). He characteristically uses added white, so-called second white, outlining the area with dilute glaze, as can be seen here on the face and limbs of the young woman, as well as on the egg being exchanged. Some additional added pigment may have been lost from the drapery, which now appears only hastily sketched. His subject matter, with the exception of one vessel in Berlin, is re-

41

lated to the funeral or to domestic scenes. The anomaly (Berlin 3175: *ARV*[2], p. 747) is not as unusual as it might seem, for Demeter and Persephone are shown, and their connections to the underworld are readily apparent.

Lekythoi such as this were frequently given as grave gifts, and the imagery of grave stelai and mourning found on many reflects this funereal use. The somber pair shown here may well be preparing for a funeral themselves. The vessels held precious scented oils, and in some cases, false interiors were created so that a smaller amount of the costly fluid could be deposited with the deceased.

BIBLIOGRAPHY: D. Kurtz, "Two Athenian White-ground Lekythoi," *Greek Vases in the J. Paul Getty Museum 4*, Occasional Papers on Antiquities 5 (Malibu, 1989), p. 128, n. 68, no. 5 (as Summa Galleries 2031).

RELATED REFERENCES: Hesperia Arts Auction, Ltd., New York, sale cat., November 27, 1990, lot 33. For Attic white-ground lekythoi, see D. Kurtz, *Athenian White Lekythoi: Patterns and Painters* (Oxford, 1975). For the Painter of Athens 1826, see *ARV*[2], pp. 745–748, 1668; *Para*, p. 413; *Beazley Addenda*[2], p. 284; J. Boardman, *Athenian Red-Figure Vases: The Classical Period* (London, 1989), p. 130ff., fig. 253. For Greek funereal practices, see D. Kurtz and J. Boardman, *Greek Burial Customs* (Ithaca, N.Y., 1971).

—K. W./J. R. G.

42

Red-figured Oinochoe (Shape III, Chous)

Attributed to the Group of Boston 10.190 [C. Watzinger]
Attic, circa 440–430 B.C.
Terracotta; H: 17.3 cm; DIAM (OF BODY): 13.65 cm
Condition: Reconstructed from fragments; some losses, especially on front over figural scene, and on legs of crouching boy at left.
Provenance: Formerly in the Falcioni collection, Viterbo; Tyszkiewicz collection, Rome; Béhague collection, Paris.

On this wine jug three nude boys wearing spiky festal crowns play with *astragals* (knucklebones). The left and center figures hold their gaming pieces in their hands, while the right-hand boy, who gestures toward his playmates, has just tossed his marker to the ground. The action may depict a variety of the game in which contestants attempt to land their pieces in holes in the ground (see van Hoorn, p. 45). The figural panel is framed by a border of descending ovolo bands, above and below, and ascending chevrons at either side.

The Attic *chous* is a specific shape of oinochoe created especially for the three-day festival of the Anthesteria, held in the spring in honor of Dionysos. On the first day, the Pithoigia, the new wine was opened and tasted. On the second, the Choes, the three-year-old male children of Athens were enrolled in their fathers' *phratries* (kinship groups). Scenes painted on many choes depict children in their festal clothing, participating in various aspects of the festival or involved in childhood pursuits. A specific miniature variety of the vase was made especially for children, and this diminutive version is often found in burials, presumably of those who died in infancy, as crawling babies are depicted on a number of them. The Fleischman chous is of the standard size used by adults for the drinking contests that took place on the festival's second day (see Hamilton, p. 121, for a differing opinion). On the third, the Chytroi, vegetables prepared in cooking pots were offered to Hermes Psychopompos, who conducted the souls of the deceased to the underworld. On this day, the spirits of the dead were free to return from the underworld and roam among the living; they were sent back at the end of the day by the exhortations of the festal participants.

The imagery found on the Fleischman chous is light in tone and reflects the gaiety and various competitions surrounding the festival. The crowns worn by the boys imitate the headdresses worn by many of the torch-racers, participants in a ritual contest of relay teams who conveyed the sacred fire from one altar to another. The boys shown here, however, are seemingly too young for such a contest. Numerous other choes depict figures playing knucklebones; none are as fine as the Fleischman chous.

The Painter of Boston 10.190, one of J. D. Beazley's "Late-Fifth-Century Pot-Painters," is known from only a small group of three vessels (one fragmentary), all of which are choes. E. Buschor made

42

the original identification of the painter, and C. Watzinger added this vase to the group. The other two vessels in the group depict the *komos* (a revel), one with children (Berlin 2658: van Hoorn; Burn). Berlin 2658 displays the same descending ovolo borders as the Fleischman chous, and the treatment of the hair with descending tendrils as seen on Boston 10.190 (see van Hoorn; Burn) is also the same, as is the side border of ascending chevrons.

BIBLIOGRAPHY: G. Koerte, *Bullettino dell'Instituto di Corrispondenza Archeologica* 48 (1876), pp. 252–253; W. Froehner, *Collection d'antiquités du Comte Michel Tyszkiewicz* (Paris, 1898), p. 15, no. 18; C. Watzinger in A. Furtwängler and K. Reichhold, *Griechische Vasenmalerei*, vol. 3 (Munich, 1932), p. 331; *Mélanges d'archéologie et d'histoire* 14 (1894), pl. 4; A. Brueckner, *Polyklets Knöchelwerfer*, BWPr 77 (1920), p. 5; B. Schröder, *Der Sport im Altertum* (Berlin, 1927), p. 82, fig. 13; G. van Hoorn, *Choes and Anthesteria* (Leiden, 1951), p. 177, no. 884; J. Dorig, "Tarentinische Knöchelspielerinnen," *Museum Helveticum* 16 (1959), pp. 32 (n. 16), 34, fig. 3; Sotheby's, Monaco, *Antiquités et objets d'art: Collection de Martine, Comtesse de Béhague provenant de la Succession du Marquis de Ganay*, sale cat., December 5, 1987, pp. 126–127, lot 147; L. Burn, *The Meidias Painter* (Oxford, 1987), pp. 86, 102, no. B3; R. Hamilton, *Choes and Anthesteria: Athenian Iconography and Ritual* (Ann Arbor, 1992), p. 205, no. 884.

RELATED REFERENCES: G. Van Hoorn, op. cit., fig. 503 (Berlin 2658), fig. 85 (Boston 10.190). L. Burn, op. cit., p. 102, no. B1 (Berlin 2658), no. B2 (Boston 10.190). For the Group of Boston 10.190, see *ARV*², p. 1318; *Beazley Addenda*², p. 363. For choes and the Anthesteria, see H. W. Parke, *Festivals of the Athenians* (London, 1977), pp. 107–124; W. Burkert, trans. by J. Raffin, *Greek Religion* (Cambridge, Mass., 1985), pp. 237–242; A. Lezzi-Hafter, "Anthesterien und Hieros Gamos. Ein Choenbild des Methyse-Malers," in J. Christiansen and T. Melander, eds., *Proceedings of the Third Symposium on Ancient Greek and Related Pottery* (Copenhagen, 1988), pp. 325–334; K. Vierneisel and B. Maeser, eds., *Kunst der Schale: Kultur des Trinkens* (Munich, 1990), pp. 442–447; L. Burn, op. cit., p. 86, n. 17 (for further references); G. van Hoorn, op. cit.

—K. W./J. R. G.

TERRACOTTA SCULPTURE AND ARCHITECTURAL DECORATION

43
Gable Sima Plaque with Lioness

East Greek, third quarter of sixth century B.C.
Terracotta; H: 44.4 cm; MAX. W: 59.7 cm; DEPTH: 14 cm
Condition: Reconstructed from four large fragments; left side of relief panel and long edge of flange in back missing.

In the construction of temples in the Mediterranean world, the facing of parts made of timber with terracotta revetments to protect exposed surfaces from the elements was practiced widely during the Archaic period of Greek art. Producible in great numbers from molds, terracotta plaques such as this one were also colorfully decorative owing to the paint applied for the patterns and over the relief figures. These precursors of ornamental marble friezes and metopes must have provided a bright and vibrant contrast to the plain walls of very large buildings.

Here, against a white background and crouching between ovolo and meander patterns in red, a snarling lioness, with her head turned back and her left forepaw upraised, crouches to the right. Her body appears to have been painted red, and the smooth mane, bordered by short, wavy tufts, black, as is the case on the less complete fragmentary plaque from the same mold in Copenhagen (see Aakerström). That she is a lioness, despite the mane, is indicated by the row of dugs along her belly (for maned lionesses, see the Berlin Painter's pelike: Alfiere et al., and a Caeretan hydria in the Louvre: Hemelrijk). The plaque was found with another from the same border showing a panther similarly posed in the opposite direction (the panther is also represented among the fragmentary plaques in Copenhagen). Although lions and panthers occur frequently as painted decorative devices on terracotta sarcophagi from Clazomenae in northern Ionia (see Cook), it has been suggested that this plaque and its companions more probably come from southern Ionia or Caria.

BIBLIOGRAPHY: Atlantis Antiquities, *Greek and Etruscan Art of the Archaic Period* (New York, 1988), pp. 10–11 (with the other panel from the same frieze).

RELATED REFERENCES: A. Aakerström, *Die architektonischen Terrakotten Kleinasiens*, vol. 1, Acta Instituti Atheniensis Regni Sueciae 11 (1966), pp. 206, 208, no. 6, fig. 67.2. N. Alfiere et al., *Spina* (Florence, 1958), pl. 3. J. M. Hemelrijk, *Caeretan Hydriae* (Mainz, 1984), pl. 87.a. R. M. Cook, *Clazomenian Sarcophagi* (Mainz, 1985). For a discussion of the comparative material from the Greek East, see A. Andren, *Architectural Terracottas from Etrusco-Italic Temples* (1939), p. LXXIXff.

—M. J-N.

43

44

Statuette Group of a Woman Feeding a Hen with Chicks

Boeotian, 500–475 B.C.
Terracotta; H: 13 cm; W: 5 cm; DEPTH: 8.3 cm
Condition: Woman's head reattached at the neck, her right arm repaired; reddish brown clay (Munsell YR 5/4); figures handmade and solid, except for woman's head, which is mold-made; rectangular base cut from a slab of wet clay; underside of base flat and smooth with small drying cracks.
Provenance: Formerly in the collection of Sir Clifford and Lady Norton.

The group was probably originally brightly colored as the fired white base pigment, though obscured and dulled by surface accretions, is preserved. The colors used to decorate these types of statuettes were left unfired and, therefore, often have not survived over time. On this piece traces of black remain on the woman's hair and on the hen's body. The woman wears a close-fitting plain garment and a cap with two peaks, which leaves a fringe of forehead curls exposed. Her arms reach toward a mother hen who shelters five chicks with her wings.

The statuette was most likely made in Boeotia, a region north of Athens centered on the city-state of Thebes and a renowned terracotta center in antiquity. It may have been manufactured at Tanagra, a city about twelve miles east of Thebes, which, beginning about 500 B.C., produced figures engaged in everyday activities such as food preparation and farming. The Fleischman statuette fits well into this group on the basis of its style, polychromy, and clay type. Although the group has decorative appeal, it probably was a grave good, as the majority of these types of statuettes have been found in tombs.

BIBLIOGRAPHY: J. Chesterman, *Classical Terracotta Figures* (London, 1974), pp. 39–40, fig. 31. F. Nicholson, *Ancient Life in Miniature: An Exhibition of Classical Terracottas from Private Collections in England*, exh. cat. (Birmingham Museum Art Gallery, 1968), p. 24, no. 73, pl. 5.

RELATED REFERENCES: R. Higgins, *Tanagra and the Figurines* (Princeton, 1986), esp. pp. 84–92.

—J. B. G.

45

Statuette Group of a Woman Carrying a Child on Her Shoulders

Boeotian, 500–475 B.C.
Terracotta; H: 15.3 cm; W: 5.25 cm; DEPTH: 4.35 cm
Condition: Intact except for minor surface chipping and staining; pale brown clay (Munsell 2Y 7/5); figures made separately by hand then attached to a small rectangular base cut from a slab of clay; imprint of a fabric weave preserved on underside of base.
Provenance: Formerly in the collection of Baron Hans von Schoen.

Like the other Boeotian terracotta group included here (cat. no. 44), the closest comparison for this type of statuette is the series from Tanagra of figures engaged in the activities of daily life. The woman's headdress is similar to that on a figure of a woman kneading dough in the National Archaeological Museum, Athens (4044: Higgins, fig. 89).

The motif of a woman carrying a child piggyback is rare in Greek art. The playful nature of the posture reflects the intimacy that developed between the child and a figure that could be either its mother or its nurse. An attendant or slave carries an infant in the same manner in an everyday scene set in the women's quarters of the house on a white-ground lekythos by the Timokrates Painter in the National Archaeological Museum, Athens (12771: *ARV*[2], pp. 743, 1668, no. 1; *Para*, p. 521; *Beazley Addenda*[2], p. 284; Rühfel), dated about 460/50 B.C. A terracotta statuette in Munich (Antikensammlungen 5441: Klein) shows four children playing around a herm, with one piggyback on the herm's shoulders. In addition to the Timokrates Painter lekythos, A. Klein also lists two terracotta statuettes as depicting women carrying a child piggyback, but these are later Tanagra figures carrying a child on one shoulder, a slightly different motif that is more common than the piggyback posture.

A separate coil of clay was used for the tie around the hair of the woman and possibly also for the bun at the back of her head. The join of the bun to her head is very skillfully smoothed, but there is part of a fingerprint preserved there at the right. A separate thin sheet of clay appears to have been added to the basic head form of the child. The child's face is quite sculptural; the eye sockets are scooped out, and he has a thin, sharp nose. The woman's face is less modeled, with paint defining the pupils, the out-

44

45

line of the eyes, the eyelashes, and the mouth. Colors are preserved: red—child's body, cross pattern on dress, woman's mouth, band around her head; black—child's hair, woman's hair, eyebrows, eyes, pupils, sleeve borders; white—woman's flesh and dress, feet, top of slab base. The paint was applied after the figures were assembled and fired, which is vividly demonstrated by splashes of the red of the child's body on the woman's neck and strokes of the white of her dress on the child's legs.

45

BIBLIOGRAPHY: Unpublished.

RELATED REFERENCES: R. Higgins, *Tanagra and the Figurines* (Princeton, 1986), esp. pp. 84–92. H. Rühfel, *Das Kind in der griechischen Kunst* (Mainz, 1984), esp. pp. 107–108, fig. 42; A. Klein, *Child Life in Greek Art* (New York, 1932), p. 27, pl. 26E. For the relationship between mothers and children, see M. Golden, *Children and Childhood in Classical Athens* (Baltimore, 1990), pp. 97–100; for the close ties between slaves and children, pp. 145–150.

—J. B. G.

MARBLE

46
Exaleiptron

Greek, circa 440–430 B.C.
Marble; H (TO TOP OF KNOB): 23 cm; DIAM (OF BODY): 13.9 cm; (OF FOOT): 9 cm; (OF KNOB): 1.8 cm
Condition: Hairline crack along edge of foot where color is particularly gray, perhaps evidence of burning; chipped area along underside of lower bowl may correspond to abraded area on shoulder; small area of encrustation along rim of body, where it fits with adjoining shoulder piece; lid cracked along one edge; square rough area on underside of foot (point of attachment to a lathe?); lathe lines visible on underside of body and interior of bowl.

The shape name of this vase, *exaleiptron,* is derived from the Greek ἐξαλείφω, meaning "to anoint." Also called plemochoe or kothon, this vessel type was used to contain perfumed oil, either as part of a toilette or as a grave offering. The sharply inward-curving lip of the vessel held in the precious fluid and prevented it from spilling. The Fleischman vase is composed of four separate elements, the stem, body, shoulder, and lid. Oftentimes, marble vessels were brightly painted, but only one small area of pigment survives on this piece—a bit of pink or red along one section of incised line on the shoulder.

There are at least nine other marble exaleiptra. Four are in Athens (NM 11362, 11365, 11368, 12292), one is in the Museum of Fine Arts in Boston (MFA 81.355: Caskey; Caskey/Beazley; Burrows/Ure), three are in Berlin (see Pernice), and one is in the Allard Pierson Museum in Amsterdam (1612: Lunsingh Scheurleer 1926; *Algemeene Gids*). There are numerous variations among profiles, the joining point between the body and shoulder elements, and lid height and knob shape. The simple shape of the knob on the Fleischman exaleiptron is nearest to the Boston piece; the Athens exaleiptra have longer, more attenuated knobs. For the profile of the body, the incised elements atop the shoulder, and the joining point between shoulder and body, the closest parallel to the Fleischman piece is Athens NM 11365. This exaleiptron and one of the Berlin examples are perhaps among the earliest of the marble varieties of this shape, since in the profile of their body and the treatment of stem and foot, they relate quite closely to contemporary Attic ceramic exaleiptra.

BIBLIOGRAPHY: Sotheby's, London, sale cat., December 14, 1990, p. 145, lot 236.

RELATED REFERENCES: L. D. Caskey, *Geometry of Greek Vases* (Boston, 1922), pp. 234–235, fig. 185a-b. L. D. Caskey and J. D. Beazley, *Attic Vase Paintings in the Museum of Fine Arts, Boston, Part I. Text* (Boston, 1931), p. 49, fig. 35 (Boston exaleiptron and two of Athens exaleiptra shown in profile drawing). R. M. Burrows and P. N. Ure, "Kothons and Vases of Allied Types," *Journal of Hellenic Studies* 31 (1911), pp. 86–88, fig. 15. E. Pernice, "Kothon und Räuchergerät," *Jahrbuch des Deutschen Archäologischen Instituts* 14 (1899), p. 71, fig. 9. C. W. Lunsingh Scheurleer, "Grieksch Steenen Vaatwerk," *Bulletin van de Vereeninging Tot Bevordering der Kennis van de Antieke Beschaving* (December 1926), pp. 7–10, fig. 1. Allard Pierson Museum, *Algemeene Gids* (Amsterdam, 1937), p. 143, no. 1371. For a discussion of exaleiptra and similarly shaped vessels, see R. M. Burrows and P. N. Ure, op. cit., pp. 72–99. For information on dates for marble vessels, see P. Zaphiropoulou, "Vases et autres objets de marbre de Rhénée," *Études Déliennes*, BCH Suppl. 1 (Paris, 1973), pp. 601–636.

—K. W.

46

47

47
Head of a Youth

Greek, late fifth century B.C.
Large-grained white marble with gray patches; H: 16 cm; W: 12.7 cm; DEPTH: 14 cm
Condition: Broken off at the neck; end of nose missing; left ear margin battered, chip from left side of lower lip; crack on right chin area; small chips and pitting of the surface, especially on left eyebrow and right cheek; surface of the marble, including broken neck surface, has a waxy, melted look consistent with an acid treatment.

This head appears to have been broken from a *herm* (a rectangular marble or bronze pillar topped by a bust with short projecting beams instead of arms and an erect penis placed on the front of the shaft). Herms were common in the cities and countryside of Greece, marking boundaries, crossroads, and sacred places, and serving as wayside shrines or public dedications. The symmetry and frontality as well as the squareness of the back of the neck on this head are characteristic of herm portraits. The top of the head is more weathered than the face, which is also typical of herms. The strands of hair are indicated by parallel grooves in a way that is usual for Archaistic herms.

The intricate hairstyle is the most notable feature of this head. The head is encircled by a twisted double cable of hair that is knotted in the middle of the front. The starting point for the braids is not immediately apparent. While similar braided hairstyles are found on other figures, notably on the Artemision Zeus and the Omphalos Apollo in the National Archaeological Museum, Athens (BR. 15161 and 45: Karouzou), the blond boy from the Acropolis Museum, Athens (689: Brouskari), and the figure known as the Conservatori Charioteer in the Conservatori Museum, Rome (1505: Jones), none of these have two plaits completely encircling the head in addition to an abundance of loose hair formed into curls, such as is seen on this head.

The treatment of the front hair, parted in the middle with small "spit curls" on either side, is reminiscent of the bronze statue of Apollo from the Piraeus (Pireaus Museum P 4645: *LIMC*). Spiral curls are arranged over the forehead symmetrically to the right and left of the part. A lock of hair has been brushed back into a large spiral curl over each temple, while a third lock of hair forms a looser spiral curl in front of each ear with smaller tendrils in front of this. At the back the hair is vertically parted, with locks brushed forward to form large spiral curls. The hair on top, badly eroded, is apparently carved as if long but with a few extra spirals escaping on the surface.

The head belongs to a small group of idealized portraits, which were probably ephebic dedications (*ephebes* were young men, eighteen to nineteen years old, who received special military and gymnastic training). It is closest in its style and carving technique to a head found at Olynthos, now in the Thessaloníki Museum (see Robinson), originally identified as a female head, but now as Apollo. Even though the Fleischman head has Apollo-like qualities, one would hesitate to identify it for certain as the god since its hairstyle, so particularized, is unlike any hairstyle on images of Apollo currently known.

BIBLIOGRAPHY: Unpublished.

RELATED REFERENCES: S. Karouzou, *National Archaeological Museum: Collection of Sculpture* (Athens, 1968), pp. 41–42, 44. M. Brouskari, *The Acropolis Museum: A Descriptive Catalogue* (Athens, 1974), p. 123. H. S. Jones, *The Sculptures of the Palazzo di Conservatori* (Oxford, 1926), pp. 211–212. *LIMC*, vol. 2, p. 239, no. 432, s.v. "Apollon" (O. Palagia). D. M. Robinson, *Excavations at Olynthus II* (Baltimore, 1930), pp. 74–78, figs. 195–196. For a related head found in Athens, see V. Machaira, "Head of a Young Athlete," in *Mind and Body: Athletic Contests in Ancient Greece*, exh. cat. (Athens, 1989), no. 223 (Athens National Archaeological Museum 468). For a similar head found in the Athenian Agora (S 2057) and a discussion of the entire group of ephebic portrait herms, see E. B. Harrison, *Archaic and Archaistic Sculpture*, vol. 11 of *The Athenian Agora* (Princeton, 1965), pp. 124–129, no. 206, pl. 55.

—J. B. G.

48
Slab with Relief of a Young Hunter

Northern Greek, second half of fourth century B.C.
Marble; H: 143.1 cm; W: 42.7 cm
Condition: Essentially intact.

The slab was part of a *naiskos* (funerary shrine), which would have had the form of a small open-fronted building. Our piece served as the monument's right-hand side wall, with the figural relief toward the inside of the shrine and the undecorated surface, which is carefully smoothed and finished above with a molding, facing outward. The top edge of the block has cuttings for the attachment of a roof piece. The figured side is carved with an anta on the edge toward the front of the naiskos, while the back edge displays *anathyrosis* (roughening of the contact surface

48

of a building block) where it abutted the rear wall.

A sunken rectangle frames the figure, seen in profile facing toward the viewer's left, of a young man dressed in a short chiton with a flat, tied belt and carrying a pair of javelins which characterize him as a hunter. His curly hair is cut short and his facial features have an idealized regularity. Slung over his shoulder and apparently attached to the javelin loop is an indistinctly carved mass. The top of this mass is rounded like a petasos pushed back over the shoulder when not needed against the hot sun (see Biesantz, no. K33), but below it merges with a vaguely carved shape that hangs down the figure's back to thigh level. There is no sign of drapery fastened around the neck in front, as there should be if the lower outline is the expected *chlamys* (a short cloak often worn by travelers or hunters); a sack or a net, difficult to parallel elsewhere, has been suggested. Detail once rendered in paint may have clarified this part of the relief.

Ambitious funerary naiskoi, assembled from several blocks and occasionally with servant figures carved in relief on the side panels, are known from Athens and its sphere of influence in the second half of the fourth century. Although the style of our piece is Atticizing and extremely accomplished, there are some hints of provincialism. The crisply carved low relief has a rhythmic, calligraphic quality which recalls the best Thessalian and Macedonian work. The way the figure faces the back wall of the naiskos, rather than the beholder, seems non-Attic. Throughout Greece, a deceased youth may be shown with a hunting dog, to symbolize the carefree pastimes of the boyish life which has been cut short, or to identify the dead person with a mourned, heroic hunter like Meleager (see Anderson). However, the characterization on our monument of the secondary figure as a hunting companion, rather than the athlete's servant usual in Attica, suggests the higher status attached to the hunt in northern and eastern Greek tradition.

A short chiton, a petasos (if this is what our figure wears over his shoulder), and paired spears are the usual attributes of male figures on Thessalian grave reliefs (see Biesantz, pp. 78–83). Each of the elements, of course, was used in real life all over the Greek world, but the literal-minded representation of substantial clothing and equipment in this combination seems to reflect a regional preference. Finally, the material of the relief is a boldly striated gray and white marble, which looks something like Pentelic but has a coarser grain and much more pronounced bands of gray discoloration. Since such marble is often seen in northern Greek monuments, including ones of modest quality, it must have a local source. Our slab, then, comes from an imposing and unusually sophisticated monument produced by a sculptor familiar with Athenian trends but probably at home somewhere further to the north.

BIBLIOGRAPHY: Unpublished.

RELATED REFERENCES: H. Biesantz, *Die thessalischen Grabreliefs* (Mainz, 1965). J. K. Anderson, *Hunting in the Ancient World* (Berkeley and Los Angeles, 1985), p. 70f. On the carved side panels from naiskoi, see B. Vierneisel-Schlörb, *Glyptothek München: Katalog der Skulpturen*, vol. 3, *Klassische Grabdenkmäler und Votivreliefs* (Munich, 1988), p. 53f., n.1.

—A. H.

49
Funerary Lion

Attic, about 330–320 B.C.
Fine-grained white marble with mica; H: 58 cm; L: 108 cm; W: 28.5 cm
Condition: Legs, tail, and end of muzzle missing; large chip on inside of right front leg, which has been reattached; surface basically intact with only minor chips; small chips from face and ends of ruff locks; lightly weathered surface, heavier on the proper right side, especially in center of body; entire surface covered with light beige encrustation.

The lion was meant to be viewed from his left side. He turns his head to face the viewer and originally crouched on his front legs. His body is carefully modeled, lean and muscular, with ribs and surface blood vessels naturalistically rendered. The mane is composed of large pointed locks lying flat with either three or four channels separating them into strands. The locks continue along the center of the back to the midpoint of the body. The first two rows of locks beside the face are parted in the center. There is a deep groove on either cheek under the eyes and a sharply incised V-shape above the eyes. The teeth are carved individually as far as the first molar with a separation between the upper and lower teeth. The surface of the mouth flanges is smooth.

The Fleischman lion is comparable to lions carved after the middle of the fourth century B.C. It is similar to a lion in the Metropolitan Museum of Art (09.221.9: Richter), a lion in Copenhagen (2448: Poulsen), and a lion in the Minneapolis Institute of Arts (25.25: Vermeule), all belonging to the small group of the largest and most dramatic sculpted lions dated just before and after 320 B.C.

The Greek sculptor in Classical times could hardly have produced a realistic representation of a lion, since he had little or no opportunity to observe the living animal. His conception of lions came from generations of artists using motifs from Syria and Mesopotamia and from actual contemplation of domestic animals such as large dogs and comparable cats.

Lions, symbols of strength and courage, were placed as guardians at the corners of family grave plots in Athens and in many places all over Greece from 440 to 320 B.C. With the exception of its use as a waterspout on buildings, the lion was chiefly a funerary animal during this period.

BIBLIOGRAPHY: Unpublished.

RELATED REFERENCES: G. M. A. Richter, *Catalogue of Greek Sculptures* (Cambridge, 1954), no. 145. F. Poulsen, *Catalogue of Ancient Sculpture* (Copenhagen, 1951), no. 238 a. C. C. Vermeule, *Greek and Roman Sculpture in America* (Berkeley, 1981), no. 84. U. Vedder, *Untersuchungen zur plastischen Ausstattung attischer Grabanlagen des 4. Jhs. v. Chr.* (Frankfurt, 1985), esp. p. 296, no. T 60 (New York lion), p. 295, no. T 56 (Minneapolis lion), p. 294, T 51 (Copenhagen lion). C. Vermeule, "Greek Funerary Animals, 450–300 B.C.," *American Journal of Archaeology* 76 (1972), pp. 49–59. C. Vermeule and P. Von Kersburg, "Appendix: Lions, Attic and Related," *American Journal of Archaeology* 72 (1968), pp. 99–101.

—J. B. G.

49

50

50
Head of a Girl

Greek, circa 300 B.C.
Marble, probably Pentelic; H: 19 cm
Condition: Most of nose and edge of right ear missing.

The little girl wears her hair in the lobed hairstyle known as the "melon coiffure." Her head is encircled by a double braid, or rather cable-twist, wound all the way around the head behind the ears, with no indication of its starting point or end. The lids of the child's large eyes are modestly lowered, and her very slightly parted lips wear a subtle smile. The surface of the face has a sfumato quality, and there is a deliberate indistinctness in the way her mouth opening is filled in behind the parted lips. On the other hand, her delicate little ears are crisply carved, and the texture of the hair is indicated with a charming freedom on the surface of the melon coiffure lobes.

The head probably comes from the figure of a child votary. The statues of the "bears," little girls who performed cult functions in the sanctuary of Artemis at Brauron in Attica, are the most famous examples of such images, and the series includes the best parallels for our piece (see Themelis). Similar representations of children, however, are found in funerary contexts, though the smiling expression might speak against such an interpretation for our head. Statues of consecrated children, boys as well as girls, are also known in connection with other cults. In style, our piece seems just one step beyond the last Athenian grave monuments, discontinued after a decree of 317 B.C. The appearance of the marble, which seems to be Pentelic, would support an attribution to Attica.

BIBLIOGRAPHY: Unpublished.

RELATED REFERENCES: P. G. Themelis, *Brauron: Guide to the Site and Museum* (Athens, 1971), p. 71 center. E. Simon, *Götter der Griechen* (Munich, 1969), p. 155f., fig. 142.

—A. H.

ΙΓΙΣΘΟΣ
ΧΟΡΗΓΟΣ

The Western Greek Colonies in South Italy and Sicily

51

BRONZE

51
Side Handle of a Hydria

South Italian, circa 550–525 B.C.
Bronze; L: 22 cm
Condition: Intact; crusty green patina over much of surface.
Provenance: Said to be from South Italy, perhaps Taranto.

This horizontal handle was one of a pair of lugs originally attached to either side of the body of a hydria for carrying the heavy filled vessel. Semicircular in cross section, it terminates in an eight-frond palmette framed by addorsed horse protomes at each end. Although the horses' necks and forelegs, which are folded up beneath them, are rendered in relief profile, they turn their fully modeled heads outward. The long strands of their manes, which hang in heavy masses on the necks, are incised, while the forelocks are rendered in ridged relief. The rounded fronds of the palmettes set between their necks are thick and fleshy, suggesting the appearance of a succulent plant.

Among the bronzes of South Italy, attachments decorated with horse protomes are popular decorative elements on tripods and vessels of various shapes. A vertical hydria handle from the collection of Norbert Schimmel, now in the Israel Museum (91.71.315: *Master Bronzes,* p. 76, no. 72), offers a fairly close parallel for the relief horse protomes, though in this instance the forelegs are simplified, and they rest on a pendant palmette that is rather flatter in design than those on the Fleischman handle. Hydria side handles of similar composition, though different in the execution of details are found on a number of vessels, including a sixth-century kalpis from Paestum (see Rolley 1986, fig. 126) and a hydria in Lecce from Rudiae (see Delli Ponte).

BIBLIOGRAPHY: Michael Ward, Inc., *Form and Ornament: The Arts of Gold, Silver and Bronze in Ancient Greece,* exh. cat. (New York, 1989), no. 3.

RELATED REFERENCES: G. Delli Ponte, *I Bronzi del Museo Provinciale di Lecce* (Galatina, 1973), p. 23, no. 22, pls. XV, XVII.

—M. T.

52
Statuette of a Youth Carrying a Diskos

South Italian, circa 525–500 B.C.
Bronze; H: 11.1 cm
Condition: Edges of diskos chipped, especially at the back; left foot bent upward; extensive corrosion obscures details of face and hair, both arms and shoulders, and, to a lesser extent, both calves, producing a mottled green and reddish brown patina.
Provenance: Said to come from South Italy.

This small, muscular youth stands with his left foot advanced (it would originally have been flat to the ground), carrying a diskos in his left hand. He has drawn his shoulders back and holds both arms, slightly flexed at the elbows, stiffly out by his sides; the hole through the closed right hand indicates that it too once contained an attribute or implement. His long hair is carefully dressed with a short fringe over the forehead and the longer locks combed back behind the ears. A beaded fillet surrounds the crown of his head, and a thick band or clip secures the tresses hanging down the back at the shoulders. The posture of the young athlete, tense and self-conscious, emphasizes the beauty of his well-trained body. The pectorals and biceps are especially developed, as would be expected in an athlete specializing in the diskos throw.

The unusual hairstyle with the barettelike clip low on the tresses in the back is found on several kouroi; most similar is the hair on the shoulders and back of the fragmentary kouros from Actium, now in the Louvre (MNB 766: Richter), and the small bronze found at Dodona and now in Berlin (Staatliche Museen 7976: Richter). However, both of these kouroi, dated by G. M. A. Richter to 575–550 and 590–570 B.C. respectively, appear on the basis of their anatomical development to be earlier in date than the Fleischman figure. Although still not accurate in the rendering of the musculature, particularly in the shape and tripartite division of the abdominal muscles, this *diskophoros* (diskos-carrier) shows a more knowledgeable rendering of the body overall and may be dated to the last quarter of the sixth century.

52

In its odd bodily proportions, with a large head and broad shoulders set over a short torso and legs, in the angular treatment of the abdominal arch, and the energetic pose with both hands holding attributes out at the sides, this diskophoros generally recalls images attributed to the workshops of Magna Graecia, such as the later bronze kouros in Frankfurt (Liebieghaus 167: Bol). The style of execution of the remaining details in the rather outmoded hairstyle and the broad face also suggests an association with the Greek colonies of South Italy, where the statuette is reported to have been found, and where it may also have been dedicated as a votive offering at one of the many local sanctuaries.

BIBLIOGRAPHY: Unpublished.

RELATED REFERENCES: G. M. A. Richter, *Kouroi* (New York, 1970), no. 74, fig. 257 (Louvre kouros); no. 45, fig. 168 (Berlin kouros). P. Bol, *Liebieghaus: Guide to the Collection of Ancient Art* (Frankfurt, 1981), pp. 53–54, fig. 64. For bronze statuettes from Magna Graecia in general, see U. Jantzen, *Bronzewerkstätten in Grossgriechenland und Sizilien,* Jahrbuch des Deutschen Archäologischen Instituts Ergänzungsheft 13 (Berlin, 1937), esp. pp. 165–166, pl. 40 (Liebieghaus kouros). For the diskos in Greek athletics, see E. N. Gardiner, *Athletics of the Ancient World* (Oxford, 1930; reprint, Chicago, 1980), pp. 154–168. For images of diskophoroi and diskoboloi, see R. Thomas, *Athletenstatuetten der Spätarchaik und der strengen Stils* (Rome, 1981), pp. 30–45, pls. x–xix; *Mind and Body: Athletic Contests in Ancient Greece,* exh. cat. (National Archaeological Museum, Athens, 1989), pp. 257–264.

—M. T.

53
Hand Mirror Decorated with the Head of Medusa

South Italian, circa 500–480 B.C.
Bronze; H: 20.2 cm; DIAM (OF DISK): 15 cm
Condition: Complete except for losses along right upper edge; handle, probably made from another material, missing; crusty green patina partially removed from mirror's back, revealing a glossy golden brown surface.
Provenance: Said to be from South Italy.

The disk of this splendid hand mirror is cast with a raised, beaded flange all around to protect the interior surfaces. On the reflective side, the polished tondo is surrounded by two incised concentric circles and framed by a kymation border. At the disk's center bottom, a handle support is attached. Two incised volutes at its top form a kind of modified Aeolic capital that suggests the support of a column beneath. The concave back of the mirror disk holds a separately made repoussé relief of a head of the Gorgon, Medusa. The comically grotesque face of the monster, with spiraling snake locks, flamelike tufted beard, and protruding tongue, has been schematized into a highly decorative motif. This image is a particularly refined version, with its symmetrical disposition of snaky curls, perfectly aligned teeth without the usual fangs, and heavy-lidded eyes that were once inlaid, probably with glass paste.

Perfectly suited to the tondo format, the Gorgon is a particularly witty choice for the decoration of the back of a mirror, as the monster's face was believed to turn to stone anyone who looked at it. Its use here is perhaps a kind of apotropaic device intended to protect the user, similar, for example, to its use on the shields of hoplites. It is not without parallels among mirrors, though none is close to the Fleischman example in quality.

This gorgoneion is most closely related to images from South Italy, where similarly decorative images appear on brightly painted terracotta roof tile ornaments (*antefixes*) and in other media (for an example of another type of terracotta antefix, in this case decorated with images of a satyr and maenad, see cat. no. 92, cover). A South Italian provenance is suggested also by the decoration on the mirror disk itself and the volute capital support at the top of the handle tang, all details that fit neatly among the South Italian hand mirrors. The decorative stylization of the grotesque image suggests a date at the end of the Archaic tradition, perhaps in the early decades of the fifth century B.C.

BIBLIOGRAPHY: Unpublished.

RELATED REFERENCES: For a similar beaded rim and impressed kymation pattern with tongues pointing out framing the mirror disk and Aeolic volutes at the top of the support, see the bronze mirror in Reggio (4779): Rolley 1986, pp. 138–139, fig. 120; R. Thomas, *Athletenstatuetten der Spätarchaik und der strengen Stils* (Rome, 1981), pl. LXII.2. For the gorgoneion as a bronze shield device, see Olympia B 110: R. Hampe and U. Jantzen, *Bericht über die Ausgrabungen in Olympia,* vol. 1 (Berlin, 1937), pp. 56–57, pl. 13. For other hand mirrors decorated with gorgoneia, on the handles rather than the backs, see P. Oberländer, *Griechische Handspiegel* (Hamburg, 1967), pp. 37, 123–124, 126, nos. 43, 175, 176, 178. For a later Greek mirror decorated with a gorgoneion, see I. Love, *Ophiuchus Collection* (Florence, 1989), pp. 109–111. For a similar gorgoneion on a South Italian terracotta antefix, see *Il Museo Nazionale di Reggio Calabria* (Rome, 1987), p. 139.

—M. T.

53

54
Group of Ornaments

West Greek, late sixth century B.C.

A. Four Frontal Female Heads
Silver, gilded
H: 4.6–3.75 cm

B. Six Frontal Bearded Male Heads
Silver, gilded
H: 4.4–1.9 cm

C. Gorgoneion
Silver, gilded
H: 3.7 cm

D. Pin Head (Perfume Dipper?)
Silver with gilding, bronze.
H: 4.05 cm

E. Four Pairs of Pins
Silver with partial gold overlays, bronze
Pair 1. H (EACH): 17.7 cm
Pair 2. Present H (EACH): 9 cm
Pair 3. Present H (a): 14 cm; (b): 17 cm
Pair 4. Present H (a): 7 cm; (b): 4.6 cm

F. Fragmentary Band with Foliate Ornament
Silver, partially gilded
Present L: 36 cm; W: 10 cm

G. D-shaped Object
Gilt bronze
H: 5.9 cm

H. Finial
Gold leaf over core material
Height 2.4 cm

I. Three Finials
Silver
Present H: 2.5–3.5 cm

54A

54B

54C

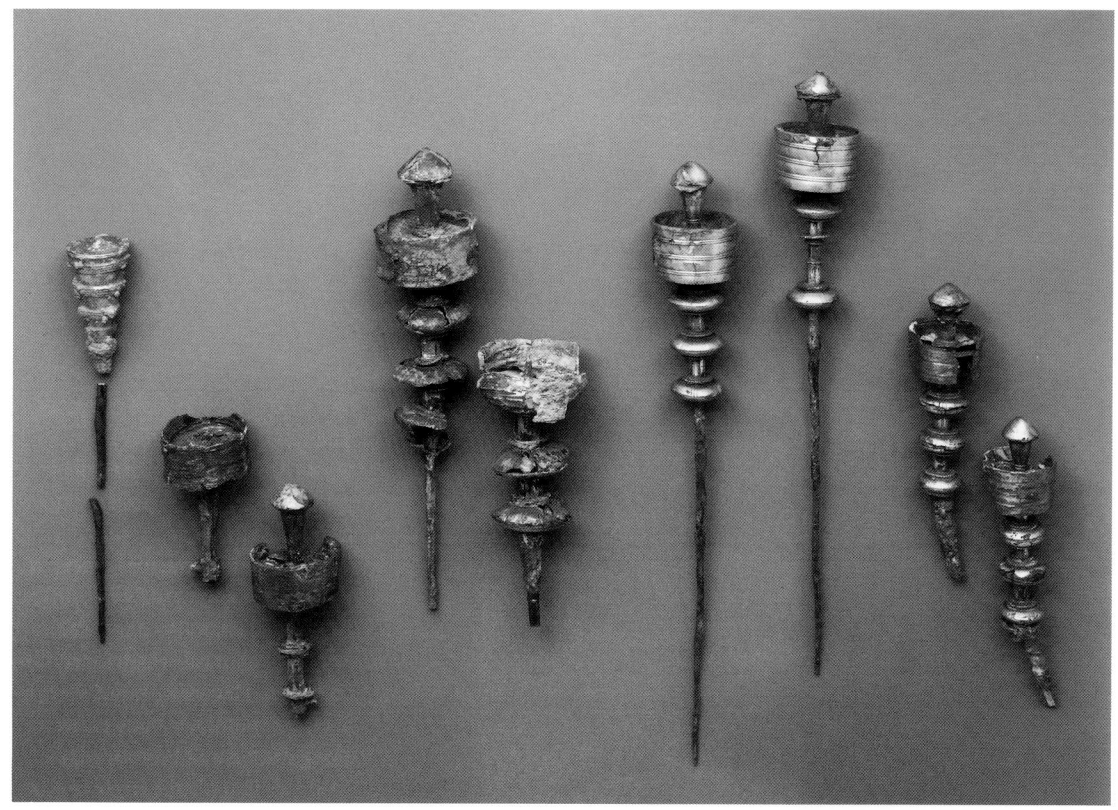

54D AND 54E

The heads, in high relief, are made of thin metal hammered into molds. They belong to three types: four heads of korai, six bearded male heads, and one gorgoneion. The korai, with full oval faces, have bulging almond eyes and fierce smiles. Their hair, in fine zigzag locks, is looped back behind the ears, then falls in loose waves to the shoulders. Each is crowned with a stephané and wears a single-strand necklace. The robust-looking, Herakles-like male heads are crowned with wreaths of pointed leaves and have short beards arranged in snail-shell curls. The gorgoneion (Medusa) is of the grimacing Archaic "lion mask" type, with furrowed brow, broad nose, and tongue protruding between pointed fangs. In this decorative rendering Medusa wears a small peaked stephané trimmed with rosettes. Originally each of the heads had a plain, fairly wide outer border. Since no trace of holes for sewing or nailing the heads to a background survives, it seems likely that this flange, a byproduct of the technique of hammering into a form, was attached with adhesive to a background of another material.

54E, DETAILS

The group also includes four pairs of long bronze pins with silver heads, parts of which are overlaid with heavy sheet gold. Each cylindrical head has an inner disk worked with a rosette, the center of which is the projecting finial. Although of the same basic type, the pairs differ slightly in detail. The pins' precise workmanship and solid construction show that they were made for actual use rather

54F

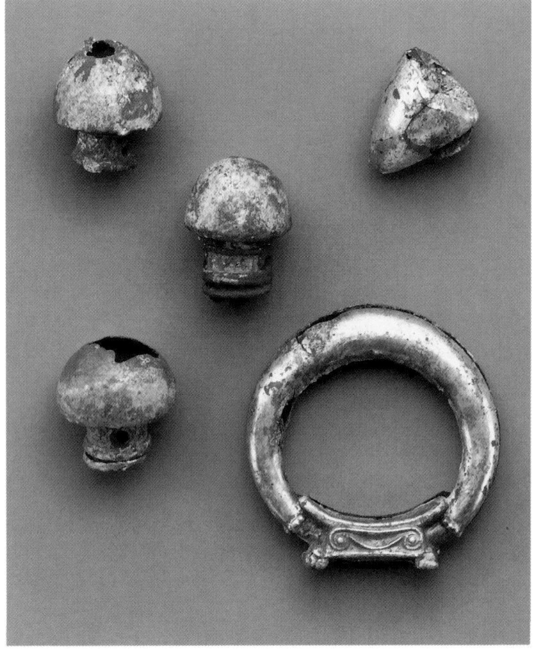

54G, H AND I

than solely for funerary purposes. The single pin has a head of different design and may have had a different function, perhaps as a perfume dipper.

The most spectacular object in the find is a wide, partly gilt silver strip to which is affixed an elaborate decoration of schematized floral forms. The background sheet is plain except for a tongue pattern along the top and bottom edges. Over this is applied a triple row of overlapping cut-out laurel leaves. On each laurel garland is superimposed a double line of small bud- or acorn-shaped finials pointing outward. Against the plain background of the middle zone large bud-shaped finials alternate with stylized blossoms made in two layers, each with a smaller, pointed finial at its center.

Also associated with the group are a stray bud-shaped finial of an odd size, three dome-shaped solid silver finials, and a D-shaped object, made in two halves, of fairly heavy gilt bronze, with its straight side in the form of a simplified, miniature Ionic capital.

The female heads have many parallels. Replicas are in the Taranto Museum and the Ortiz collection (see *Ortiz Collection*); related examples from another mold are in two private collections (see *Classical Art*). Similar heads, made into pendants, are part of a necklace from Ruvo (see Langlotz). The male head has a replica in the Ortiz collection (see *Ortiz Collection*). The pins are of a well-known South Italian type found in burials at Taranto (see Loiacono/de Juliis). Fragments of wide bands with schematized floral ornament are also in the Taranto Museum and in a private collection, but like ours they are badly damaged and their purpose is uncertain. Since ours, at least, was found in association with the sets of pins, it is likely that such bands are articles of female adornment. All seem to be bent on a slight curve which may be original. Because of their fragile projecting elements, they cannot have been belts worn at the waist, at least of a living person. Possibly they come from *polos*-shaped ceremonial crowns like those sometimes worn by priestesses or brides; the loosely attached decorations seem designed to shimmer with a wearer's movements. The ornament has a dense and fanciful richness hardly seen again until the Age of Alexander.

BIBLIOGRAPHY: Unpublished.

RELATED REFERENCES: *The George Ortiz Collection: Antiquities from Ur to Byzantium*, exh. cat. (Berne, 1993), no. 123. André Emmerich Gallery, *Classical Art from a New York Collection*, exh. cat. (New York, 1977), no. 130n (a-c). E. Langlotz, *The Art of Magna Graecia* (London, 1965), color pl. VII. D. Loiacono and E. de Juliis, *Il Museo Nazionale di Taranto* (Taranto, 1985), p. 325, no. 384.

—A. H.

55A

55
Pair of Reliefs

Early Hellenistic, early third century B.C.
Silver with gilding

A. Gorgoneion

H: 6.4 cm; W: 8.6 cm
Condition: Hollow in back and broken off all around.

Medusa, shown in sensitively rendered low relief with chased details, has a rounded, youthful face set frontally on a broad aegis decorated with a scale pattern and snakes knotted around its edge. This type of Medusa had a long life, adorning, for example, the back of the late Hellenistic Tazza Farnese (see *LIMC*, no. 223). It appears in the painted relief of an early Hellenistic tomb in Naples (*LIMC*, no. 220) and is also related to the softer-looking, more flat-faced examples among the frontal female heads on Canosan askoi. The imbricated background is seen on early third-century terracotta miniature shields from the Tomb of the Erotes at Eretria (*LIMC*, no. 190).

55B

B. Skylla

H: 6.2 cm; W: 7.1 cm
Condition: Broken off all around; separate backing plate attached.

Skylla is represented here as a mermaidlike hybrid, winged and holding an oar. Seen frontally, her human upper half is almost childishly plump. Her two tails emerge from a foliate apron and coil symmetrically to either side, where they end in the small dog heads that identify her as the legendary sea monster. Skylla, whose name means "puppy," was a predatory creature that personified the terrors of the sea. In the *Odyssey* she is described as making a sound like a puppy's bark. In later art she has a woman's upper half, often with an apronlike fringe of canine protomes above her single or double fishtail.

Our type of decoratively stylized Skylla appears in gilded terracotta appliqués from Taranto (see *Gli ori di Taranto*, nos. 35, 36). More loosely related is the Skylla on Canosan relief pilgrim flasks (see *Enciclopedia dell'arte antica*, vol. 2, fig. 464). Skylla is winged on, for example, the box mirror in Berlin (Pergamonmuseum 8391: Schefold/Jung). The workmanship of the tails and Skylla's large-headed, rather babyish appearance recall the style of the creatures from Poseidon's retinue shown in relief on the famous silver-gilt box from Ruvo in the Taranto Museum (see *Gli ori di Taranto*, p. 58f., no. 8). The Skylla emblema from the somewhat later silver treasure in the Metropolitan Museum of Art (98.11.22: Bothmer) displays a similar treatment of the tails but a very different representation of Skylla's human part.

BIBLIOGRAPHY: Unpublished.

RELATED REFERENCES: *LIMC*, vol. 4, p. 285ff., no. 223, s.v. "Gorgo, Gorgones" (S.-C. Dahlinger, I. Krauskopf). E. M. De Juliis et al., *Gli ori di Taranto in eta ellenistica*, exh. cat. (Milan, 1984). K. Schefold and F. Jung, *Die Sagen von den Argonauten, von Theben und Troia in der klassischen und hellenistischen Kunst* (Munich, 1989), fig. 307. D. von Bothmer, "A Greek and Roman Treasury," *Metropolitan Museum of Art Bulletin* (Summer 1984), cover, no. 94.

—A. H.

TERRACOTTA VASES

Phlyax Plays

The three bell kraters that follow (cat. nos. 56, 57, 58) all show scenes from phlyax plays. The word "phlyax" seems to have been used in Magna Graecia to designate both the actors in a certain type of rustic comedy (Pollux IX. 149) and the actual performance. Hesychius derives the word from φλυαρεῖν (to talk nonsense); F. Radermacher (*Zur Geschichte der griechische Komödie*) suggested that it might also have a connection with φλέω in the sense of "tumid" or "swelling," a not inappropriate designation for the actors in phlyax plays, who often wear well-padded costumes.

A phlyax play might consist of a parody of some well-known tragic theme, or deal, in a burlesque or farcical fashion, with the deeds of gods and of heroes, or with episodes from everyday life. Such performances might well have been at an almost impromptu level, as the simple staging required for their production would seem to indicate. Later on, in the early third century B.C., the poet Rhinthon, who was of Syracusan origin but worked at Taras, seems to have given literary form to this type of comedy, of which he was regarded as the originator, although its roots go back a long way. Recent work, in particular by O. Taplin, C. W. Dearden, E. Csapo, and other scholars, has shown that some at least of the so-called phlyax vases of Magna Graecia are direct reflections of Attic comedies of the early fourth century B.C. This seems reasonably clear in the case of a vase like the one in Würzburg (H 5697), on which a scene from the *Thesmophoriazusae* of Aristophanes is represented; it is rather less obvious in some of the other examples cited.

On all three of the Fleischman bell kraters a simple form of stage is represented; this consists of a low wooden platform supported by two or three posts. These stage settings tend to become more elaborate as time goes on; posts are replaced by columns, sometimes with curtains draped between them, and more flights of steps are used to link the stage with ground level.

Phlyax vases make their first appearance in Magna Graecia at the end of the fifth century B.C. The main period of their production lies between about 380 and 340 B.C., though a few go down to the last years of the century. In general, phlyax vases are fairly uniform in treatment: the typical costume for males, which looks back to that of the padded dancers on Corinthian vases of the sixth century, consists of close-fitting tights (i.e. simulating nudity), with padding on the belly and the rump, and a large phallus. Over this basic costume may be worn a tunic, sometimes shown in added white, or a cloak.

A mask, appropriate to the role being played, will also be worn. Female players tend to wear normal dress but also wear masks that identify their role (the mask types cited below are based on the categories established by T. B. L. Webster and J. R. Green in *Monuments Illustrating Old and Middle Comedy*, BICS Suppl. 39 (1978), pp. 13–26.

56
Red-figured Bell Krater

Name Vase of the Choregos Painter [A. D. Trendall]
Apulian, circa 380 B.C.
Terracotta; H: 37 cm; DIAM (OF MOUTH): 45 cm; (OF FOOT): 17 cm
Condition: Reconstructed from fragments; seven pairs of rivet holes for ancient repairs, mostly just below the rim.

This vase has a phlyax scene on the obverse and a genre scene on the reverse; there is a band of laurel below the rim and, encircling the vase beneath the pictures, a band of meanders in groups of three, interspersed by saltire squares. Below the handles are superposed palmette fans, with spiraling side-scrolls. Immediately in front of the door on the stage stands Aigisthos, identified by the inscription ΑΙΓΙΣΘΟΣ above his head. He wears an elaborate long-sleeved garment, richly embroidered with a variety of patterns and caught up at the waist by a studded girdle; behind his back falls a black-bordered cloak, fastened at the throat with a brooch, and over his chest is a crossed shoulder-cord, decorated with black dots. His left hand grasps two spears; with his right he touches the pilos on his head. There seems to be no good reason to think of him as wearing a mask; his costume corresponds closely to those worn by characters in tragic scenes on South Italian vases. The pilos, cloak, and two spears are typical attributes of the traveler and suggest that Aigisthos has come from afar.

Facing Aigisthos, with his right hand raised in an admonitory gesture while his left grasps a crooked stick, is a white-haired phlyax actor (Mask L), inscribed ΧΟΡΗΓΟΣ (choregos), who wears tights and a short, padded tunic. Beside him on an upturned basket, serving as a sort of *bema* (podium), stands a second phlyax wearing a similar costume, with a tunic patterned with dot-clusters and having a dot-stripe lower border. ΠΥΡΡΙΑ[Σ] is inscribed above his head, probably to indicate that he was a "red-head," perhaps a Thracian slave, since they were apparently noted for their red hair. His mask is Type N and is shown in fully frontal view, a comparatively rare phenomenon on phlyax vases. His right hand is stretched out in a deictic gesture, his left is akimbo. Beside him, leaning on a crooked stick held in his left hand, is another phlyax, inscribed ΧΟΡΗΓΟΣ, like the first. His drapery is similar to that of his counterpart, but his feet are bare and he wears a mask with black hair (Type P).

The interpretation of this scene raises considerable problems. The two figures inscribed "choregos" might stand for the two leaders of the semichorus, but, as O. Taplin reminds us (*Comic Angels*, pp. 57–58), members of the chorus are not usually present on comic vases, and it is therefore more likely that they represent the backers of the play.

The fact that the two choregoi are wearing typical comic costume and that they are standing on an actual stage suggests that they are performing a part in the play itself, which might well be concerned, like the *Proagon (Preview)* of Aristophanes, with a comic version

56 SIDE A

56 SIDE A–B

SIDE B

of the preparations for a stage production. In that case the figure of Aigisthos, who seems to have come on stage through the half-open door to the left, might well stand for Tragedy. Pyrrhias could then represent Comedy, in the characteristic pose of the declaimer, and the two choregoi, the older supporting Aigisthos, the younger, Pyrrhias, may be arguing which of the two they will ultimately back. Otherwise, we may have here a parody of the scene, as represented in both the *Choephoroi* of Aeschylus and the *Elektra* of Sophocles, in which Aigisthos arrives at the palace. It could have been inspired by a lost comedy like that referred to by Aristotle in the *Poetics* (1453.a. 36–39) in which the bitterest enemies (e.g., Orestes and Aigisthos) walk off at the end as good friends.

The reverse depicts a typical genre scene. In the center seated on a chair (*klismos*) is a young woman wearing a chiton, with a piece of black-bordered drapery covering the lower part of her body. Her hair is bound up in a sphendone; she wears an earring and necklace composed of black beads. A small bird is perched on the index finger of her left hand; her right arm rests in a languid pose on the arm of the chair, her slippered feet on a footstool. Standing behind her, with her head turned away to the left, is a girl attendant, wearing a long chiton with a black stripe down the side and grasping the handle of a large fan with both hands. To the right, in a relaxed, almost frontal pose, resting his weight on his right leg, with his left crossed in front of it, stands a nude youth; his left arm is enveloped in a piece of black-edged drapery, from beneath which emerges a stick. Perched on his extended right arm is a small feline (perhaps a marten or some form of cat) which looks down at the bird on the hand of the seated woman. Hanging above, in the center, is a black-dotted fillet with tasseled ends.

In pattern-work and style this vase is very closely comparable to two other bell kraters—one in Milan (Museo Civico Archeologico A 0:9:2841: *RVAp*, Suppl. II, p. 7, no. 1/123), the other in Cleveland (89.73: Trendall, *Bulletin; RVAp*, Suppl. II, Postscript, p. 495, no. 1/125). All three are the work of a single artist, who has been named the Choregos Painter after the Fleischman vase. Particularly characteristic of his pattern-work are: (1) the palmettes with side-scrolls below the handles, (2) the laurel wreath above the pictures and, below, meanders in groups of three separated by saltire squares, and (3) the ovoli at the handle roots. The treatment of the phlyakes is remarkably similar, especially in the rendering of their costumes and masks, as is that of the stage, on which the graining of the wooden floor is clearly shown. The large fan and the pose of the girl who holds it find a close parallel on a bell krater in Newcastle-upon-Tyne (19.1524: *RVAp* I, p. 27, no. 1/121), which is probably also by the Choregos Painter, who is a follower of the Sisyphus Painter and a contemporary of the Ariadne Painter in the early years of the fourth century B.C.

BIBLIOGRAPHY [F93]: A. D. Trendall, "Farce and Tragedy in South Italian Vase-painting," in T. Rasmussen and N. Spivey, eds., *Looking at Greek Vases* (Oxford, 1991), p. 165, fig. 67; idem, "A New Early Apulian Phlyax Vase," *The Bulletin of the Cleveland Museum of Art* 79, 1 (1992), pp. 1–15, figs. 7, 8, 11; idem, *RVAp*, Suppl. II, no. 1/124, pl. 1.3–4; O. Taplin, *Comic Angels* (Oxford, 1993), p. 55ff., pl. 9.1. M. Schmidt, "Tracce del Teatro comico attico nella Magna Grecia," *Vitae Mimus* (Incontri del Dipartimento di Scienze del l'Antichità dell' Università di Pavia) 6 (1993), pp. 37–38.

RELATED REFERENCES: On the Cleveland bell krater, see A. D. Trendall, *Bulletin*, op. cit., pp. 1–15, figs. 1, 4, 5, 9, cover. On phlyax vases in general, see H. Heydemann in *Jahrbuch des Deutschen Archäologischen Instituts* 1 (1886), pp. 260–313; A. D. Trendall, *Phlyax Vases*, 2nd ed. = BICS Suppl. 19 (1967), pp. 1–7; J. R. Green, "Theatre Production, 1971–1986," *Lustrum* 31 (1989), pp. 66–85; O. Taplin, op. cit., pp. ix-xii; M. Schmidt, op. cit., pp. 27–41, with bibl.; K. J. Maidment, "The Later Comic Chorus," *Classical Quarterly* 29 (1935), p. 3ff.; C. W. Dearden, "Phlyax Comedy in Magna Graecia: A Reassessment," in *Studies in Honour of T. B. L. Webster*, vol. 2 (1988), pp. 33–41; idem, "Epicharmus, Phlyax and Sicilian Comedy" in *Eumousia: Studies in Honour of Alexander Cambitoglou* (1990), p. 55ff.; A. D. Trendall, "Farce and Tragedy in South Italian Vase-painting," op. cit., pp. 151–182. For an illustrated list of comic masks, see T. B. L. Webster and J. R. Green, *Monuments Illustrating Old and Middle Comedy*, BICS Suppl. 39 (1978), pp. 13–26; A. D. Trendall, "Masks on Apulian Red-figured Vases," in *Studies in Honour of T. B. L. Webster*, vol. 2 (1988), pp. 136–154. On stages and staging, see S. Gogos in *ÖJh* 54 (1983), pp. 59–86; ibid., 55 (1984), pp. 27–53; L. Massei, "Note sul λογεῖον fliacico," in *Studi Classici Orientali* 23 (1974), pp. 54–59, figs. 1–3. For the Choregos Painter, see A. D. Trendall, "Farce and Tragedy in South Italian Vase-painting," op. cit., p. 154ff.; idem, *Bulletin of the Cleveland Museum of Art* 79, 1 (1992), pp. 1–15; idem, *RVAp*, Suppl. II, pp. 7–8, 495 ; O. Taplin, op. cit., p. 55ff. For the animal on the reverse, see A. Ashmead, "Greek Cats," *Expedition* 20, 3 (1978), p. 38ff.; D. von Bothmer in *Antiquities from the Collection of Christos G. Bastis* (1987), p. 283, no. 165.

—A. D. T.

57
Red-figured Bell Krater

Attributed to the Rainone Painter [A. D. Trendall]
Apulian, circa 370 B.C.
Terracotta; H: 27.3 cm; DIAM (OF MOUTH): 30 cm; (OF FOOT): 12.5 cm
Condition: Reconstructed from large fragments; section of rim just above temple pediment restored; fractures on reverse and some restoration on lower part of reverse.

The obverse represents a scene from a phlyax play taking place on a simple wooden stage consisting of a floor supported by a post at either end. To the left is an elaborate door between Ionic columns; there is part of a pediment above, with an acroterion, suggestive of a temple rather than an ordinary building. Two white-haired phlyakes, each of whom wears a piece of black-bordered drapery across the lower part of his body, to which a phallus is attached, are standing beside a large basket (*cista*), decorated with parallel rows of simple ornamental patterns. The phlyax on the left has just lifted the elaborately decorated lid of the basket to reveal an ithyphallic child with a ramlike head in added white, with golden brown highlights on the hair. His extended left arm is grasped at the wrist by the second phlyax, who turns his head slightly to the left with a somewhat puzzled look. Both phlyakes wear the mask of the old man (Types M and L).

What this scene is intended to represent is not easy to discover. The presence of the cista, opened to reveal the child, suggests a parody of the story of Erichthonios. When Hephaistos attempted to violate Athena, led on by a malicious joke by Poseidon who told him that she was about to visit his smithy in expectation of having him make love to her, she tore herself away, and the semen of Hephaistos struck her above the knee. This she wiped off with a piece of wool, which she threw onto the ground, accidentally fertilizing Ge (Earth), who happened to be there on a visit. Athena took care of the infant and named him Erichthonios—probably from the Greek words for wool (*erion*) and earth (*chthon*). The child was placed in a sacred basket, guarded by serpents; curiosity led the daughters of Cecrops to look inside the basket and, stricken with madness, they hurled themselves from the Acropolis and perished. If what we have here is a parody of this story, the building to the left could be the Erechtheum, the precinct of Athena, where the child was left; as the mystical chest is opened it reveals not Erichthonios, however, but a comic version of him with the head and features of a ram to remind us of the wool, which helped to impregnate Earth. Perhaps it is this unexpected discovery that causes the puzzled look on the face of the second phlyax. Such a parody would be completely in keeping with the tradition of the phlyax plays.

The reverse represents the usual three draped youths, who so frequently appear on the reverses of Apulian bell kraters. All three are completely enveloped in their himatia, which are, however, draped in slightly different fashion for each of them. As the obverse design is highly stylized, it is to the reverse that we must turn to help us in attributing the vase, and this at once indicates that the vase is the work of the Rainone Painter, who decorated several other bell kraters with almost identical youths. Particularly characteristic is the one in the center, with his left arm akimbo and covered by the drapery with an almost sleevelike effect. The name vase of the Rainone Painter also has a phlyax scene on the obverse—a parody of the story of Antigone, which lends some support to the suggestion that the scene on our vase is also a parody. The Rainone Painter was a follower of the Tarporley Painter and must have been active in the second quarter of the fourth century; our vase must come fairly early in his extant work, since the frieze of stopped meanders is confined to the area immediately below the pictures and does not encircle the vase, as is the regular practice on his later vases.

BIBLIOGRAPHY: *RVAp* I, no. 4/224 a (as Vidigulfo, Castello dei Landriani 261); *RVAp*, Suppl. II, p. 15.

RELATED REFERENCES: For the story of Erichthonios, see Apollodoros III.14.6, and the notes by J. G. Frazer in the Loeb edition, vol. 2, p. 90ff.; Hyginus, *Fab.* 166 (where the child is part serpent). For the treatment of the myth on Attic vases, see K. Schefold, *Die Göttersage in der klassischen und hellenistischen Kunst* (Munich, 1981), pp. 48–56; *LIMC*, vol. 4, p. 928ff, nos. 1–16 (U. Kron); M. Schmidt, "Die Entdeckung der Erichthonios," *AM* 83 (1968), pp. 200–212. For the Rainone Painter, see *RVAp* I, pp. 95–96. A. D. Trendall, *Red-figured Vases of South Italy and Sicily* (London, 1989), p. 76.

—A. D. T.

57 SIDE A

57 SIDE B

58

Red-figured Bell Krater

Attributed to the Cotugno Painter [A. D. Trendall]
Apulian, circa 370–360 B.C.
Terracotta; H: 34 cm; DIAM (OF MOUTH): 38 cm; (OF FOOT): 15.3 cm
Condition: Good, with little retouching.

The obverse of this vase represents a scene from a phlyax play enacted on a simple stage, similar to the one on the preceding vase (cat. no. 57) but supported by three posts instead of two. To the left stands a slave (Mask: Type B) wearing tights; his right arm is akimbo and he grasps a stick in his left hand. He looks toward an aging woman, who is wearing a mask of Type U and has short white hair, a wrinkled face and brow, a stubby nose, and thick lips. She wears a peplos with a long overfall, on which the fold-lines are clearly designated, and holds with both hands a long piece of drapery which billows out above her head. Coming up to her from behind, with his right arm outstretched and a lecherous look in his eye is a regal figure, wearing a crown and carrying a bird-topped scepter; these attributes suggest that he is intended to represent Zeus. His mask is that of the old man—Type G, with scanty white hair, straggly beard, and a large visible ear. He is wearing tights; across the lower part of his body and over his left shoulder is draped a short cloak with a dot-stripe border. The scene is clearly a comic version of one of the many amorous adventures of Zeus, but there is nothing that enables us

58 SIDE A

58 SIDE B

SIDE A–B

specifically to identify the woman. The reverse depicts a genre scene—a draped youth standing between two women, both wearing peploi with overfalls. The one to left holds a thyrsos in her right hand, while she gestures toward the youth with her left; the other holds in both hands a long fillet, with tasseled ends, decorated with a series of small circles.

The Fleischman bell krater adds a third to the two already attributed to the Cotugno Painter, who takes his name from the former owner of a vase in Bari (Museo Archeologico 8014: *RVAp* 1, p. 266, no. 10/45). All three kraters are very similar in shape and decoration—the feet have reserved bands at the top and bottom, as well as at the join of stem to foot; there are bands of ovoli below the laurel wreath around the rim and immediately above the pictures on both sides, as well as around the handle roots. Below the figured scenes runs a continuous band of meanders, interspersed here, as on the Bari krater, with saltire squares. Below the handles are superposed palmette fans, with spiraling side-tendrils. The treatment of the phlyax scenes is also very consistent; all show the events as taking place on a low wooden stage supported by three posts, and the grouping of the figures on this and the Bari vase is identical, as are the costumes worn by the two female figures. The reverses are also very similar; the youth in the center on our vase is repeated on the other two, with slight variations, and the entire scene is very like that on Bari 8014,

59 SIDE A

FRONT

SIDE B, DETAIL

except that there the woman is placed in the center and flanked by two youths, whereas our vase reverses their positions. The Fleischman vase is in a much better state of preservation than the other two, on both of which substantial portions of the obverses are missing. The Cotugno Painter is associated with the Judgment Painter and his circle.

BIBLIOGRAPHY: *RVAp,* Suppl. II, Postscript, p. 564, no. 10/46a (as on the Swiss market).

RELATED REFERENCES: On the Cotugno Painter, see *RVAp* I, p. 266, nos. 10/45, 46 (Bari, Museo Archeologico 8014 and Taranto, Museo Archeologico Nazionale 107937). For additional information on Bari 8014, see E. M. de Juliis, *Il Museo Archeologico di Bari,* pl. 40.1; *Aspetti del Quotidiano* (Bari, 1986), p. 9 and cover; *Dov' è Atene* (Bari, 1988), p. 66, fig. 5; *LIMC,* vol. 4, p. 533, no. 185, pl. 325.2, s.v. "Helene."

—A. D. T.

59
Red-figured Askos

Apulian, circa 360–350 B.C.
Terracotta; H: 17 cm; W: 16 cm
Condition: Reconstructed from large fragments; a few minor abrasions.

The askos is a comparatively rare shape in Early Apulian vase painting. The Felton Painter decorated one with a phlyax scene running around its entire body (Ruvo, Museo Jatta 1402: *RVAp* I, p. 175, no. 7/68), and another, closer in shape and style to the Fleischman askos (Museo Jatta 957: *RVAp,* Suppl. I, p. 39, no. 11/133a), has been placed in the Meer Group, one of the groups of minor vases which follow on from the Felton Painter and are contemporary with the work of the Iliupersis Painter and his circle. Our vase belongs to U. Rüdiger's Type A, sometimes referred to as a "duck" askos. It is set on a low foot, with a flaring mouth at the front, and a strap handle that connects it to the base of the pointed knob (often likened to a duck's tail) at the back. Beneath the handle are two rows of reserved triangles, set base to base. Either end of the vase is decorated with superposed palmette fans, with spiraling side-scrolls.

The main decoration consists of a single phlyax on either side, although the two may be seen as combining to form a single picture. On one side is an older man, with white hair and beard (Mask: Type E), wearing a padded tunic over his close-fitting tights; in his upraised right hand he brandishes a stick, as he runs to the right in pursuit of his slave, who is running as hard as he can over ground represented by a double row of pebbles. The slave's costume is similar to that of his master; his expression is one of fear and anxiety; his mouth is open and his left arm outstretched. He wears a mask of Type N.

"Master and slave" was clearly a very popular subject for phlyax plays; they appear in a variety of scenes on the vases, though this is the only one which shows them separately.

Askoi of Type A sometimes have a figured scene only on the end below the mouth, the sides being decorated with scrollwork or floral patterns; a good example is in Basel (Antikenmuseum Z 303: *BICS* 9 [1962], p. 23, no. 130 bis, pl. 1, fig. 4), which shows a grotesque, pot-bellied, naked male figure running to the left with a shopping basket in his right hand; it has not yet been attributed, but must belong to the same general area as our vase.

BIBLIOGRAPHY: *RVAp,* Suppl. II, p. 74, no. 11/133b, pl. XII.5–6.

RELATED REFERENCES: For the shape, see U. Rüdiger, "Askoi in Unteritalien," *RM* 73 (1966), pp. 1–9; M. G. Kanowski, *Containers of Classical Greece* (St. Lucia, Queensland, 1983), pp. 30–32.

—A. D. T.

60
Red-figured Stemless Cup

Name Vase of the Painter of the Fleischman Phlyax Cup [A. D. Trendall]
Apulian, circa 360–350 B.C.
Terracotta; H: 5.8 cm; MAX. W: 23.9 cm; DIAM (OF BOWL): 16.1 cm
Condition: Intact.

This stemless cup has a low disk foot, decorated on the underside with four reserved concentric rings, and a deep interior with a wide offset lip zone decorated with an ivy wreath, the triple stem of which is incised and bears widely spaced pairs of leaves and berry clusters in added white. In the tondo on a band of dotted ovoli stands a phlyax (Mask: Type B), with a spray in his right hand and a small wreath in his left, facing Dionysos, who is seated on a chair, holding in his right hand a phiale containing three white objects, probably eggs. Dionysos has a piece of star-patterned drapery over the lower part of his body and wears a wreath of white-leaved ivy. The picture is encircled by a band of

60

ovoli, the black element having a reserved center, unlike those immediately below the picture, where the center is solid black.

The exterior is decorated on either side with a scene containing two figures. On side A is seated a draped woman; in her right hand she holds a phiale, decorated on the outside with a row of black dots, with three small white objects in it (eggs?). She faces a small nude figure of Eros, who is kneeling in front of her; he has a white circlet around his head and holds a white fillet with both hands. Side B shows similar figures, with their positions reversed—a kneeling nude boy, with left arm outstretched, beside a seated woman who turns her head to the left to look at him. She holds a phiale, again decorated with a row of black dots, in her right hand and a mirror in her left. She wears a chiton, with a piece of drapery across the lower part of her body and a radiate stephané on her head.

The cup was originally thought of as Proto-Paestan, but the drawing of the women's drapery, especially in the area above the waist, the treatment of the border of the short cloak over the lower part of their bodies, the rendering of the faces and, in particular, the radiate diadems in their hair, find very close parallels on (1) a mug in a private collection in Naples (see *RVAp,* Suppl. II, p. 50, no. 8/130d, pl. VII.1) and on (2) an unpublished lekanis once on the Paris market (see Hôtel des Ventes), on the lid of which are depicted seated women, kneeling satyrs and erotes, all of whom closely match those on this stemless cup. On the latter, one of the seated women holds a phiale decorated with a row of black dots, exactly as on the Fleischman vase. All three vases would appear to be the work of the same painter, one of the minor artists in the workshop of the Iliupersis Painter, who were active around the middle of the fourth century B.C. He might be given the name of the Painter of the Fleischman Phlyax Cup.

BIBLIOGRAPHY: Unpublished.

RELATED REFERENCES: Hôtel des Ventes d'Auxerre, *Civilisations Antiques Orient rituel et sacré,* April 17, 1988, pamphlet, p. 4 (no. 129). For vases by the Iliupersis Painter and from his workshop, see *RVAp* I, p. 184ff.; *RVAp,* Supp. II, p. 50.

—A. D. T.

61

Red-figured Patera with Anthropomorphic Handle

Attributed to the Circle of the Lycurgus Painter [A. D. Trendall]
Apulian, mid-fourth century B.C.
Terracotta; H: 47 cm; DIAM (OF BOWL): 26 cm
Condition: Good; little loss of added white; ocher paint on handle abraded in places.

Although terracotta casts made from metal handles are fairly common in Apulia, especially from tombs at Canosa, complete paterae, consisting of a shallow bowl decorated in red-figure and attached to a handle in the form of a nude youth with upraised hands, are rare; only two other examples are so far known, both rather later in date. One of these was once on the New York market (*RVAp,* Suppl. II, p. 284, no. 27/72–1, pl. LXXIV.4); the other is in the Allard Pierson Museum in Amsterdam (3573: *RVAp,* II, p. 1031, no. 30/97). The handle on our vase consists of a nude youth with long, curling hair; both his arms are upraised to support the attachment to the patera; there are two vent holes behind, one in the middle of his head, the other in the small of his back. His body is covered with ocher-yellow paint, either to simulate gold, or as a preparation for gilding.

The shallow bowl of the patera is decorated internally with a frontal head of Medusa in the tondo; it is encircled by a wreath of white vine leaves, with grapes in clusters of three, around a band of wave pattern. The Gorgon's head is shown fully frontal; the eyes, nose, and mouth are carefully drawn, and the hair is shown as a mass of curls, enlivened by brushed glaze. Above the brow is a black-dotted diadem, with three leaves in added white and yellow set in the hair above it. She has white earrings and two pairs of snakes are entwined around her head; one pair is knotted at her throat, the other emerges on either side of the head, partly concealed beneath the hair. Their mouths are open; their bellies are white and their bodies are decorated with a row of black and white dots. The underside of the patera has a low, reserved base ring and a border of veined laurel leaves, with white berries on a long stem between them.

This vase is not easy to place, since the design is highly stylized and fully frontal faces are compara-

61

tively rare in Apulian vase painting; those which regularly appear on the necks of large volute kraters are normally turned slightly to the left or to the right. A Gorgon head on an oinochoe in Bochum (S 1010: *RVAp* II, p. 1065, no. 15/37b), from the Group of the Dublin Situlae, is comparable, though here the snakes are entwined in the hair, as well as encircling the neck. Closer perhaps are the frontal female heads (1) in the tondo of a dish formerly on the London market (see Sotheby's; *RVAp*, Suppl. II, p. 110, no. 16/23b) and attributed to the Lycurgus Painter, on which the eyes show the same touch of added white as on our vase and on several others by the same painter (e.g., *RVAp* I, nos. 16/5–6), and (2) on the underside of the base of the situla in New York (Metropolitan Museum of Art 56.171.64: *RVAp* I, no. 16/17, pl. 151.1). The laurel wreath with veined leaves and white berries that encircles the back of our vase is repeated on *RVAp* I, nos. 16/2 and 12. These comparisons suggest that our vase should probably be associated with the Lycurgus Painter around the middle of the fourth century B.C. A rather later version of the same subject is found on an oinochoe in Taranto (Museo Archeologico Nazionale 61501: *RVAp* II, p. 954, no. 28/356, pl. 373.6).

BIBLIOGRAPHY: Unpublished.

RELATED REFERENCES: Sotheby's, London, sale cat., May 20, 1985, no. 382, pl. 40.1. For bronze handle prototypes, see P. Amandry, "Manches de patère et de miroir grecs," *Monuments et mémoires: Fondation E. Piot* 47 (1953), p. 48ff.; U. Jantzen, *Griechische Griff-Phialen*, BWPr 114 (1958); G. Schneider-Herrmann, "Apulische Schalengriffe verschiedner Formen," *Bulletin Antieke Beschaving* 37 (1962), pp. 40–51 (lists of extant examples). For terracotta casts, see S. Mollard-Besques, *Catalogue raisonné des figurines et reliefs en terre-cuite grecs, etrusques et romains*, vol. 4 (Paris, 1986), p. 95f., esp. no. D 3859, pl. 91b (from Canosa); *Oudheidkundige Mededelingen uit het Rijksmuseum van Oudheiden te Leiden* 63 (1982), pp. 95–97.

—A. D. T.

62
Red-figured Volute Krater

Attributed to the Underworld Painter [A. D. Trendall]
Apulian, circa 330–320 B.C.
Terracotta; H: 63.5 cm; DIAM: 26 cm
Condition: Good; slight abrasions of the added white on architrave of naiskos, at lower right-hand corner of plinth, and on meander frieze; also on faces of masks on obverse of volute handles.

Both sides of this vase depict funerary scenes—on the obverse are a woman and youth with offerings at a *naiskos* (funerary shrine), in which is a seated youth, on the reverse there are similar figures on either side of a stele. The naiskos and the youth inside it are shown in added white to simulate the stone of the actual grave monument. The naiskos stands on a low plinth, decorated with a white scroll pattern, and Ionic columns support the undecorated architrave and pediment. The seated youth holds in his upraised right hand a comic mask of an old man with white hair and beard (Type L or M), below him is an open scroll. To the left of the naiskos stands a woman, wearing a sleeveless tunic, caught up at the waist by a black ribbon girdle; she holds a looped fillet in her right hand and a mirror in her left. On the other side stands a nude youth; his left arm is enveloped in a piece of drapery; his upraised right hand supports an open box. The reverse represents a similar pair beside a stele with a patera as the crowning element. The woman, who closely resembles her counterpart on the obverse, holds a bunch of grapes in her right hand and a cista in her left. The youth to the right of the stele is nude, save for a piece of drapery over his left arm.

The neck of the obverse is decorated with the head of a black-bearded and balding satyr facing left; it is set amid scrolling tendrils and flowers. On the reverse is a palmette fan flanked by leaves and scrolls. Below the handles are superposed palmettes, the lower of triangular shape, with side-scrolls and smaller fans.

The handle volutes are decorated with female heads in low relief, wearing diadems; on the obverse their hair is golden yellow and their faces white, on the reverse they have black hair and red faces.

The combination of a naiskos scene on the obverse with a stele scene on the reverse is popular with the painters of larger Apulian vases (e.g., volute kraters, amphorae, loutrophoroi) from the second quarter of the fourth century onward. This combination occurs on several of the less pretentious vases of the Underworld Painter (e.g., *RVAp* II, p. 535ff., nos.

62 SIDE A

62 SIDE B

SIDE A–B

18/291–293, 299, 301–302; *RVAp,* Suppl. II, no. 302a), to whose hand this vase may be attributed. The draped woman, wearing a black girdle, tied with a loop at the waist, each end of which terminates in a row of dots, is almost a "stock figure" in his funerary scenes (cf. the reverses of Munich, Antikensammlungen 3296, 3297: *RVAp,* pls. 195.2, 194.2, 198.2), as is the nude youth with drapery covering one arm (see *RVAp,* Suppl. II, pl. XLI.2). The floral setting for the satyr's head on the obverse finds close parallels on other vases by the Underworld Painter (see *RVAp,* nos. 18/289, 290, pls. 198.1, 199.1), though the satyr's head itself with its black beard and receding hair is less usual in this context.

Masks associated with figures in naiskoi are rare in Apulian vase painting; nearest to our vase is the volute krater in Trieste (Civico Museo di Storia d'Arte S 383: *RVAp* I, p. 410, no. 15/70), which shows a partly draped youth in a naiskos, holding a long-haired female mask (Type SS) in his right hand; a *kithara* (lyre) is suspended above in the top left-hand corner. A second volute krater (Lecce, Museo Provinciale Castromediano 3544: *RVAp* I, p. 410, no. 15/69) shows a half-draped youth in a naiskos standing in front of a hound with a kantharos above him and the mask of a white-haired old man suspended in the top left-hand corner. These two vases suggest a connection between the man in the naiskos and the comic stage; on the former, in view of the kithara suspended above, he may have been a composer as well as an actor; on the latter the comic mask is comparatively incidental and is perhaps an indication that the man was an actor, though the hound and the kantharos are more in keeping with the normal pursuits of youth. That the mask on the Fleischman krater is held in the young man's hand suggests a closer connection with the stage, and this is reinforced by the presence of the open scroll. We may reasonably conclude that we have here the representation of a young actor holding in his hand the mask of one of the roles he once played, with perhaps the text on the scroll below, which is marked by a series of parallel lines of dots, presumably to imitate writing.

BIBLIOGRAPHY: *RVAp,* Suppl. II, p. 162, no. 18/293b, pl. XLI.3; O. Taplin, *Comic Angels* (Oxford, 1993), p. 93, pl. 23.23.

RELATED REFERENCES: For the Underworld Painter, see *RVAp* II, pp. 531–540; *RVAp,* Suppl. I, pp. 83–86; *RVAp,* Suppl. II, pp. 161–165, and the bibliographies there given; A. D. Trendall, *Red-figured Vases of South Italy and Sicily* (London, 1989), pp. 90–92, 273ff. (bibl.). For vases depicting grave monuments, see H. Lohmann, *Grabmäler auf unteritalischen Vasen,* Archäologische Forschungen 7 (Berlin, 1979). For the heads on handle volutes, see, L. Giuliani, "Vierfältige Lockenköpfe," *Kanon: Festschrift Berger* (= *Antike Kunst,* Beiheft 15 [1988]), pp. 159–164.

—A. D. T.

63
Situla

Attributed to the Workshop of the Konnakis Painter [A. D. Trendall]
Apulian (Gnathia ware), circa 360–350 B.C.
Terracotta; H (WITHOUT HANDLE ATTACHMENTS): 21.9 cm; (WITH ATTACHMENTS): 25.5 cm; DIAM (OF MOUTH): 21 cm
Condition: Intact; extremely well preserved.

The *situla* is essentially a bucketlike container, equipped with handles; it could be used both for mixing and for carrying wine and water. The three forms in which it appears in South Italian pottery are clearly derived from metal prototypes. The first, of which the Fleischman vase is a good example, has a wide, open mouth with a projecting rim, a slightly concave, cylindrical body, which curves outwards as it approaches the base, normally supported by three low feet. This shape is rare in Apulian red-figured ware, only five examples being so far known, but it is frequently represented in actual use on the vases themselves.

The Fleischman situla represents a phlyax actor moving to the right on an uneven ground line consisting of two rows of dots; his head is turned back to the left, with the eyes looking slightly upward. He wears tights of a deep reddish brown color, and over them a short-sleeved tunic in added white, with a cloak, enlivened by an ocher-shaded wash, draped across the lower part of his body and over his left arm, which is completely concealed beneath it. On his feet are open sandals; the toes are clearly visible, the nails shown as white dots. The legs and right arm have vertical white lines on them, presumably to indicate the seams of the tights. His right hand is splayed out. Around his head is a wreath with white leaves and small berries; on either side of his head, just above the ears, a short spray sticks upward, with a white ribbon looped around the ear on his left side. He wears a mask (Type E), on which the facial details are highlighted; the wide-open mouth clearly reveals his teeth and is outlined in white, to suggest a mustache and a short white beard. The reverse is undecorated.

In style the vase is close to the work of the Konnakis Painter, and it may well be by his own hand. We may note as characteristic of his style—the isolation of the figure against the black background, the absence of filling ornaments or framing, and the ivy trail with incised stems and white leaves, with rounded bottoms, separated by berries in clusters of three. Also characteristic is the careful polychromy, which finds good parallels on his vases (see Forti, pls. X–XII, XXXIII a, c). The Konnakis Painter, one of the founding fathers of the Gnathia style, was active in the second quarter of the fourth century B.C., and this vase should be dated around 360–350.

BIBLIOGRAPHY: Unpublished.

RELATED REFERENCES: For the shape and prototypes in metal, see B. Schroeder, *Griechische Bronzeeimer,* BWPr 74 (1914); J. Kastelic, *Situlenkunst* (Vienna, 1964); A. D. Trendall, "Two South Italian Red-figure Vases in a Private Collection in Sorengo,"

63

Numismatica e antichità classiche 19 (1990), pp. 118–120. For Gnathia in general, see L. Forti, *La Ceramica di Gnathia* (Naples, 1965); J. R. Green, "The Gnathia Pottery of Apulia," in M. E. Mayo and Kenneth Hamma, eds., *The Art of South Italy: Vases from Magna Graecia*, exh. cat. (Virginia, Museum of Fine Arts, 1982), pp. 252–259; idem, "Some Painters of Gnathia Vases," *BICS* 15 (1968), pp. 34–50; idem, "Gnathian Addenda," *BICS* 18 (1971), pp. 30–38; idem, *Gnathia Pottery in the Akademisches Kunstmuseum Bonn* (Mainz, 1976); idem, "Some Gnathia Pottery in the J. Paul Getty Museum," *Greek Vases in the J. Paul Getty Museum* 3, Occasional Papers on Antiquities 2 (Malibu, 1985), pp. 115–138; T. B. L. Webster, "Towards a Classification of Apulian Gnathia," *BICS* 15 (1968), pp. 1–33.

—A. D. T.

64
Calyx Krater

Proto-Paestan, circa 360 B.C.
Terracotta; H: 37.5 cm; DIAM (OF MOUTH): 38.5 cm; (OF FOOT): 15 cm
Condition: Reconstructed from fragments; minor abrasions of the added white and surface of rim.

On the obverse of this calyx krater is a lively Dionysiac scene showing three figures moving to the left over gently undulating ground, depicted by curving lines in added white, from which spring a few plants. Leading the procession is a youthful Dionysos, naked save for a piece of drapery over both arms and flowing behind his back; he has long, curling hair, with a white ivy wreath around his head, and in his left hand he holds a kithara, with the plectrum in his right. He turns his head back to the right to look at the white-haired phlyax who follows him, a blazing torch in each hand. The phlyax wears the mask of the old man (Type L or M). Behind him comes a bearded satyr whose face is seen in three-quarter view; his chest and belly are covered with shaggy hair, and he is playing a flute, one reed of which is held in each hand. He carries piggyback a small Eros, who wears a spiky diadem; his hands grasp the satyr's hair above the brow, and his legs come down on either side of his head. The wings of Eros are outspread; they are enlivened by a white edge at the top and black or black-on-white dots on the feathers.

The reverse is more banal and represents a nude, bearded satyr bending forward over his left foot, which rests upon a rock pile; in his left hand he holds a string of beads and with his right gesticulates towards a maenad, who stands beside him, wearing a chiton with a black stripe down its left side and a dot-stripe border. She holds a thyrsos in her left hand, her right arm is extended, with the hand drooping languidly down.

On stylistic grounds the vase may be assigned to the Group of Louvre K 240, which provides a direct link between Sicilian red-figured ware of the second quarter of the fourth century B.C. and the first truly Paestan vases from the workshop of Asteas. The vases in this group mostly depict Dionysiac scenes, often with theatrical con-

64 SIDE B

64 SIDE A

nections, indicated by the presence of a comic mask or a phlyax actor (as on this vase); very close parallels to these scenes may be seen on the earlier vases attributed to Asteas (cf. *RVP*, pls. 17, 20–22). Some of the vases in this group were found in Sicily (e.g., Lipari, Museo Regionale Eoliano 9604, 9558, 927, Gela, Museo Archeologico 8255–6, Syracuse, Museo Archeologico Regionale 29966: *RVP*, nos. 91, 92, 99, 102–103), others on the mainland (*RVP*, nos. 95, 100 and probably 96, 97), but as yet there is no clear evidence when the move from Sicily to the Paestan area took place. The general treatment of the Fleischman krater would place it nearer the vases from Sicily, but unfortunately its find spot is not recorded. Many parallels with other vases in this group can be found: for the dot-stripe border to the drapery (see *RVP*, p. 44ff., nos. 94–98), for the lyre (nos. 96, 97), for the use of flutes (nos. 91, 94, 96, 97), for the ground lines with small plants (nos. 91, 94, 98), and for similar processions (nos. 94, 96–100). Closer still is the papposilen carrying an Eros piggyback on no. 98; also the white-haired phlyakes on Lipari 927 (no. 99) and on the two fragmentary vases Gela 8255 and Syracuse 29966 (nos. 103, 104). We may note also that the saltire squares that accompany the frieze of meanders below the pictures are the same as those on Louvre K 241 (no. 98); they are closely modeled on the type favored by the Dirce Painter (cf. *RVP*, pls. 1, 2), another immediate forerunner of Paestan.

The simpler design of the reverse is also in keeping with those on the other vases in this group (cf. Lipari 9604, 9558: *RVP*, nos. 91, 92; Taranto, Museo Archeologico Nazionale 50240: no. 100), all of which show a maenad in the company of a satyr or youth.

We may also note the presence of a dot-stripe border on the piece of drapery over the arms of Dionysos and on the lower border of the chiton worn by the maenad on the reverse. This appears frequently on other vases in the Group of Louvre K 240 and becomes one of the most characteristic features of Paestan drapery (see *RVP*, p. 12).

The vase is probably datable to the second quarter of the fourth century B.C., not long before the middle of the century.

BIBLIOGRAPHY [F94]: *RVP*, p. 46, no. 101, pl. 13b–c; idem, *Red-Figured Vases of South Italy and Sicily* (London, 1989), p. 199, ill. 341; O. Taplin, *Comic Angels* (Oxford, 1993), p. 33, n. 11.

RELATED REFERENCES: For the Group of Louvre K 240, see *RVP*, pp. 42–48; A. D. Trendall, "Two South Italian Red-figure Vases in a Private Collection in Sorengo," *Numismatica e antichità classiche* 9 (1990), pp. 117–124. For Eros carried piggyback, see Louvre K 241: *RVP*, op. cit., pl. 12d; *LIMC*, vol. 3, p. 914.

—A. D. T.

65
Squat Lekythos

Attributed to Asteas [A. D. Trendall]
Paestan, circa 350–340 B.C.
Terracotta; H: 45.5 cm; DIAM (OF BASE): 18.3 cm
Condition: Reconstructed from large fragments; some of added color on laver and on palmette on which Hesperid is seated missing; some retouching.

The squat lekythos in its smaller form, with a comparatively narrow body, decorated with a single figure or a female head, is common in Paestan vase painting; this larger form, with a much squatter body, better adapted for mythological compositions, appears far less frequently. The subsidiary decoration is extensive; on the neck, white ivy, ovoli, and tongues descend in bands to the shoulder, which is decorated with pairs of leaves in added white, separated by paired berries on stems, and meeting in a central five-petaled flower (cf. Naples 2873: Volkommer). Beneath the handle are superposed palmette fans with spiraling tendrils, accompanied by palmettes and terminating in florals or "Asteas flowers." Immediately below the handle join is a relief plate, perhaps cast from metalwork, with a crouching panther, surrounded by spiraling tendrils, above a leaflike pattern and an inverted palmette.

The scene on this vase depicts the Garden of the Hesperides, not, as more commonly shown, as the setting for the twelfth labor of Herakles, but in its own right as an idyllic fantasy. In the center is the tree bearing the golden apples; around its trunk is coiled the serpent, with a red crest, white belly, and golden brown scales, feeding tranquilly from a phiale held out in the hand of one of the Hesperids. On the other side a second Hesperid is plucking an apple from the tree to add to the two she is already holding in her left hand. These two Hesperids are similarly dressed; they wear peploi with overfalls bordered with a checkered pattern, which is also used for the lower border; a dot-stripe runs down the side of the garment, which is decorated with rows of dots and stars. The Hesperid on the left also wears a short mantle in added red. Both wear bracelets and the one on the left has her hair done up in a sphendone, the other in a red headband, with a ponytail emerging at the back. On the left a half-naked Hesperid, the lower portion of whose body is covered by a piece of drapery with a dot-stripe border, sits upon an enclosed white palmette fan—a typical device in Paestan scenes (e.g., *RVP*, nos. 2/37, 53, 62, 76, 137)—looking into a mirror held in her right hand, while she adjusts her long hair with her left. To the far right another Hesperid, dressed like the apple-gatherer, but with wave-pattern borders on the lower hem and overfall of her peplos, leans over a washbasin, shown in added white, holding out a beaded wreath towards a small bird (now partly vanished) perched on the rim. Above her to the right, an elderly, balding satyr (shown bust-length), with white hair and beard, and a mass of white hair on his chest, looks down from behind rising ground, an egg in his right hand, a thyrsos in his left, and a panther-skin over his left shoulder. At the foot of the tree are two birds in added white, a dove to the left and a duck to the right; above, on both sides of the tree, hang a looped red fillet and a wreath with a fillet wound through it.

The vase may be attributed to Asteas and makes a remarkable pendant to his signed squat lekythos in Naples, which depicts a similar scene but also shows the youthful Herakles in the garden, with busts of Pan, Hera(?), Hermes, and Donakis above. The central scene is very similar, with the serpent coiled around the golden-apple tree, being fed from a phiale held by a Hesperid named Kalypso on the left—here she is seated on a spiraling tendril—while a second Hesperid in a similar pose to the one on our vase plucks a third apple to add to the two already in her left hand. The two birds are also shown at the bottom of the tree, though on the Naples vase the dove perches on the Hesperid's foot. The Fleischman lekythos looks to be a little earlier than the one in Naples—the figures of the Hesperids are less mature, and they look closer to the women on some of the earlier vases attributed to Asteas (e.g., *RVP*, no. 2/76 for the Hesperid on the far left; no. 2/86 for the face of the Hesperid to the left of the apple tree; no. 2/87 for the pose with the bent leg). The serpent also finds a parallel on the Medea lekythos in Bochum (S 1080: *RVP*, no. 2/143). No less typical of Asteas are the decorative patterns, especially the so-called Asteas flower, which here appears in conjunction with a floral springing from a calyx at the end of a spiraling tendril (*RVP*, no. 2/134 shows exactly the same combination; cf. also the florals on nos. 2/102, 106, 109). The rendering of the drapery, with its clearly defined fold-lines, the dot-stripe, checkerboard, and wave-pattern borders, is also very much in the manner of Asteas, as is the treatment of the women's hair and headdresses. The two other large squat lekythoi (Paestum 4794 and Bochum S 1080: *RVP*, p. 109, nos. 2/142, 143), attributed to Asteas, also provide good parallels for the shape, pattern-work, and treatment of a mythological subject, though both seem slightly later in date. Our vase should be placed fairly early in the second half of the fourth century, circa 350–340 B.C.

BIBLIOGRAPHY: Unpublished.

RELATED REFERENCES: R. Vollkommer, *Herakles in the Art of Classical Greece* (Oxford, 1988), pp. 16–19, p. 65, no. 472 (where the squat lekythos by Asteas [Naples 2873] is erroneously connected with a satyr-play). For Paestan ware in general and Asteas in particular, see *RVP*, chaps. 4–6; idem, *The Red-figured Vases of South Italy and Sicily*, (London, 1989), pp. 196–202. For the typical squat lekythos shape from the Asteas-Python workshop, see *RVP*, p. 86, no. 135; p. 99ff. (Naples 2873); p. 109ff., nos. 142, 143 (Paestum 4794; Bochum S 1080); p. 132, nos. 209a, 210 (Paestum 32129; Meer, priv. coll.). For a good later example, see Louvre N 3148: *RVP*, p. 260, no. 1025. For the Asteas flower, see *RVP*, p. 18, fig. 4; for other combinations of floral and Asteas flower (which seem confined to Asteas), see *RVP*, nos. 2/102, 106, 109, 134. For the Garden of the Hesperides, see *LIMC*, vol. 5, p. 396f.

—A. D. T.

65

65

LIMESTONE

66
Lion's Head Waterspout

66

West Greek, mid-fifth century B.C.
Limestone; H: 17.5 cm; DEPTH: 18.5 cm
Condition: Broken off in back; slight damage on top; lower jaw broken and mended, with some losses; ear tops scuffed.

The lion has a narrow, bony face with a strongly marked median line. In style it recalls the much larger lion's head waterspouts from the south side of the great temple built at Himera in Sicily after the victory over the Carthaginians in 479 B.C. (see Mertens-Horn). Our piece is not, however, made in one with a sima, but was carved as a separate plug and had water draining through it from above rather than behind. It is unlikely to have been a fountain spout, set into the wall of a spring house, because it has not been eroded by continuously trickling water. The head must have served, perhaps as a repair, in a small, atypical construction (see Langlotz for two small, separately made waterspouts of later date). It differs from early lions at Himera and elsewhere in having no mane on the neck behind the ruff; much of the smooth rear section would have been sunk into the background.

BIBLIOGRAPHY: Unpublished.

RELATED REFERENCES: M. Mertens-Horn, *Die Löwenkopf-Wasserspeier des griechischen Westens im VI und V Jahrhundert vor Christus im Vergleich mit den Löwen des griechischen Mutterlandes* (Mainz, 1988), no. 23A. E. Langlotz, *The Art of Magna Graecia* (London, 1965), pl. 157.

—A. H.

67

67
Sculptural Fragment: Two Women at a Fountain

Tarantine, late fourth–early third century B.C.
Limestone; H: 17 cm; W: 23.5 cm
Condition: Broken off at top and bottom; head of woman at right missing.

Two women flank a fountain from the lion's head spout of which water gushes. The woman at the left had some further object, probably a water jar, fixed by a bronze pin to the top of her head. The piece is not a relief but is neatly finished at the back; the "background" is the upright wall of the fountain in front of which the two women stand.

Such sculptures, carved from soft, fine-grained local limestone, were produced at Taranto in the late fourth and the first half of the third century B.C. Their style is a local interpretation of mainstream Greek trends. It is clear that they decorated small funerary buildings, but their positions can in most cases only be inferred. Cutout groups like ours, quite frequent among the finds, are usually said to be pedimental, but it would make no sense to finish backs that were not intended to be seen, especially as in some such pieces details of considerable interest appear at the back (see Carter, no. 313, pl. 52 b–c). A few at least of the figure groups may have served as acroteria of small, ornate buildings, such as are sometimes reflected in Apulian vase painting.

Water carriers, often identified as the Danaids, were a favorite subject in South Italian funerary art, appearing in vase painting as well as on the Tarantine reliefs. The implied theme may be purification through toil. Another Tarantine limestone fragment with a woman at a fountain, close in style and scale to ours, is on loan to the Basel Antikenmuseum (see Carter, no. 172, pl. 28).

BIBLIOGRAPHY: Unpublished.

RELATED REFERENCES: J. Coleman Carter, *The Sculpture of Taras,* Transactions of the American Philosophical Society, n.s. 65, pt. 7 (Philadelphia, 1975). E. Keuls, *Water-Carriers in Hades: A Study of Catharsis through Toil in Antiquity* (Amsterdam, 1974).

—A. H.

Etruria

68

BRONZE AND LEAD

68
Finial in the Form of a Lion's Head

Etruscan, late sixth century B.C.
Bronze, with details added in white paste and iron; H: 8 cm; L: 9.5 cm
Condition: Edge of neck broken away all around; large area of mane missing above right ear; hole in left ear.

The lion's head and neck are hammered in one piece from very thin bronze sheet. The open jaw is almost completely filled by the centrally grooved tongue. The upper and lower incisors are embossed in the round, the other teeth and the gums modeled in low relief. Stylized skin-folds fanning out from the upper lip contract the sides of the muzzle into curving ridges and continue around the back of the mouth and along the sides of the lower jaw. On top of the muzzle crescent-shaped ridges running back from the nose diverge to frame a raised lozenge-shaped projection between the eyebrows, each of which is ornamented with a punched circle filled with an incised whorl. The whites of the eyes are inlaid in paste, their irises formed by bronze disks inset in this and held in place by split-pins of iron, which also serve to represent the pupils. The mane consists of striated, flamelike tufts of hair arranged in superimposed rows which decrease from four on top of the neck to two underneath it. The ears are folded back; on top of the front of each is a whorl similar to those above the eyes.

That the head was an ornamental protome, not a fragment of a complete lion, is proved by the regular, cylindrical shape of the neck. The use of the lion protome as a decorative motif was introduced into Etruscan art from the East during the orientalizing period. An Assyrian vase hammered from sheet bronze in the shape of a lion's head and dating from the late eighth century B.C. has been found at Veii (see Brown, p. 12f., pl. VIb). Early Etruscan examples are provided by repoussé bronze lion's heads which were originally attached below the feet of the kouros figures on the chariot from Monteleone in the Metropolitan Museum of Art, which dates from about 540 B.C. (see Emiliozzi); while more elaborately worked lion's heads with inlaid eyes appear as decorative bosses on the small round shields of hammered bronze made as decoration for Tarquinian tombs in the second half of the sixth and the early fifth century B.C. (see Brown, p. 101f., pl. XLI). But all these lack the depth of the present head. Of comparable dimensions (diameter: 9 cm, length: 11 cm) is a repoussé lion's head with inlaid eyes from the Tomba del Guerriero in the Osteria necropolis of Vulci (Rome, Villa Giulia 63580: Cristofani). A series of nail holes around the edge of the neck of this piece indicates that the head was attached to a wooden object, probably the end of a chariot pole. Although the rear of the neck, where the nail holes would have been situated, is missing on the present head, it is probable that it, too, originally decorated the end of a wooden pole of some kind.

There seems to be no close stylistic parallel for the Fleischman bronze, but the rare motif of the whorls above the eyebrows and on the ears occurs on gold lion's head pendants of the late sixth century B.C., two in Berlin (Antikenmuseum, Charlottenburg GI 416, 417: Cristofani/Martelli) and one in Paris (Cabinet des Médailles, Collection de Luynes 502: von Mercklin), as well as on a fragmentary bronze lion of the second half of the fifth century in St. Petersburg (Hermitage V493: *Die Welt*).

BIBLIOGRAPHY: Unpublished.

RELATED REFERENCES: W. L. Brown, *The Etruscan Lion* (Oxford, 1960). A. Emiliozzi in *Antichità dall'Umbria a New York*, exh. cat. (Perugia, 1991), p. 113ff., figs. 112, 115. M. Cristofani, ed., *La civiltà degli Etruschi* (Milan, 1985), p. 300f., no. 11.21, 9. M. Cristofani and M. Martelli, *L'oro degli Etruschi* (Novara, 1983), no. 157. E. von Mercklin, "Etruskischer Bronzelöwe in der Ermitage," *Scritti in onore di B. Nogara* (Vatican, 1937), p. 278ff., pl. XXXI.2–5. *Die Welt der Etrusker* (Berlin, 1988), p. 272f, fig. and no. D 2.28.

—S. H.

69
Statuette of a Nude Youth

Etruscan, late sixth century B.C.
Bronze; H: 9 cm
Condition: Left hand missing; smooth dark brown and green patina.
Provenance: Formerly in the d'Avray and Perrier collections.

The statuette is cast solid together with its flat rectangular base, which is pierced by two holes for mounting. Like a Greek kouros, the youth stands with his left foot slightly advanced. The left forearm is extended horizontally and the miss-

69

ing hand probably held out an offering; the right arm is slightly bent with the hand stretched out flat, the palm turned inwards. The feet are large and the powerful legs long compared with the torso. The shoulders are wide and sloping, the clavicles rendered in relief as an arc with a notch in the center. He looks straight ahead. His face is round with a low, receding forehead, widely arched brows, almond eyes, a short, narrow-bridged nose and a large, slightly smiling mouth. The forehead is covered by a two-tiered fringe of hair, incised with parallel strokes; and a wavy mass of hair hangs down the back of the neck. The crown of the head is left smooth, as if covered by a close-fitting cap. The big ears are only summarily modeled.

The smooth, rounded bodily forms, which are interrupted by few anatomical details, the swelling curves of the muscular limbs, and the broad shoulders are characteristic of the Ionian style of Middle Archaic Etruscan sculpture (see, for example, the fine statuette in the British Museum: Richardson; Haynes 1985, no. 35). The massive anatomy is balanced by the springy readiness of the pose and the alert expression. The statuette is said to come from Vulci, a provenance that accords well with its style.

BIBLIOGRAPHY: Unpublished.

RELATED REFERENCES: E. Richardson, *Etruscan Votive Bronzes* (Mainz, 1983), figs. 461, 462.

—S. H.

70

Statuette of an Archer (from the top of a candelabrum)

Etruscan, late sixth–early fifth century B.C.
Bronze; H (INCLUDING BASE): 11.5 cm
Condition: Bow missing; green patina with patches of encrustation.
Provenance: Formerly in the d'Avray and Perrier collections.

The statuette is cast solid together with its base. The molded edge of the circular base is decorated with a tongue pattern. The young archer stands with his left foot slightly advanced, his left arm at his side, the fist closed, and his right forearm raised to hold the now-missing bow. He is dressed in pointed boots, tight leggings with an incised rhomboid pattern, and a close-fitting upper garment the whole of which, including the elbow-length sleeves, is covered with rows of zigzags. Over this is draped a doe-skin, its forelegs knotted on his left shoulder, its head hanging in front of his groin, its short tail over his buttocks, and its hind legs down his legs, their pointed hooves reaching to mid-calf. The skin is dappled with incised almond-shaped spots, suggesting that it is that of a fallow deer. Slung from a bandolier passing over the archer's right shoulder is a curved bow-case which hangs under his left arm. The case is incised with a close crisscross pattern. On his head he wears a tall conical cap with a budlike peak curving over forward and a long pointed neck-guard which falls down his back almost to the waist. Decorated all over with a zigzag pattern, the cap is lined with leather or some other soft material, a long lappet of which hangs free in front of each ear. He has a cheerful face with large, level, almond-shaped eyes, a short nose, a slightly smiling mouth, and a heavy chin. A fringe of hair frames his low forehead.

Costume, cap, and bow-case are characteristic of Scythian archers and Amazons as depicted on Greek vases, by which the Etruscan artist was probably inspired. But the deerskin denotes a hunter, and it is likely that the bronze represents an Italic divinity or hero connected with the chase. Closely related to the present bronze in type are a statuette from Valle Fuino in the Vatican (Museo Gregoriano Etrusco 12056: Ciotti), ascribed to a workshop of Vulci; a less well-preserved piece in Geneva (Musée d'Art et d'Histoire MF 1017: Richardson); and one in Paris (see de Ridder). A similar, but bearded figure from Contarina (Adria, Museo Nazionale 15844: Cristofani) has been mistakenly described as wearing a lion's skin and identified as Herakles.

The present statuette is said to have been found at Vulci, a provenance with which its style agrees.

BIBLIOGRAPHY: Unpublished.

RELATED REFERENCES: U. Ciotti in *Antichità dall'Umbria in Vaticano*, exh. cat. (Perugia, 1988), p. 104ff., fig. 6.5. E. Richardson, *Etruscan Votive Bronzes* (Mainz, 1983), p. 361f., Type XI, fig. 867. A. de Ridder, *Les Bronzes antiques du Louvre*, vol. 1 (Paris, 1913), no. 223. M. Cristofani, *I bronzi degli Etruschi* (Novara, 1985), p. 282, fig. 95.

—S. H.

70

71

Statuette of a Youth Putting on His Cuirass (from the top of a candelabrum)

Etruscan, 490–470 B.C.
Bronze; H (INCLUDING BASE): 10 cm; (OF FIGURE): 8 cm
Condition: Intact except for damage to right elbow, missing right ear, and abrasion of nose; decoration of cuirass very worn; greenish brown patina with patches of azurite on cuirass and top of base.
Provenance: Formerly in the Pitt-Rivers collection.

The statuette is cast solid together with its spool-shaped base, the upper rim of which is edged with beading above an egg-and-dart pattern. The youth stands with his left leg forward and both bare feet firmly on the ground. He is dressed in a brief, short-sleeved tunic of light material, the hem of which is decorated with dots and hangs in an undulating edge over his buttocks and groin, leaving the genitalia uncovered. Over this tunic he wears a cuirass with a double row of flaps (*pteryges*) along its lower edge. The cuirass is ornamented with rows of punched circles and semicircles. With both elbows raised, the youth has seized in each hand one of the two shoulder-flaps of the cuirass, pulling them forward so as to fasten them in front of his chest. The flaps are incised with crisscrossed lines. He looks straight ahead with his wide-open almond eyes. His oval face has a small mouth, the lips of which are slightly pulled up at the corners, and a firm chin. Finely striated and circled by a fillet, his hair hangs in

71

a broad mass over his nape and shoulders; a fringe falls on his low forehead and temples, leaving the ears free.

Three candelabra excavated at Spina were surmounted by figures of warriors donning their cuirasses (see Hostetter). But these statuettes, which are less delicately modeled, show the stage preceding that illustrated by the present bronze, the warriors being still engaged in fitting the open cuirasses around the torso. Here the subsequent act of pulling the shoulder-flaps upward and forward to fasten them on the closed cuirass is represented—and with great skill. The form of the base and the facial type of the youth suggest that the statuette was probably made in a workshop of Vulci.

BIBLIOGRAPHY: Unpublished.

RELATED REFERENCES: E. Hostetter, *Bronzes from Spina* (Mainz, 1986), p. 54ff., nos. 28–30, pls. 34–37.

—S. H.

72

Statuette of a Nude Youth Brandishing a Weapon

Umbrian, early fifth century B.C.
Bronze; H (INCLUDING TANGS): approx. 27 cm; (OF FIGURE): 22.2 cm
Condition: Intact except for missing, separately cast weapon; smooth greenish brown patina.

The statuette is cast solid. The casting tangs under the feet were retained in order to fix it in a base, suggesting that it was a votive figure. Of slim and elongated build, the youth strides forward vigorously with his left leg advanced; in the pierced fist of his raised right arm he held a weapon, presumably a spear. To balance the movement of throwing, he swings his left arm forward with fist closed. Clavicles, pectoral and abdominal muscles, and buttocks are schematically, but successfully modeled in ridges and smoothly flattened planes. He has a large oval face with strongly arched brows, big almond eyes with emphatic lids, a long straight nose, a small smiling mouth, and a massive chin. His hair, which hangs in a thick smooth fringe over forehead and temples, is tied with a ribbon, over which, behind his small ears, it is rolled up.

A nude youth brandishing a weapon is a frequent subject of Etruscan and Italic art of the Middle and Late Archaic period (circa 550–470 B.C.), and a fine Etruscan example is the bronze statuette in the British Museum (Bronze 518: Richardson). The closest parallel to the present example is an Umbrian kouros in Palestrina (Museo Archeologico Prenestino 1502: Quattrocchi), which E. Richardson thinks may be by the same hand (private communication).

Umbrian bronze workers of the Late Archaic period copied Etruscan prototypes, but developed their own characteristically mannered style in doing so. The best Umbrian bronzes, among which the Fleischman statuette must be numbered, achieve a nice balance between recognizably human proportions and extreme attenuation.

BIBLIOGRAPHY: Unpublished.

RELATED REFERENCES: E. Richardson, *Etruscan Votive Bronzes*, vol. 2 (Mainz, 1983), fig. 471. G. Quattrocchi, *Il Museo Archeologico Prenestino* (Rome, 1956), figs. 43–45.

—S. H.

72

73

73
Foot of a Cista in the Form of a Winged Youth

Etruscan, early fifth century B.C.
Bronze; H (INCLUDING BASE): 15.2 cm
Condition: Intact; slightly encrusted light green patina.

The foot, which is cast solid, consists of a feline paw surmounted by an attachment plate in openwork relief. From the paw, which has strongly developed fleshy pads under the claws, rises an elongated, flat Ionic capital with an incised triangle and a beaded abacus. On this, but extending beyond it on either side, rests a row of symmetrically opposed cut-out waves. Over their crests skims a nude winged youth with his knees bent in a position known as "Knielauf," an Archaic schema for rapid movement. His legs are in profile to the right, his body and head frontal, the transition between the two directions being reflected in a slight twist in the detailed musculature of the torso. He wears soft pointed boots to which small wings are attached at the heels. His sinewy arms with their finely modeled hands are stretched out sideways in front of his upswept, sickle-shaped wings, the feathers of which are indicated by incision. The youth has a delicately modeled oval face with sharply defined almond eyes, a long, thin nose, and a full, slightly smiling mouth. His hair, which is crowned by a beaded fillet, falls in a close striated fringe on forehead and temples and hangs in a broad flap behind his shoulders.

The foot was once one of a set of three belonging to a cylindrical *cista* (box for cosmetics); the back of the relief is slightly curved to fit the box, and the back of the leg is recessed behind the capital to support it. From the curvature of the relief the diameter of the cista can be estimated to have been about 20.5 centimeters. A nude winged youth hurries over the waves of the ocean on an engraved Etruscan mirror (now lost, see Gerhard), and a comparable figure appears on a terracotta antefix from Temple B at Pyrgi. The latter has been plausibly identified as Usil, the Etruscan sun-god (see Vacano), and it is likely that the figure on the cista foot represents the same divinity.

F. Jurgeit lists three similar cista feet, as well as four closely related feet on a rectangular box found in the La Boncia tomb at Chiusi, but all these are smaller than the present foot and cruder in workmanship. Jurgeit ascribes them to a Chiusine workshop of the early fifth century B.C., but suggests that this type of cista foot may be a Vulcian invention. It is therefore significant that the Fleischman example, much larger and more sensitive in its anatomical modeling than the Chiusine examples, is said to have been found at Vulci.

BIBLIOGRAPHY: Unpublished.

RELATED REFERENCES: E. Gerhard, *Etruskische Spiegel*, vol. 1 (Berlin, 1840), pl. 120. O. W. v. Vacano, "Gibt es Beziehungen zwischen dem Bauschmuck des Tempels B und der Kultgöttin von Pyrgi?" *Akten des Kolloquiums "Die Göttin von Pyrgi," Tübingen 1979*, Biblioteca di Studi Etruschi 12 (Florence, 1981), p. 154ff., pl. 17c. F. Jurgeit, *Cistenfüsse: etruskische und praenestiner Bronzewerkstätten*, vol. 2, pt. 1 of *Le Ciste prenestine* (Viterbo, 1986), pp. 35–37, 100–103, pls. XV c, d, e, XVI a.

—S. H.

74

74
Statuette of a Seated African Boy (from the top of a candelabrum)

Etruscan, fifth century B.C.
Bronze; H (INCLUDING BASE): 5.7 cm
Condition: Nose and front of base damaged; light green patina with areas of black.

The statuette is cast solid together with the base. The upper edge of the spool-shaped base is decorated with beading. The nude boy sits with his right leg bent and flat on the ground and his left knee drawn up to his chest. He supports his left arm on his left thigh and cradles his bent head in his hand. His right hand rests on his left upper arm. His sad face is carefully characterized as that of a black African: low, furrowed forehead, thick brows, large protruding, sharply defined eyes, short broad nose, and full lips. The dense curls of his short, caplike hair are rendered by circular punchmarks with a central dot. His hunched back and his prominent shoulder blades and ribs realistically portray the boy's poor physical condition, while dejection is manifest in his attitude and facial expression.

The form of the base clearly indicates that the statuette was the decorative finial of a candelabrum. Etruscan candelabra were frequently made in matching pairs (see cat. no. 75 below), and a candelabrum in the antiquarium of the Villa Giulia, Rome (24417), which is surmounted by a similar but less well-preserved statuette, may well be the companion piece of the one to which our statuette belonged. The Etruscans were familiar with the natives of Africa as a result of their extensive maritime trading and piracy, and sometimes, undoubtedly, they took African slaves and employed them as domestic servants. Heads of blacks are occasionally represented in Etruscan plastic vases of the fourth century

B.C. (see Harari) and on terracotta antefixes from Pyrgi and Cerveteri (see *Image of the Black*), but they are rare in Etruscan bronze sculpture. The present statuette is a sympathetic rendering of an unusual subject.

BIBLIOGRAPHY: Unpublished.

RELATED REFERENCES: M. Harari, *Il gruppo Clusium* (Rome, 1980), pls. XLVIII, LVII. F. M. Snowden, *The Image of the Black in Western Art,* vol. 1 (Cambridge, Mass., 1976), figs. 181, 183.

—S. H.

75
Pair of Candelabra

Etruscan, first half of the fifth century B.C.
Bronze; H (a): 138.2 cm; (b): 136.7 cm
Condition: Intact; crusty light green patina.
Provenance: Formerly in the collection of Basia Johnson.

Cast solid in several separate pieces, the candelabra were made as a pair, but differ slightly in size and details. Each is supported on a tripod base formed by three lion's legs on round pedestals, the junctions between them decorated by pendant palmettes with a drop at the end of each petal. The upper part of the base is ornamented in low relief with an elaborate pattern of smaller palmettes combined with studded volutes. The tapering shaft rises above a series of four moldings: a double reel, a large torus with beading around the circumference, another double reel, and a collar of ovolo and beading. The slender shaft is fluted, except at the lower end, where it is smooth and incised with four overlapping rows of feather pattern. It is surmounted by a disk shaped like an inverted saucer, decorated with a tongue-and-dart pattern and a smaller ovolo molding. Superimposed on this is a double reel, a wider disk with a beaded edge, another double reel, and a crowning disk from which project four curved and faceted branches ending in lotus-shaped prickets.

75

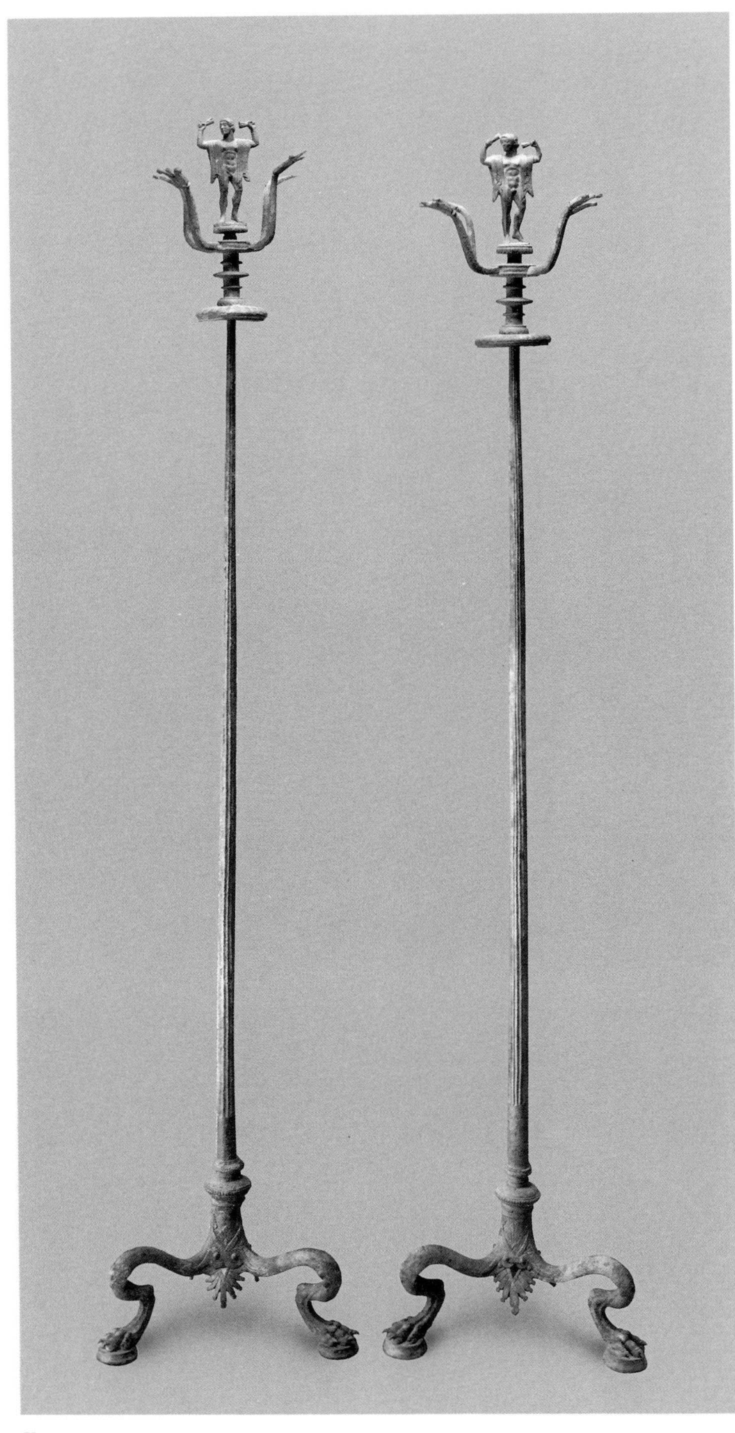

75

Rising from the center of the crown is a circular base with an ovolo molding under beading, on which stands a statuette of a dancing youth. His only garment is a short mantle draped symmetrically over his back and shoulders and hanging in front in stacked folds, the zigzagging edges of which end in swallowtails at the height of his thighs. The youth is poised with his weight on the right leg, the left flexed and touching the ground with the toes only. In each hand he holds aloft a pair of *crotala* (castanets) with which he accompanies his dancing. He has turned his head to his right. His finely incised hair is brushed down from the crown and falls in a thick striated fringe over his forehead and temples; behind the ears it curls inward to form a roll. He has large almond-shaped eyes under arching brows, a straight nose, an unsmiling mouth, and a strong chin.

The elaborate musculature of the dancers' bodies, particularly that of the chest and abdomen, is inspired by Greek prototypes of the late sixth and early fifth centuries B.C. and can be compared with the anatomy of the youths on pseudo-red-figured vases of the decade 470–460 B.C. by the Vulcian vase painter Praxias and his followers (see Brendel). The reputed Vulcian provenance of the candelabra is further supported by the fact that they fall into the category of Type C 1 of A. Testa's classification of candelabra; many C 1 candelabra have been found at Vulci. For the style of the figures we may compare a nude dancer with crotala on a candelabrum in the Museo Gregoriano

Etrusco of the Vatican (12396: Testa, p. 41f., no. 10), which may be from the same workshop.

Candelabra are typically Etruscan, the Greeks having used oil lamps for illumination. They seem to have been employed in pairs, as illustrated in the fresco from the Tomba Golini at Settecamini near Orvieto (see *Pittura Etrusca*), and a number of pairs have, in fact, been found in excavations (see, for example, Martelli; Hostetter; Testa, p. 50ff., nos. 13, 14). In its rich decoration and excellent preservation, the present pair is a particularly fine example.

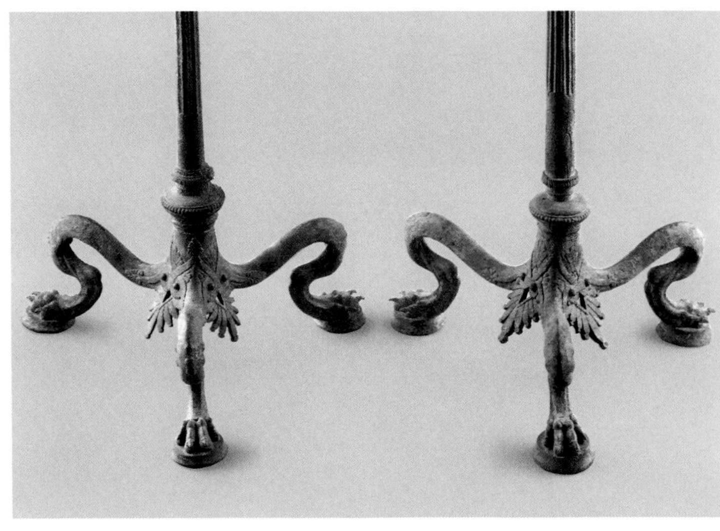

75

BIBLIOGRAPHY: Unpublished.

RELATED REFERENCES: O. Brendel, *Etruscan Art* (Harmondsworth, 1978), fig. 200. A. E. Feruglio, ed., *Pittura Etrusca a Orvieto* (Rome, 1982), p. 53. A. Testa, *Candelabri e Thymiateria* (Rome, 1989). M. Martelli in *Prima Italia*, exh. cat. (Brussels, 1981), p. 200ff. (from Montepulciano). E. Hostetter, *Bronzes from Spina*, vol. 1 (Mainz, 1986), p. 60ff., nos. 31, 32. For the classification of candelabra, see Testa, op. cit., p. 34ff., Appendix A, pp. 201–204.

—S. H.

76
Statuette of a Striding Warrior

Umbrian, mid-fifth century B.C.
Bronze; H (INCLUDING TANGS): 31 cm; (OF FIGURE): 26 cm
Condition: Lower end of crest and separately made javelin and shield missing; light green crusty patina with blue patches.

The statuette is cast solid; the long triangular casting tangs under the feet have been retained to fasten it to a base. A slim elongated figure, the warrior wears full hoplite armor and strides forward in an attitude of attack with left leg stiffly advanced, body and head upright and facing to the front. In his raised right hand he brandished a weapon, which the extended index and middle fingers show to have been a javelin. On his left arm he held out a shield. He is barefoot but his legs are protected by plain long greaves. He wears a short chiton which hangs below his cuirass in three semicircular lappets incised with concentric folds, one covering his genitals and the other two his buttocks. Below a twisted, ropelike belt, his cuirass ends in two tiers of leather tabs (*pteryges*), alternately plain and incised with crosses. Above the belt the cuirass is decorated with a jumble of punched ornaments: tongue patterns, checkerboards, and spirals. The long U-shaped shoulder flaps, which are also punched with spirals, are fastened in front with incised crossing laces. On his head the warrior wears a tall crested Attic helmet, the crest-holder incised with hatched triangles, the rest of the helmet punched with tongue patterns and spirals. The long hinged cheek pieces are turned up. The sharply outlined, bulging eyes, the straight, jutting nose, and the long mustache, which covers the upper lip and curves down on to the spade-shaped beard, combine to give the face an expression of fixed ferocity.

Like that of many other Etruscan and Umbrian votive bronzes of the fifth century B.C., the warrior's panoply is based on Attic prototypes of the early part of that century, but such warriors are normally

76

76

clean-shaven. Other bearded examples are, however, provided by a helmeted javelin-thrower of Etruscan style in the British Museum (Bronze 452: Haynes 1985, no. 98; Richardson) and an Umbrian spear-wielding hoplite in the Ashmolean Museum, Oxford (Fortnum B 6: Richardson). The latter is closely related to the present statuette in its attenuated proportions, its attitude, and in many of the details of the armor. They are probably products of the same workshop.

BIBLIOGRAPHY: Unpublished.

RELATED REFERENCES: E. Richardson, *Etruscan Votive Bronzes* (Mainz, 1983), fig. 406 (British Museum), figs. 421, 422 (Ashmolean).

—S. H.

77

Statuette of a Youth Carrying a Diskos

Etruscan, late fifth–early fourth century B.C.
Bronze; H (INCLUDING BASE AND LUG FOR SUSPENSION RING): 12.5 cm; (OF FIGURE): 9.6 cm
Condition: Intact except for index and second fingers of right hand; green, powdery patina still covering base and suspension ring has been largely removed from figure, leaving it reddish brown.
Provenance: Formerly in the Pomerance collection.

The statuette is cast solid together with the base and the lug for the suspension ring, which was cast separately. The upper rim of the spool-shaped base has a beaded molding. The sturdy body of the nude athlete is carefully modeled. He stands in repose with both feet firmly on the ground, his left leg slightly advanced with the knee relaxed, his weight on his right leg, so that the right hip is pushed up and outward. He raises his muscular arms a little away from his body with the elbows slightly bent. In the palm of his overlarge left hand he balances the diskos, while with his right hand he seems to be pointing to his right, the direction in which his head too is turned. The head is disproportionately large, the neck short and massive. Characterized by level brows, large eyes, a straight nose, and a sullen mouth, his serious face is framed by a thick cap of hair, composed of finely incised flame-like locks, which covers both ears.

The statuette probably decorated a stand for utensils, a ring of hooks for the suspension of which would have been attached below the base. Figures of diskos-carriers and other athletes were popular

77

as ornaments for Etruscan utensil stands and candelabra (see Haynes 1985, no. 133); they were probably inspired by palaestra scenes on imported Greek pottery. Although the present statuette has competently adapted a Classical Greek prototype, its squat proportions and large head are, nevertheless, characteristically Etruscan.

The facial structure, the treatment of the hair, and the shape of the base all point to a Vulcian workshop.

BIBLIOGRAPHY: D. von Bothmer, ed., *Ancient Art from New York Private Collections,* exh. cat. (Metropolitan Museum of Art, New York, 1961), no. 154, pl. 54; *The Pomerance Collection of Ancient Art,* exh. cat. (Brooklyn Museum, 1966), no. 120; R. S. Teitz, *Masterpieces of Etruscan Art,* exh. cat. (Worcester Art Museum, Worcester, Mass., 1967), no. 58; *Master Bronzes,* no. 178.

—S. H.

78
Patera Handle in the Form of a Nude Winged Girl

Etruscan, second half of fourth century B.C.
Bronze; H (INCLUDING BASE): 21 cm
Condition: Intact; smooth reddish brown patina.

Except for the wings, the whole of the handle is cast solid in one piece, including a lug for a suspension ring under the base. The girl stands on a triangular base, the edges of which are decorated with a tongue pattern between beaded moldings. Her left leg is relaxed in an easy contrapposto pose. She is nude except for soft slippers with high tongues. Of slim, athletic build, she has narrow hips and small, firm breasts. She rests her right hand on her head and holds a large alabastron in the left. Her serious face has a triangular forehead, sharply outlined, level eyes with lightly punched circles for the irises, a straight nose, and a small full-lipped mouth. Parted in the center, her hair is drawn back from the brow and encircled by a ribbon, visible at the back, where the hair is swept up under and over it to fall in a free-hanging tail. She wears a thick plaited-chain necklace from which hang a central bulla and two semilunate pendants framing her breasts. A bracelet with three pendants adorns her left upper arm; above it protrudes a flat rectangular object, the nature of which is uncertain. Her outspread wings were cast together as a separate piece and are attached to her back by a rivet; the feathers are modeled and incised on the front only. From the girl's head rises a curving attachment plate to which a bowl-

78

78

shaped patera was originally joined by soft solder. A rosette projects forward on either side of the base of the plate, the back of which is decorated in low relief with an antithetic pair of finlike shapes with hatched ribs and a cusped upper edge.

A nude winged and bejeweled girl holding a scent bottle is probably a Lasa, a mythical being most often represented in Etruscan art as an attendant of Turan, the goddess of love, but also associated with other divinities. Lasae, who are sometimes wingless, have been compared with the Nymphs of Greek mythology; they, too, appear in various contexts and cannot be connected with any specific activity.

The main uses of paterae seem to have been to pour water over the hands before sacrificing and over the body when bathing. A nude figure at toilet or after athletic exercise was thus an appropriate subject for a patera handle and handles in this form are frequent; but draped figures of both sexes also occur.

Closely related to the present bronze are a handle in the form of a winged Lasa in the Metropolitan Museum of Art (Rogers Fund 1919, 19.192.65) and another in the form of a nude girl in the Museum of Fine Arts, Boston (H. L. Pierce Fund, 98.679: *The Gods Delight,* no. 49). For a draped figure in the same attitude, compare the girl wearing a chiton and holding an alabastron on the British Museum patera (51.8–6.3: Haynes 1985, no. 178).

The treatment of the hair here recalls that of Praenestine bronze figures influenced by Tarantine prototypes of the second half of the fourth century B.C. (for example, the figures of Dionysos on the "Cista Ficoroni" in the Villa Giulia, Rome [see Haynes 1985, no. 172] and on the "Ciste Napoléon" in the Louvre [see Haynes 1985, no. 173]), a date which agrees well with the graceful proportions and stance of the figure.

BIBLIOGRAPHY: Sotheby, Parke Bernet, New York, sale cat., December 11, 1980, lot 181, color pl.; Haynes 1985, p. 320; *The Gods Delight,* no. 48.

RELATED REFERENCES: For a discussion of Lasae, see A. Rallo, *Lasa: Iconografia e esegesi* (Florence, 1974).

—S. H.

79
Statuette of a Nude Youth

Etruscan, last quarter of fourth century B.C.
Bronze; H: 19.7 cm
Condition: Legs below knee, left arm, tips of last two fingers of right hand missing; right ear damaged; pale gray-green patina.

Perhaps cast hollow over a partial core. The break in the left shoulder shows what appears to be core material. The youth stands with his weight on the left leg, the right leg relaxed, his right hand resting on his lowered hip, and his head turned a little upward and to the left, an attitude which gives the figure a spiral twist. In his left hand he probably held an object. Although competent and fluid, the modeling is somewhat summary, with coarsely incised detail. The bullet-shaped head on a strong neck has blank oval eyes, a short nose, and a small mouth. The striated hair, which is swept back from the low forehead and temples and upward at the nape, appears to be rolled around a band (invisible), exposing the barely detailed ears. Fine incised lines mark the pubic hair. A votive inscription of two lines of unequal length runs from the figure's right armpit down the flank and the inside of the right thigh to the knee and from the right pectoral muscle to the pubic region. It reads: *ecn : turce : avle : havrnas : tuthina : apana : selvansl [:] tularias.*

The Lysippian proportions of the slim, muscular body with its long thighs and the comparatively small head suggest that the artist was inspired by Greek sculptural prototypes of the second half of the fourth century B.C. rather than by Polykleitan athletes of the second half of the fifth century with their more compact volumes and self-contained stance. The type of a youth standing with hand on hip and either nude or draped in a short mantle is known from a number of Etruscan bronze statuettes of the fourth century B.C.; they represent either athletes such as the diskos-carrying youth in Basel (Antikenmuseum KÄ.512: Dohrn) and the elongated youth from the Stafford Collection (*Master Bronzes,* no. 180), or dedicants such as two votive figures in the British Museum (see Haynes 1985, nos. 154, 155), both inscribed with a dedication to Selvans.

The inscription on the present bronze may be translated as follows: "Avle Havrnas gave this [for *or* in the] tuthina of the father to Selvans of the boundaries." The meaning of *tuthina* is still debated, but the word probably refers to a rural district or village; *apana* means "of the father," and the dedicatory character of the inscription is clear. The family name Havrna or Havrnas (in the form Havrenies or Harenies) is known from inscriptions incised on a set of banqueting vessels found together in a tomb at Bolsena (see Brunn; Pandolfini; *The Etruscans*), but now divided between the British Museum (Bronze 651 and 655: Haynes 1985, no. 167) and the Museo Gregoriano Etrusco of the Vatican (see Helbig). It is probable, therefore, that the present statuette was made for a member of this family from Bolsena. Selvans is an important Etruscan divinity (related to the Latin Silvanus), concerned with the protection of the fields and their boundaries. A stone cippus inscribed with the god's name was discovered at the sanctuary of Pozzarello near Bolsena (see *Testimonia Linguae Etruscae;* Colonna 1964), and Bolsena is also the provenance of a votive statuette of a draped youth bearing the name of Selvans inscribed on his body and mantle (Rome, Villa Giulia 59459: Colonna 1971). Clearly, Selvans was worshipped in the region of Bolsena, and the suggestion that the Fleischman statuette comes from that area has much to recommend it.

BIBLIOGRAPHY [F56]: C. de Simone, *Secondo congresso internazionale Etrusco, Atti,* vol. 3 (Florence, 1985), p. 1316f., pl. 1; idem, *Studi Etruschi* 55 (1987–88), p. 346, no. 128; G. Colonna, *Atti e memorie della Accademia Petrarca di Lettere, Arti e Scienze,* n. s. 47 (1985),

79

p. 184ff.; L. B. van der Meer, *The Bronze Liver of Piacenza* (Amsterdam, 1987), p. 61f.; L. and G. Bonfante, "Problems in Decipherment," *Bibliothèque des cahiers de l'Institut de Linguistique de Louvain* 49 (1989), p. 189ff., 197, fig. 13; *The Gods Delight,* no. 47; A. Morandi, *Epigrafia di Bolsena Etrusca* (Rome, 1990), p. 85f., no. 29; L. Bonfante, "Un bronzetto votivo di Bolsena (?)," *Archeologia Classica* 43, 2 (1991), p. 834ff.; G. Colonna, "Le iscrizioni votive etrusche," *Atti del Convegno Internazionale ANATHEMA: Regime delle offerte e vita dei santuari nel Mediterraneo antico, 15–18 guigno 1989 = Scienze dell'Antichità, Storia, Archeologia, Antropologia* 3–4 (Rome, 1989–90), pp. 875–903; M. Bentz, *Etruskische Votivbronzen des Hellenismus,* Biblioteca di Studi Etruschi 25 (Florence, 1992), p. 200.

RELATED REFERENCES: T. Dohrn, *Die etruskische Kunst im Zeitalter der griechischen Klassik* (Mainz, 1982), p. 30f., pl. 13. H. Brunn in *Bullettino dell'Istituto di Correspondenza Archeologica* (Rome, 1857), p. 34f. M. Pandolfini, "Rivista di epigrafia etrusca," *Studi Etruschi* 44 (1976), p. 243ff., nos. 46, 49, 51. *The Etruscans: Legacy of a Lost Civilization from the Vatican Museums,* exh. cat. (Memphis, Tenn., 1992), p. 134ff. W. Helbig, *Führer durch die öffentlichen Sammlungen klassischer Altertümer in Rom,* 4th ed., vol. 1 (Tübingen, 1963), no. 674. M. Pallottino, ed., *Testimonia Linguae Etruscae* (Florence, 1968), p. 114, no. 900. G. Colonna, "Rivista di epigrafia etrusca," *Studi Etruschi* 32 (1964), p. 161ff; ibid., *Studi Etruschi* 39 (1971), p. 336, pl. LXX. For the meaning of *tuthina,* see G. Colonna, "Le iscrizioni votive etrusche," op. cit., p. 887f., nn. 65, 66. For the meaning of *apana,* see C. de Simone, op. cit., p. 347; L. Bonfante, "Un bronzetto votivo de Bolsena," op. cit. For Selvans, see M. Cristofani, ed., *Dizionario della Civiltà Etrusca* (1985), p. 269, s.v. "Selvans"; L. B. van der Meer, op. cit., p. 58ff.

—S. H.

80
Votive Statuette of Herakles

Etruscan, late fourth–early third century B.C.
Bronze; H: 24.3 cm
Condition: Thumb and front part of fingers of left hand, object/s held in it, free-hanging end of lion's skin, and penis missing; right foot and ankle, front of left foot restored; smooth greenish black patina.

The statuette was cast solid in two pieces, the missing part of the lion's skin, which hung free from the figure's left forearm, having been cast separately and attached with three pins, of which one is preserved.

The youthful hero stands in a relaxed attitude with his weight on his right leg and his left bent and set slightly back with the heel raised from the ground. He rests his right hand on his hip and extends his left forearm; in the palm of his left hand he probably held the apples of the Hesperides (see Hill). His head is erect and turned a little to his left, his gaze fixed on the distance. The modeling of the torso is anatomically detailed, while the muscles of buttocks and legs are treated in broad faceted planes. On his head he wears the mask of the lion's skin, the lower jaw of which covers his ears and frames his face like a high collar. The animal's front paws are knotted on the hero's chest while the rest of the skin is draped over his back and left forearm. The hairs of the pelt are incised in rows of fine short strokes, the dorsal ridge being marked by a herringbone pattern.

First found in Cypriote statuary, the type of Herakles wearing the lion's skin over his head and fas-

80

80

tened around his body becomes a favorite subject of Attic black-figure vase painting in the sixth century B.C. This Archaic tradition of Herakles wearing the skin as a garment rather than holding it as an attribute survives into the Hellenistic period in Etruria, where the hero continued to be immensely popular (see Cristofani). In Greece the Macedonian dynasty revived the type to celebrate their descent from Herakles (see, for example, coins: Francke/Hirmer; and the marble portrait head of Alexander the Great from the Kerameikos, Athens: Ninou). The present votive statuette reflects the powerfully modeled, yet fluid and elongated physical types created by Lysippos, the only sculptor permitted to portray Alexander the Great. The calm pose and distant gaze of the statuette and the attribute of the Hesperidean apples suggest that he is represented here as a redeemer rather than a fighter.

BIBLIOGRAPHY [F63]: S. Reinach, "L'Héracles de Polyclète," *Revue des études anciennes* (1910), p. 6f.; idem, *Répertoire de la statuaire grecque et romaine,* vol. 4 (Paris, 1913), p. 128, no. 3 (fig leaf later removed); Sotheby's, London, sale cat., December 8, 1970, p. 21, no. 34; J. P. Uhlenbrock, *Herakles: Passage of the Hero Through 1000 Years of Classical Art,* exh. cat. (New Rochelle, N.Y., 1986), no. 32; *LIMC,* vol. 5, pt. 1, p. 203, no. 44, s.v. "Herakles."

RELATED REFERENCES: D. K. Hill, *Catalogue of Sculpture in the Walters Gallery of Art* (Baltimore, 1949), p. 46, no. 95, pl. 22. M. Cristofani, *I bronzi degli Etruschi* (Novara, 1985), nos. 94, 97, 98. P. R. Francke and M. Hirmer, *Die griechische Münze* (Munich, 1964), figs. 169, 172, 173. K. Ninou, ed., *Alexander the Great: History and Legend in Art* (Thessaloníki, 1980), p. 69.

—S. H.

81
Statuette of a Man

Etruscan, early third century B.C.
Bronze; H (EXCLUDING CASTING TANGS): 31.6 cm
Condition: Index finger of left hand broken off and reattached; inset patches missing in drapery on left shoulder and behind right thigh; smooth grayish green patina.
Provenance: Said to be from Orvieto.

F. Roncalli (see *Antichità dall'Umbria*) states that the statuette was cast hollow over a partial core, which would not be surprising in view of its size; this seems to be confirmed by what appear to be the remains of iron chaplets in the recesses on the left shoulder and behind the right thigh. The casting tangs under the feet were retained for mounting the votive statuette on a base. A casting fault at the top of the left calf is masked by a patch. The nipples are inlaid in copper.

Barefoot and wearing a short semicircular mantle draped around his waist and over his left shoulder and arm, the figure stands with his weight on his left leg, his right leg relaxed and placed slightly to the side, a pose which imparts an S-curve to his slim, smooth torso. Both arms are bent, the right raised a little from the body, and both the large, delicately modeled hands are spread in a gesture of prayer. Compared with his youthful body, his large head appears unexpectedly mature. Deep folds in the right side of his neck emphasize the turn of his head to the right and upward. The face is vigorously modeled with large, deep-set eyes, of which iris and pupil are indicated, under the furrowed brow. Slight depressions slanting away from the nostrils and from the corners of the full-lipped mouth enliven the modeling of the broad cheeks. Parted in the center, the hair, which covers the ears, ends in sickle-shaped clusters of short striated locks, of which a symmetrical pair rises prominently above the center of the forehead.

As H. Hoffmann points out, the serious and pathetic expression of the face, the arrangement of the locks above the forehead, and the studied twist of the head are clearly inspired by portraits of Alexander the Great. Another votive statuette with features modeled on those of Alexander the Great is in the Staatliche Antikensammlung, Munich (3003: Richardson) and is closely related to the present piece.

A votive inscription is engraved on the mantle in two parts, the first running along its edge from the left wrist to the corner of the garment below and reading: *Vel Matunas turce* (Vel Matunas dedicated). The second part follows the slanting folds of the drapery at the back, running from below the left elbow to the right buttock, where, to avoid the patch at this point, the last three letters are written boustrophedon above: *lur : mitla : cvera* (precious/sacred gift to Lur). Vel is a frequent personal name. Matunas is the family name of a noble South Etruscan gens, recorded both at Cerveteri and at Tarquinia. However, the forms of some of the letters of the inscription are typical for the region of Orvieto-Volsinii-Bolsena. The name Lur occurs in inscriptions from Bolsena and Perugia and is thought to be that of a divinity. It seems likely therefore, despite the votary's South Etruscan descent, that his dedication was made in Central Etruria, in the neighborhood of Volsinii; and this conclusion is strengthened by the reported provenance of the statuette from Orvieto. Situated not far from Orvieto was, we know, the "Fanum Voltumnae," the federal sanctuary of the twelve Etruscan cities, whose inhabitants and leaders met there at regular intervals for common worship, political counsel, and the celebration of games. We may conjecture that it was the universal prestige of this pan-Etruscan sanctuary which induced a South Etruscan prince to dedicate a statuette of himself in the likeness of Alexander the Great to a divinity of this distant region.

BIBLIOGRAPHY [F225]: H. Hoffmann in *Master Bronzes,* no. 187; O. Brendel, *Etruscan Art* (Harmondsworth, 1978), p. 411; F. Roncalli, "Etrusco CVER, CVERA = greco ΑΓΑΛΜΑ," *La Parola del Passato* (1983), p. 290, no. 5; *Wealth of the Ancient World,* exh. cat. (Fort Worth, 1983), no. 36; Sotheby's, New York, *The Nelson Bunker Hunt Collection, The William Herbert Hunt Collection,* auct. cat., June, 19, 1990, no. 37; F. Roncalli in *Antichità dall'Umbria a New York* (Perugia, 1991), p. 289, no. 6.12; M. Bentz, *Etruskische Votivbronzen des Hellenismus,* Biblioteca di Studi Etruschi 25 (Florence, 1992), p. 194f.

RELATED REFERENCES: E. Richardson, "The Types of Hellenistic Votive Bronzes from Central Italy," *Eius Virtutis Studiosi: Classical and Postclassical Studies in Memory of Frank Edward Brown (1908–1988)* (Hannover and London, 1993), pp. 283–284, fig. 2. For letterforms typical of Orvieto-Volsinii-Bolsena, see F. Roncalli, *Antichità dall 'Umbria,* op. cit., p. 291.

—S. H.

81

82
Statuette of a Boy Running with a Tray of Food

Italic, probably third century B.C.
Bronze; H: 8.3 cm
Condition: Right foot and object held in right hand missing.

The figure runs forward, carrying aloft a round tray covered with strips of food; the missing object in his right hand may have been a skewer with pieces of meat. His hair is cut short and he wears a kiltlike garment decorated with incised dots. His jaunty, if rushed, attitude is that of a proud waiter.

Figures similarly clad and occupied appear in a cooking scene in the Tomba Golini I at Orvieto (see Steingräber), and, still more strikingly, on a cista in Brussels with inscriptions in Praenestine dialect (see Bordenache). In the latter, the young men preparing food wear cloths tied around their waists like our figure. They brandish dishes, including round platters, and a sophisticated range of cooking implements. One figure hurries forward with a kebab, crying *ASOM FERO* (I'm bringing the grilled meat!).

The proportions of the figure's boyish head with its round face and pudding-bowl haircut recall some Etruscan pieces. However, the lively but completely unpretentious quality of our figure, without showy surface detail, seems alien to mainstream Etruscan bronze work, which makes a display of technical virtuousity and tends to value rhythmic composition, based on traditional forms, above direct observation of reality. The ingenuous style of the statuette sets it even more decisively apart from the later, sophisticated figures of grotesque slaves with trays of food from the Vesuvian cities. The piece can probably be ascribed to a central Italian workshop, active in early Hellenistic times, somewhere on the fringes of Etruscan art.

BIBLIOGRAPHY: Unpublished.

RELATED REFERENCES: S. Steingräber, ed., *Etruscan Painting* (1986), p. 278f., no. 32. G. Bordenache Battaglia, *Le Ciste prenestine,* vol. 1 (Rome, 1979), no. 12. L. Bonfante, *Etruscan Life and Afterlife: A Handbook of Etruscan Studies* (Detroit, 1986), p. 261, fig. VIII-42, n. 89.

—A. H.

83
Mirror with Relief-Decorated Cover

Etruscan, third century B.C.
Bronze; DIAM (INCLUDING HINGE): 15.1 cm
Condition: All but a small fragment of swing-handle of cover missing; wrongly assembled with reflecting surface of mirror reversed; hinge pin modern; large piece of ancient fabric adheres to reflecting surface of mirror; crusty green patina.

The mirror and its cover, apart from the relief, are both cast and decorated on their concave undersides with a series of concentric circles in relief, formed on a lathe. The two parts of the hinge are connected to them by rivets. Of slightly smaller diameter than the cover it decorates, the relief was formed by hammering sheet bronze into a negative mold or matrix, after which it was finished by chasing from the outside, filled with lead and fixed on the cover with soft solder. A small swing-handle attached at the bottom of the relief served to open the mirror.

The scene on the cover is framed by palm fronds which rise symmetrically from two small lion's masks in whose mouths the swing-handle was fixed. A large acanthus separates the lion's masks, a garlanded bull's head (*bucranium*) the tops of the palm fronds. Within this frame a man and a woman face one another. The man, on the left, stands with his left foot set on a small outcrop of the horizontally stippled rocky ground. From his left hand, which rests over his raised thigh, dangles a stout gnarled stick. The elbow of his bent right arm is also supported on this thigh and with

83

the index finger he gesticulates didactically at his female companion. Barefoot, but wearing an anklet on his left leg, he is dressed in a short, belted tunic (*exomis*), the characteristic garment of workmen and beggars, which was fastened on the left shoulder only. He is bearded and wears a pilos on his curly hair. The woman stands listening to him intently with her left leg crossed over the right, her right arm wrapped round her waist, and her left hand, in which she holds a spindle, raised to her chin in a gesture of perplexity. Her long chiton is pleated and belted, its crinkly overfold hanging around her hips. She wears a bracelet on her left wrist and a necklace; and her centrally parted hair is twisted into a roll at the sides and tied in a low chignon at the back. Behind her hangs her mantle, draped over some invisible support. At her feet sits a large dog with a collar; it looks up at the man, placing its right forepaw against his shin. The space between the dog's head and the bucranium at the top of the relief is filled by a frontal masklike face, the hair covered with the horned head of a goatskin, the forelegs of which are knotted under the chin.

The man's dress, the dog's familiarity, and the woman's uncertainty identify the scene as the return of Ulysses to Ithaca, where Penelope meets him but has yet to be convinced that the stranger disguised as a beggar is her long-lost husband, though his faithful dog Argos has recognized him instantly. The Etruscan artist has skillfully telescoped the Homeric sequence of events into a single out-of-doors scene, using for it figures based on fifth-century Greek prototypes. A non-Greek addition is the head clad in a goatskin, an image of Juno Sospita, the purely Italic goddess who protected cities, women, and marriages.

The ten known mirror reliefs with this scene have recently been studied by E. H. Richardson and I. Jucker, who was the first to identify the frontal head as that of Juno Sospita. In number five of Jucker's list of these reliefs the head of Juno Sospita is replaced by a rosette; other reliefs show other variations in detail, such as the form of the framing wreath or the absence of the acanthus. It is thus evident that more than one matrix existed for shaping the reliefs, though they were probably all made in a single South Etruscan workshop. The present well-preserved example may be Jucker's number four, which comes from Tarquinia, and which she describes as "lost?".

BIBLIOGRAPHY: Unpublished.

RELATED REFERENCES: E. H. Richardson, "A Mirror in the Duke University Classical Collection and the Etruscan Version of Odysseus' Return," *RM* 89 (1982), p. 27ff. I. Jucker, "Bemerkungen zu einigen etruskischen Klappspiegeln," *RM* 95 (1988), p. 5ff.

—S. H.

84
Weight in the Form of a Double Head

Etruscan, third–second century B.C.
Bronze filled with lead; H (INCLUDING SUSPENSION RING): 6.8 cm; WEIGHT: 332.5 g
Condition: Intact; olive green patina.

The weight is cast hollow, the open underside showing the lead filling. There is a cast hole in the hair on one side. The two male half-heads are set back-to-back on a single neck, the bottom of which is decorated with a shallow cavetto molding. They share a pair of pointed horse's ears that rise between the coarsely grooved tresses of the hair. The ears characterize both creatures as satyrs, though one has wilder features than the other. The former's triangular forehead, from which the hair recedes, is scored with deep horizontal furrows and high-arching wrinkles fanning out from the bridge of the nose. The frowning brows are hatched with a herringbone pattern. His large eyes, which have incised irises with dots for pupils, are of unequal size and set at different levels. The short snub nose is displaced to the right of the face, the sullen mouth and cleft chin slightly to the left of center. The other face is rather more regular and human. Chin, mouth, and nose are aligned. The eyes and eyebrows are treated in the same way as those of his companion. The smooth forehead is framed by a short fringe and thick clusters of locks falling over the temples and cheeks.

Bronze human heads filled with lead and fitted with a suspension

84

ring are a familiar form of weight for steelyards in the Roman period, and there can be little doubt that this bronze is an Etruscan predecessor. What is evidently another Etruscan specimen has been recently excavated near Chianciano (see Rastrelli). Half-heads, often those of a maenad and a silenos, are similarly combined in a group of small Etruscan bronze vases of the third and second centuries B.C. (see Haynes/Menzel), which may have suggested the form of this unusual weight. A similar weight, probably by the same hand, is in the possession of the Galleria Serodine, Ascona (information from Ariel Herrmann).

BIBLIOGRAPHY: Unpublished.

RELATED REFERENCES: A. Rastrelli, "Scavi e Scoperte nel Territorio di Chianciano Terme: L'Edificio sacro dei Fucoli," *La civiltà di Chiusi de del suo territorio, Atti del XVII Convegno di Studi Etruschi ed Italici, Chianciano Terme, 28 maggio-1 giugno, 1989* (Florence, 1993), pls. XX, XXI. S. Haynes and H. Menzel, "Etruskische Bronzekopfgefässe," *Jahrbuch des Römisch-Germanischen Zentralmuseums Mainz* 6 (1959), p. 122f., pls. 56, 57.

—S. H.

85
Plastic Vase in the Form of a Female Head

Etruscan, second century B.C.
Bronze; H: 9.1 cm
Condition: Small part of rim broken off at front, otherwise intact; separately cast stopper with its attachment chain and main suspension chains missing; smooth brown and green patina with patches of encrustation.

The vase is cast, apart from its flat base which was cut from hammered sheet metal and soldered in position. Set on a sturdy neck, the woman's head is tilted slightly upward. She has an oval face, a triangular forehead, sharply outlined almond eyes with incised iris and pupil, a long, straight nose, and full, parted lips. Her wavy hair is divided down the center and tied by a fillet, only visible at the front, where two rising locks emerge from it and fall to either side. On the temples the hair is twisted around the fillet in a roll covering the tops of the ears; at the back it is brushed up from the nape and tucked in. She wears earrings in the form of an inverted pyramid ending in a drop. Chains for a handle were originally attached to the vessel by the vertical lugs on either side of the cylindrical molded mouth; the stopper would have been fastened to the handle by a third chain.

The bronze belongs to a large and varied group of portable containers in the form of a head, which were probably used for cosmetics. Dating from the third and second centuries B.C., they seem from their variety and wide distribution to have been made in more than one Etruscan city. When, as in this and the great majority of cases, the head is female, it may represent Turan, the Etruscan goddess of love, or Lasa, a nymphlike supernatural being often appearing in scenes of female toilet (see, for example, no. 78 above). The typology and interpretation of the group as a whole have been discussed by S. Haynes and H. Menzel. The pres-

85

ent example differs from most of the others in the simple, naturalistic treatment of the hair, the front of which resembles that of Diana on a second-century B.C. terracotta antefix from Nemi in the Villa Giulia Museum, Rome (see *Enea nel Lazio*).

BIBLIOGRAPHY: Unpublished.

RELATED REFERENCES: S. Haynes and H. Menzel, "Etruskische Bronzekopfgefässe," *Jahrbuch des Römisch-Germanischen Zentralmuseums Mainz* 6 (1959), p. 110ff.; see also Haynes 1985, nos. 193, 194. *Enea nel Lazio, archeologia e mito*, exh. cat. (Rome, 1981), p. 25f., no. A 26.

—S. H.

TERRACOTTA VASES AND ARCHITECTURAL DECORATION

86
Pithos with the Blinding of Polyphemos

Etruscan, 650–625 B.C.
Terracotta; H: 104 cm
Condition: Reconstructed from fragments.

This large pithos is wheel-made of a red impasto fabric that has been burnished to provide a lustrous surface for the decoration applied in a cream-colored slip. There are four sets of white rays encircling the vase, one descending from the mouth and three ascending from the foot. These are separated from each other and from the picture zone by sets of parallel lines and, just below the picture, a guilloche pattern. The lid is decorated in the same technique with open and filled triangles and herringbone patterns on the stirrup handles.

The widest zone of the pithos, just below the double-loop handles, has been decorated with a scene of

86 SIDE A, DETAIL

Odysseus and two of his men poking out the eye of Polyphemos, the Cyclops, just before their escape from his cave. The giant Cyclops, although no larger than the Greeks, is seated on a four-legged stool to our left; he grasps the large pointed stick as it moves toward his eye. In front of him is a large amphora, perhaps an allusion to the wine that Odysseus had him drink before attempting this daring maneuver. The three Greeks who wield the stick are differentiated in their clothing, but there is no indication which is Odysseus. We might assume he leads the charge.

Odd plants frame this scene; behind Polyphemos are large conical blossoms hung with fillets and to our far left a tall, narrow plant that looks like a tree bearing lollipops. On the opposite side of the vase at the far left are a pair of horses, one white, the other dotted, but each with curiously long, curling attachments to its bridle and a very long tail. At the right is a lion with its prey, a small deer, hanging from its mouth. Behind the lion is another pair of conical blossoms.

Nearly contemporary versions of this scene are found on the well-known proto-Argive krater in Argos and the proto-Attic amphora in Eleusis (see *LIMC*), and later in many representations throughout the Mediterranean. Polyphemos can be shown with one, two, or three eyes, but in this scene is usually shown lying down or seated on the floor of his cave. An Etruscan wall painting from the Tomb of the Ogre in Tarquinia is inscribed with the Etruscan names of the Cyclops

86 SIDE B, DETAILS

86 SIDE A

87

and Odysseus: Cuclu and Uthste.

The size, shape, and style of decoration of this vase (but not its subject) is nearly identical to a pithos in the Kelts collection in San Diego (see Hesperia Arts), which shows two confronted winged horses and has a lid nearly identical to the pithos in the following entry (cat. no. 87). A lidded pithos in the Ortiz collection is similarly related (see *Ortiz Collection*); one might think of them all as being from the same workshop in southern Etruria. They have a stylistic relationship with a similarly painted impasto tripod krater from the Regolini-Galassi tomb in Cerveteri (now in the Louvre, MNB 1781: Martelli). A more distant cousin is the Polledrara amphora in the British Museum (H230: Pryce).

BIBLIOGRAPHY: Unpublished.

RELATED REFERENCES: Hesperia Arts, New York, auct. cat., November 27, 1990, no. 26. *The George Ortiz Collection: Antiquities from Ur to Byzantium*, exh. cat. (Berne, 1993), no. 186. M. Martelli, ed., *La ceramica degli Etruschi* (Novara, 1987), no. 43. F. N. Pryce, *CVA* Great Britain 10, British Museum 7, pl. 9.3. For Etruscan and other parallels of this narrative, see *LIMC*, vol. 6, pp. 154–159, no. 17 (proto-Attic amphora), no. 26 (Tomb of the Ogre), s.v. "Kyklops" (Icard-Gianolio), and *LIMC*, vol. 6, pp. 954–957, s.v. "Odysseus" (Touchefeu-Meynier); p. 973, s.v. "Odysseus/Uthuze" (Camporeale). On this fabric, but citing only vases with geometric and vegetal ornament, see M. A. Del Chiaro, "An Early Etruscan Red Impasto Vase," *Elvehjem Museum of Art, University of Wisconsin-Madison Bulletin* (1981–83), pp. 29–31; R. L. Gordon, Jr., "Evidence for an Etruscan Workshop," *Muse* (Annual of the Museum of Art and Archaeology, University of Missouri) 5 (1971), pp. 35–42.

—K. H.

87
Pithos with Geometric Decoration

Etruscan, 650–625 B.C.
Terracotta; H: 70 cm
Condition: Reconstructed from fragments.

This pithos, like the one above (cat. no. 86), is of a red impasto fabric with a burnished surface and decoration added in a cream-colored slip. The vase has no handles, but the lid has stirrup handles positioned around a tall central knob.

The decoration on both the lid and the body of the vase consists solely of geometric ornament. Starting at the bottom of the vase, there are solid white rays, crosshatched rays, a running wave pattern, two rows of alternating checked and red triangles, and finally solid white rays descending from the mouth. Each decorative frieze is separated from its neighbors by a set of three lines. The lid is likewise decorated with sets of parallel lines, interrupted near the bottom by a guilloche pattern, and at the shoulder of the lid by a series of crosshatched rays.

The lid is nearly identical to the lid of an unpublished pithos in the Kelts collection in San Diego; for this and other parallels of the type, see the previous entry.

BIBLIOGRAPHY: Unpublished.

RELATED REFERENCES: See previous entry.

—K. H.

88–89
A–B Pair of Plates
C–D Pair of Plates

Attributed to various Caeretan workshops [J. Szilágy]
Etruscan, Plate A: 660–640 B.C.; Plate B: 680–660 B.C.; Plates C and D: 680–670 B.C.
Terracotta; DIAM (A): 37 cm; (B): 35 cm; (C): 29 cm; (D): 28 cm
Condition: All four nearly complete, reconstructed from fragments; all except panther plate have two holes near rim for hanging.

These plates, of red impasto with white engobe, are of a shape called "spanti" by the Etruscans. To the underside of each of these plates has been applied a single scheme of decoration in red paint, with variations in content. On each, the raised foot and the rim are decorated with concentric circles and other geometric ornament, and in the band between the two is a procession of animals.

On plate A, three panthers with extended tongues, curling tails, and filler ornament between their legs fill the relatively large figure zone with repeating curvilinear patterns. On plate B, the five speckled birds with patches of herringbone decoration beneath them are confined to a narrower zone. In addition to concentric circles the subsidiary decoration on this plate has near the rim a continuous wave pattern, and around the edge of the foot solid bands of decoration in a whirligig pattern. Although the painter cannot be identified, he seems to be in the circle of the Crane Painter.

More truly a pair than plates A and B, plates C and D are very similar in size and exhibit identical decoration; both are of the so-

88–89A

88–89B

88–89C

88–89D

called Heron Class of subgeometric pottery from southern Etruria. On each, the underside of the foot is divided into eight sections, while five very stylized herons occupy the figure zone between solid bands and concentric circles at the foot and rim. In addition to plates, the heron decoration appears on amphorae, kraters, oinochoai, and ollas in a fairly discrete geographic area in southern Etruria. Its introduction into Etruria may have been in the form of similarly shaped Phoenician red-slip plates (also decorated with concentric circles) found at Pithekoussai (see Leach, p. 306). (The author wishes to thank J. Szilágy for his assistance with this entry.)

BIBLIOGRAPHY: Unpublished.

RELATED REFERENCES: S. Leach "Subgeometric 'Heron' Pottery: Caere and Campania," in J. Swaddlin, ed., *Italian Iron Age Artefacts in the British Museum* (London, 1986), pp. 305–330. On the Etruscan name for such plates, see G. Colonna, "Nomi etruschi di vasi," *Archeologia Classica* 35–36 (1973–74), pp. 144–146. On the fabric, see J. Szilágy in *Secondo congresso internazionale Etrusco, Atti* (Rome, 1989), pp. 626–627, pl. 11b; M. Martelli, ed., *La ceramica degli Etruschi* (Novara, 1987), nos. 33, 41. On Greek antecedants, see M. Martelli, "Prima di Aristonothos," *Prospettiva* 38 (1984), pp. 2–15. On the Crane Painter, see R. Dik, "Un'anfora etrusca con raffigurazioni orientalizanti da Veio," *Mededelingen van het Nederlands Instituut te Rome* 42 [n.s. 7] (1980), pp. 15–30, pls. 3.2–3, 4.2. On the herons, see J. Szilágy, op. cit., pp. 620–621. On the Heron Class, see E. Rystedt, "An Etruscan Plate," *Medelhavsmuset Bulletin* 11 (1976), pp. 50–54; S. Leach, *Subgeometric Pottery from Southern Etruria* (Göteborg, 1987); S. Leach, op. cit., pp. 305–330. For a plate by the same hand as the heron plates, see R. Dik, "Un'anfora orientalizzante etrusca nel museo Allard Pierson," *Bulletin Antieke Beschaving* 56 (1981), pp. 50, 70, fig. 16. For similar birds in later Etruscan bucchero, see M. Robertson, *Greek, Etruscan and Roman Vases in the Lady Lever Art Gallery, Port Sunlight* (Merseyside, England, 1987), p. 53, no. 72.

—K. H.

90

Amphora

Attributed to the Tityos Painter [J. R. Guy]
Pontic, circa 530–510 B.C.
Terracotta; H: 35 cm; DIAM (OF MOUTH): 13 cm; (OF FOOT): 12 cm
Condition: Intact with well-preserved surface and added colors.

The broadly sloping shoulder of this neck amphora shows, between black panels at the base of the handles, the death of Medusa, surrounded on side A and side B by her Gorgon sisters. All were among the monstrous children of Phorkys and his sister Keto. On side A one winged Gorgon runs to our left, her head frontal and "dumbbells" in her hands. To our right Medusa, already beheaded, slumps to the ground as two winged horses emerge from her neck. These represent her children, Pegasos and Chrysaor, but here perhaps both should be named Pegasos, since this appearance is appropriate only for Pegasos, Chrysaor typically being shown in human form. On side B two additional Gorgons are shown frontally. Like their sisters on the other side, they carry dumbbells and wear heavy black dresses and pointed shoes. Their awkward frontal faces with ponderous masses of kinky red hair and wide open mouths with protruding tongues intensify their threatening visages as the nightmare monsters of Greek mythology.

Above them on each side of the neck are the heraldically confronted bodies of two panthers that meet in a single frontal head with a gaping mouth and widespread eyes. Behind each panther a branch grows up. Below the Gorgons is a decorative band of alternately upright and inverted palmettes between horizontal lines.

Around the lower half of the vase is a continuous frieze of animals with leafy plants scattered throughout. In the center of side A a deer poised on its rear legs browses in a tree, while to our right a panther, with a frontal head like those on the neck, and to our left a lion, move away from the deer and around the sides of the vase to threaten two rams butting heads on side B.

The foot, of inverted echinus (convex curved) shape, the cylindrical handles, and the echinus mouth are all painted black. The top of the lip is reserved, the inside black. The zone above the foot is filled with ten black rays. Interior detail is incised on all figures. The variety of colors is enhanced by the golden brown of the dilute glaze used for the palmette frieze and the plants and by the well-preserved added red and white used on the figures throughout.

This version of the death of Medusa with two Pegasoi may be a local Etruscan variant since it occurs on at least one other Etruscan vase. Perseus, the slayer of Medusa, is usually present and his absence here may indicate his relatively unimportant role in this telling of the story. Equally absent, perhaps hidden in the hunting bag of the missing Perseus, is the snakey-haired head of Medusa, reputed to have had the ability to turn to stone anyone who saw it. The objects which the Gorgons carry, the dumbbells, remain unexplained in our knowledge of antiquity. The attitude of the Gorgons on side B seems to suggest, however, that these things are either weapons or symbols of the frightful potency the Gorgons embody.

BIBLIOGRAPHY: Unpublished.

RELATED REFERENCES: For Pegasoi on vases, see R. Bianchi-Bandinelli and A. Giuliano, *Etruschi ed italici prima del dominio di Roma* (Milan, 1973), fig. 206 (Archaeological Museum, Florence); Hesperia Arts, New York, auct. cat., November 27, 1990, no. 26 (Kelts collection pithos, see cat. no. 86 above); J. Sieveking and R. Hackl, *Die königlische Vasensammlung zu München* (Munich, 1912), nos. 843, 854. For dumbbells in Gorgons' hands, see a relief block in Istanbul from the Archaic temple of Apollo at Didyma and a neck amphora in Reading (47.VI.1): *CVA* Great Britain 12, Reading 1, pls. 36–37. On the Tityos Painter, see *Enciclopedia dell'arte antica*, vol. 7, p. 887, s.v. "Tityos, Pittore di" (P. Bocci); L. Hannestad, *The Followers of the Paris Painter* (Copenhagen, 1976), pp. 17–31, 56–60, pls. 12–24; P. Ducati, *Pontische Vasen* (Rome, 1968). For another vase by this painter, see Münzen und Medaillen A. G., Basel, *Auction Catalogue 56* (1980), no. 49.

—K. H./J. R. G.

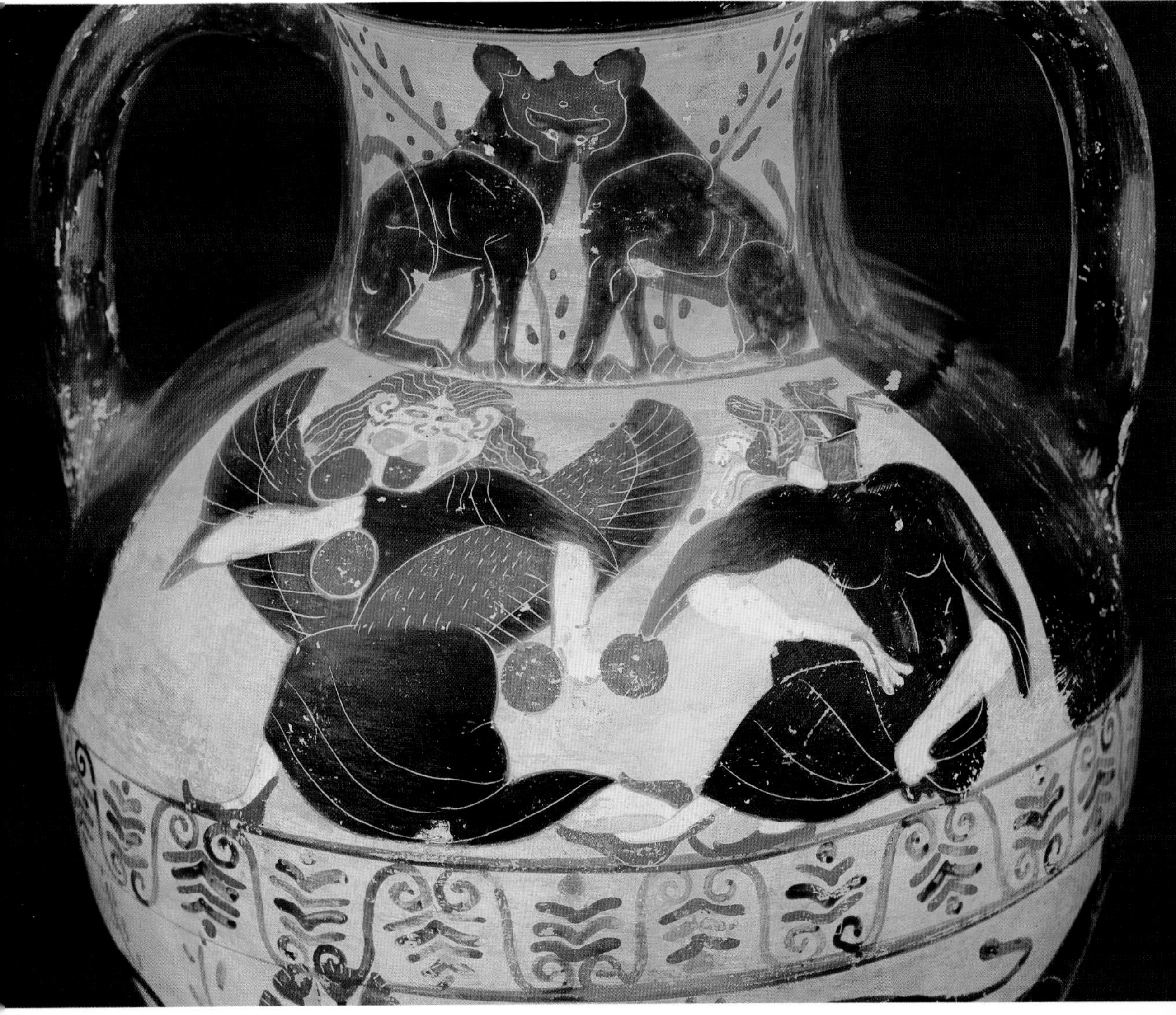

90 SIDE A, DETAIL

SIDE A–B

90 SIDE A

SIDE B

91
Painted Wall Panel

Etruscan, 520–510 B.C.
Terracotta; H: 88 cm; W: 52.5 cm;
DEPTH: 4.5 cm
Condition: Reconstructed from fragments; bottom of panel, including feet and right shin of figure, lower part of right side, and large pieces of left side missing.

The clay is a light reddish brown with black micaceous inclusions. The background is painted a creamy white. Finely outlined in black, the figure is filled in with ocher, red, and blackish brown. Incised preliminary drawing is visible in places.

At the top of the panel is an ornamental frieze enclosed above and below by black, white, and red stripes and a row of black dots. The frieze consists of black S-scrolls inclined at an angle to one other, with an ivy leaf in black outline outside each junction between them, and a red palmette filled with black and outlined with black dots in each intermediate space.

The rest of the panel is occupied by the standing figure of a draped youth, the crown of whose head overlaps the lower border of the frieze. While his sturdy legs and his abdomen are shown in profile to the right, his chest is frontal, and his head turned back in profile to the left. Both arms are bent with elbows outward, the slim right hand extended in front of the chest, the left holding aslant a long, crooked stick with a forked top. The youth has red flesh and blackish brown hair. His long nose and receding forehead together form an almost unbroken line. The mouth is small and slightly pursed, the chin firm. Represented frontally under a finely arched brow, the eye is painted in white with a black iris. The curly hair hangs loosely over the temple and in front of the ear, behind which it falls in a thick mass over the back of the neck. The youth wears a short ocher-colored tunic of thin crinkly material fastened on the right shoulder by three small round buttons, whence pleats rendered by three parallel wavy black lines descend diagonally to the chest. Draped around his torso and over his left shoulder is a black mantle, the ends of which hang in stacked undulating folds down to his knees. The lower edge of the mantle is decorated with a line of small white circles under a white zigzag, and each end is bordered with a broad red stripe outside a white one. The zigzag line of division between them suggests weaving technique, as does the less conspicuous dentillation of the inner edge of the white stripe where it meets the fabric's black ground.

In profile and physique the youth can be compared with figures in tomb paintings of the late Archaic Ionicizing period at Tarquinia, such as the Tomba delle Iscrizioni (see Steingräber, no. 74), Tomba dei Giocolieri (no. 70), Tomba della Fustigazione (no. 67), Tomba delle Olimpiadi (no. 92), Tomba dei Bacchanti (no. 43), and Tomba del Barone (no. 44), all of which are dated to the last quarter of the sixth century. Some of these tombs were probably decorated by East Greek immigrant artists and are related in style to painted pottery produced in Etruria by master craftsmen transplanted from Northern Ionia (see Rizzo).

The youth's poise and dignity are enhanced by the frontality of his torso and arms, while the unusual form of his staff indicates that he holds a specific office. In Etruscan wall painting a forked staff is represented only rarely, but it occurs frequently in Greek vase painting, where it characterizes the bearers as officials charged with the physical education and supervision of athletes. In the small frieze from the Tomba delle Bighe at Tarquinia, which shows games held in honor of the dead (see Steingräber, no. 47), three men draped in bordered mantles each hold a forked staff (one of which is wavy), while addressing nude athletes. Dating from about 490 B.C., the paintings of this tomb are strongly marked by Attic influences.

The fact that the youth's head is turned to the rear suggests a formal connection with another figure on a neighboring panel. Further confirmation that our panel was not an isolated piece is provided by the interruption of the pattern at both ends of the frieze. The juxtaposition in the frieze of isolated ivy leaves and dotted palmettes alternating above and below oblique double spirals cannot be directly paralleled, but the individual elements occur both in Etruscan wall painting and on "Pontic" vases: the dotted palmettes with spirals are found in the Tomba delle Leonesse (Steingräber, no. 77) and on "Pontic" oinochoai in Basel (see Hannestadt) and Bonn (see Ducati), while ivy leaves with spirals are found on

91

the ornamented columina of a number of Tarquinian tombs (see Steingräber, color pl. 198). But we meet the most closely related combination of these motifs on a panel from the votive deposit from the Vignaccia site at Cerveteri, which was sold in 1902 to the University of California by the Archpriest of Cerveteri (Berkeley, Museum of Anthropology 8/2990: Matteucig). Made for the decoration of the walls of sacred and profane buildings and tombs, such panels have rarely survived in situ. All the then known examples of Caeretan terracotta panels were published by F. Roncalli in 1965; he further discussed the subject in his contribution to S. Steingräber's book on Etruscan painting. A number of additional fragments that have appeared on the art market have been acquired by museums (see Christiansen; Del Chiaro; Charles Ede Ltd.). It is probable that all these pieces, including the Fleischman panel, originally came from Cerveteri.

BIBLIOGRAPHY: Unpublished.

RELATED REFERENCES: S. Steingräber, *Etruscan Painting* (New York, 1986). M. A. Rizzo, "Ceramografia e pittura parietale in età orientalizzante e arcaica," in *Pittura Etrusca al Museo di Villa Giulia* (Rome, 1989), p. 179f. L. Hannestadt, *The Followers of the Paris Painter* (Copenhagen, 1976), pl. 34a, b. P. Ducati, *Pontische Vasen* (Berlin, 1932), pl. 17b. G. Matteucig, "A Painted Terracotta Plaque from Caere," in *Hommages à A. Grenier* (Brussels, 1962), p. 1151ff., pl. 224. F. Roncalli, *Le lastre dipinte da Cerveteri* (Florence, 1966). Idem, "Etruscan Tomb-painting and Other Types of Etruscan and Italic Painting," in S. Steingräber, op. cit., p. 74ff. J. Christiansen, "En Etruskisk Afrodite," *Meddelelser fra Ny Carlsberq Glyptotek* (1988), p. 47ff. M. Del Chiaro, "Two Etruscan Painted Terracotta Panels," *GettyMusJ* 11 (1983), p. 129ff. Charles Ede Ltd., *Antiquities*, Catalogue 145, no. 3 (now in the Villa Giulia, Rome). For representations of trainers on Attic vases, see J. Jüthner, *Die athletischen Leibesübungen der Griechen*, vol. 1 (Vienna, 1965), p. 182f., pls. 5, 15–17, 21, 23, and vol. 2, pls. 23a, 44b, 56b, 84a, b.

—S. H.

92

Antefix in the Form of a Maenad and Silenos Dancing

South Etruscan, early fifth century B.C.
Terracotta; H: 54.6 cm
Condition: Maenad's left hand, silenos' left ear, left arm and shoulder, left hip, leg, and foot, and right end of base missing; maenad's right elbow and castanets chipped.

The two-figure group is moldmade and joined at the rear to the semicircular end of a cover-tile which reached to the back of their knees, and of which fragments are preserved. An arched buttress joined the top of the cover-tile to the back of the figures at waist level, where a stump of it survives. There is a hole in the crown of each head. The maenad and the silenos stride toward the right side-by-side, each with an arm around the other's shoulder. The barefooted maenad's legs are in profile, her body in three-quarter view, and her head, which she averts from that of the silenos, almost frontal. The forms of her short body and sturdy limbs are clearly visible under the stylized folds of her thin, clinging chiton; over her back is draped a light mantle, the ends of which fall forward from her shoulders in zigzags. In her right hand, which is raised to her waist, she holds castanets. Her round face is framed by crisply waved hair under a stephané of cavetto form. She has widely arched brows, almond eyes, a straight nose, and pouting lips. The silenos turns his broad face back toward her. From his wavy hair, which is crowned with an ivy wreath, protrude pointed horse's ears. His large almond eyes bulge beneath the deeply furrowed, plunging brow. He has a snub nose and a broad, fleshy mouth framed by a wavy mustache and a crinkly spade-shaped beard with a small additional lappet of hair hanging over it from the lower lip. In his left hand he holds a drinking horn. His right hoof appears between the girl's feet. A fragmentary antefix in the Metropolitan Museum of Art (38.11.6: Richter), which is probably from the same mold, shows that the silenos' back was covered by a panther-skin, one paw of which hung down between his left leg and that of the maenad.

The original polychrome painting is exceptionally well preserved. The front of the base is decorated with a pattern of red crenellations along the bottom and black crenellations along the top, the ground between being painted white. The maenad's skin was white, her hair black. Her chiton is red and scattered with rosettes composed of four white dots; around its lower edge runs a wide border of alternating black and white stripes, the uppermost white and serrated along its top. Her mantle is striped in white, red, and black, the last

92

dotted with white. On the maenad's neck was painted a black band, of which a short section on the right, including one vertical bead, survives. The silenos has ocher flesh, black hair, black irises, and red lips; his pubic hair is indicated with thin black lines.

A number of fragmentary examples of antefixes of this type, as well as of molds for producing them, have come to light at Civita Castellana (Falerii) (see Andrén; Sprenger/Bartoloni), finds which clearly prove their local manufacture. But the votive deposit of Campetti at Veii has yielded the head of a silenos of identical type and made of Veian clay (see Vagnetti 1971), which led P. J. Riis to suggest that this type of antefix was invented at Veii. The lower half of an antefix of this type with a provenance from Veii is in a private collection in Switzerland (see Jucker), and similar fragments have recently been excavated in Rome (see Cristofani). Other fragmentary antefixes belonging to the same series are without provenance, one in the Cleveland Museum of Art and another (mentioned above) in the Metropolitan Museum of Art (see Teitz). Modelers evidently traveled with their molds from place to place to renew the terracotta decoration of temples, wherever this might be required (see Vagnetti 1966). The present antefix, one of the largest known examples of this type, is said to come from Cerveteri.

BIBLIOGRAPHY: *Wealth of the Ancient World*, exh. cat. (Fort Worth, 1983), no. 16, frontispiece; Sotheby's, New York, *The Nelson Bunker Hunt Collection, The William Herbert Hunt Collection*, auct. cat., June 19, 1990, no. 16.

RELATED REFERENCES: G. Richter, *Handbook of the Etruscan Collection* (New York, 1940), p. 22, fig. 56. A. Andrén, *Architectural Terracottas from Etrusco-Italic Temples* (Lund, 1939/40), p. 99ff., pls. 32, 33, figs. 111, 114. M. Sprenger and G. Bartoloni, *The Etruscans* (New York, 1983), nos. 137, 138. L. Vagnetti, *Deposito Votivo di Campetti a Veii* (Florence, 1971), p. 27f., pl. 3.3. P. J. Riis, *Etruscan Types of Heads* (Copenhagen, 1981), p. 44ff., fig. 29, pl. Veii 13, K, L. I. Jucker, *Italy of the Etruscans*, exh. cat. (The Israel Museum, Jerusalem, 1991), no. 332. M. Cristofani, ed., *La grande Roma dei Tarquinii* (Rome, 1990), p. 62f., no. 3.4, color pl. v. R. S. Teitz, *Masterpieces of Etruscan Art* (Worcester, Mass., 1967), pp. 26, 48f., nos. 10, 34. L. Vagnetti, "Nota sull'attivita dei coroplasti etruschi," *Archeologia Classica* 18 (1966), p. 110ff.

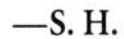

—S. H.

92, BACK VIEW

The Hellenistic World after the Death of Alexander

93

BRONZE AND LEAD

93
Weight with Head of a Beardless Male

Hellenistic, mid-third century B.C.
Lead; H: 5.7 cm; W: 4.8 cm; DEPTH: 0.4–0.8 cm; WEIGHT: 160 g
Condition: Intact; crusty whitish surface.
Provenance: Said to come from Asia Minor.

The weight, a rectangular plaque without a border, is stamped in medium-low relief with a beardless male head, seen in profile facing our left. He has short, thick hair and wears a large circular diadem instead of the usual headband with the ends hanging down over the neck. The back is undecorated and somewhat irregular, probably from the process of adjusting the weight.

The head, the portrait of an eastern Hellenistic ruler, is similar to some dies in the silver tetradrachm coinage of Philetairos of Pergamon (274–263 B.C.), which bear an image of Seleukos I Nikator wearing the same unusual circular diadem (see Leu; Jenkins). A bronze head, believed to be Seleukos I, now in Naples, has the same deep-set eyes and the curved, energetic mouth (see Richter; Fittschen). This remarkable weight-portrait displays the head of the strong, decisive founder of the Seleucid dynasty, after whom several cities in the East are named. As the weight is said to come from Asia Minor, Seleucia Pieriae or Seleucia ad Calycadnum Ciliciae could be the place of origin and the mid-third century B.C. a possible date.

In its weight of 160 grams the piece is equal to ten silver tetradrachms of the third century B.C. It appears to have been cast in lead from a bronze weight-standard with more clearly drawn portrait features.

BIBLIOGRAPHY: Unpublished.

RELATED REFERENCES: Leu sale 48, Zurich, 1989, no. 221. G. K. Jenkins, *Ancient Greek Coins* (Fribourg, 1972), p. 224, no. 553. G. M. A. Richter, *Portraits of the Greeks,* vol. 3 (London, 1965), figs. 1867, 1868; K. Fittschen, *Griechische Porträts* (Darmstadt, 1988), pl. 84.3–4.

—L. M.

94
Mina Weight with Elephant

Hellenistic, second part of third century B.C.
Bronze; H: 11.2 cm; W: 11.5 cm; DEPTH: 0.8 cm; WEIGHT: 534.5 g (MINA OF SELEUCIA PIERIAE)
Condition: Intact.
Provenance: Found near Adana (together with cat. no. 95).

The square weight has a hole, evidently for suspension, pierced in the center top. Side A is framed by a plain molding and, within this, an astragal. An Indian elephant, shown in medium-low relief and seen in profile facing the viewer's left, stands on a plain ground line. In the field in front of his head is a small anchor. The piece is inscribed in relief letters: ΣΕΛΕΥΚΕΙΟΣ (Seleucian [meaning the city of]) above the elephant and MNA (mina) below. The reverse has a net pattern and a raised, roughly rectangular area that is evidence of the weight of the piece being adjusted.

This piece is remarkable for several reasons: (1) this type of weight is common in lead but extremely rare in bronze, (2) it is unusual to find the Seleucid anchor in the field on such a weight, (3) the elephant is unusually well designed as are the script and the carved border, and (4) the piece lacks a date. The fact that the weight is bronze combined with the missing date indicates that the weight is of early manufacture. Lead elephant weights from Antioch and Seleucia are known from the period between the beginning of the Seleucid era in 312 B.C. and the founding of Seleucia Pieriae itself in 109 B.C. E. Seyrig

94

has stressed that the first known dated elephant weight (Louvre 3303: Michon; de Ridder) is marked 126 of the Seleucid era (= 187/186 B.C.). Our bronze weight is, thus, considerably earlier, which is also indicated by the early form of the *alpha*.

A very similar, but less well-preserved piece, in lead with a weight of 1057 grams, in the American Numismatic Society has been attributed to Antioch (see Lang). It may, however, come from the same die as our bronze weight. Another lead double mina of Seleucia with the elephant marching to the right is of somewhat lighter weight (1035 grams) and later date (see Qedar).

The inscription is an enigma. It seems inconceivable that an early official weight, not to mention a weight-standard, from the Seleucid empire would contain an error in the inscription, i.e., ΣΕΛΕΥΚΕΙΟΣ (Seleucian) instead of ΣΕΛΕΥΚΕΙΟΝ, the usual inscription, meaning ([the weight] of the inhabitants of Seleucia). M. Rostovtzeff proposes an explanation for the anomaly: "ΣΕΛΕΥΚΕΙΟΣ means ΣΕΛΕΥΚΕΙΟΣ ΧΑΡΑΚΤΗΡ" (Seleucian means Seleucian mark).

The weights in this region and period vary considerably. This may be the reason for the addition of the rectangular metal piece to our specimen. It seems, however, that a mina between 500 and 600 grams is the most frequent weight. Thus, the figure of 534.5 grams fits into this scheme.

BIBLIOGRAPHY: Unpublished.

RELATED REFERENCES: E. Seyrig, *Scripta Varia*, Bibliothèque Archéologique et Historique 125 (Paris, 1985), pp. 375–379, 402–407. *Dictionnaire des antiquités grecques et romaines d'après les textes . . .*, vol. 4, p. 556, s.v. "Pondus" (E. Michon). A. de Ridder, *Les Bronzes antiques du Louvre*, vol. 2 (Paris, 1915), no. 3303. M. Lang in *Museum Notes of the American Numismatic Society* (1968), p.2, no. 5. S. Qedar in Münzzentrum, Cologne, *Gewichte aus drei Jahrtausenden*, vol. 4, sale 49, 1983, no. 5078. M. Rostovtzeff, *Social and Economic History of the Hellenistic World* (Oxford, 1953), p. 453, no. 2, pl. 55.2. For the well-known lead double mina (1069 grams) of Antioch, depicting an elephant, see E. Babelon and J.-A. Blanchet, *Catalogue des bronzes antiques de la Bibliothèque Nationale* (Paris, 1895), no. 2246.

—L. M.

95

96

95
Tetarton Weight with Zebu

Hellenistic, first half of second century B.C.
Bronze; H: 7.2 cm; W: 7.2 cm; DEPTH: 0.5 cm; WEIGHT: 128 g (TETARTON = QUARTER MINA OF SELEUCIA PIERIAE)
Condition: Intact except for battered lower right corner.
Provenance: Found near Adana (together with cat. no. 94).

The weight is a square plaque. Its obverse has a plain molded frame and shows, in medium-low relief, a walking zebu seen in profile facing the viewer's left. The piece is inscribed in relief letters. Above the figure ΣΕΛΕΥΚΕΙΟΝ (of the inhabitants of Seleucia). Below: TETAPTON (tetarton). The reverse has a very regular net pattern within a narrow raised border.

The zebu, a humped Indian breed of bull known as a "Brahman" in the West today, had much the same appearance as the modern variety, and, like the elephant, was an appropriate symbol for an eastward-looking city founded by one of Alexander's successors. A closely related lead tetarton of Antioch in Syria, with a zebu walking toward the right and the same net pattern, is in the Bibliothèque Nationale in Paris (see Seyrig; Babelon/Blanchet; Michon).

BIBLIOGRAPHY: Unpublished.

RELATED REFERENCES: E. Seyrig, *Scripta Varia,* Bibliothèque Archéologique et Historique 125 (Paris, 1985), pp. 367–416, no. 3. E. Babelon and J.-A. Blanchet, *Catalogue des bronzes antiques de la Bibliothèque Nationale* (Paris, 1895), no. 2247. *Dictionnaire des antiquités grecques et romaines d'après les textes* . . . , vol. 4, p. 555, fig. 5776, s.v. "Pondus" (E. Michon). For weights from Seleucia (including a tetarton with zebu walking toward the left [location not indicated]), see Michon, op. cit., p. 548ff.

—L. M.

96
Tetarton Weight with Ship's Prow

Hellenistic, first half of second century B.C.
Bronze; H: 7.3 cm; W: 7.1 cm; DEPTH: 0.7 cm; WEIGHT: 109.8 g (TETARTON = QUARTER MINA OF SELEUCIA PIERIAE)
Condition: Rather crusty olive green and black patina.

The weight is an approximately square plaque. On the obverse a ship's prow is shown in relief within a molded frame. The vessel is a war-

97

ship with rams and bow ornament. An unclear trace in the field above may be an anchor. The piece is inscribed ΣΕΛΕΥΚΕΙΟΝ (of the inhabitants of Seleucia) above the ship and ΤΕΤΑΡΤΟΝ (tetarton) below. The reverse is decorated with a precise network pattern.

The piece is similar in lettering, square network design, and framing to the other bronze weights from Seleucia in the Fleischman collection (see cat. nos. 94, 95).

BIBLIOGRAPHY: Unpublished.

RELATED REFERENCES: For a weight with a ship's prow and the same network pattern, but with a Tanit-sign in the center, see S. Qedar in Münzzentrum, Cologne, *Gewichte aus drei Jahrtausenden*, vol. 4, sale 49, 1983, no. 5073.

—L. M.

97

Half-mina Weight with Anchor and Dolphin

Hellenistic, 151/150 B.C.
Lead; H: 8.5 cm; W: 8.7 cm; DEPTH: irregular, approximately 0.6 cm; WEIGHT: 232.3 g (HALF-MINA OF SELEUCIA PIERIAE)
Condition: Upper left corner missing; lower right corner bent.

The obverse depicts a large anchor, framed by a plain, narrow border. Two dolphins flank the anchor, while a third is shown below it at the right. The piece is inscribed [ΣΕΛΕ]ΚΕΙΟΝ / ΒΞΡ ΕΠΙ / ΕΥΔΩΡΟΥ / ΗΜΙΜΝΑΙΟΝ (Of the Seleucids / 162 / under / Eudoros / half mina [162 of the Seleucid era = 151/150 B.C.]). The reverse is decorated with a rhomboid network pattern, whereas earlier pieces have a square net design.

Syrian and Phoenician weights are dated according to several eras. The most frequently used are the eras of the Seleucid empire (beginning 312 B.C.), of Tyre (beginning 127/126 B.C.), and of Seleucia Pieriae itself (beginning 109/108 B.C.). The shape, design, and script of our piece suggest a late Hellenistic origin. The anchor is a common Seleucid emblem, widely used even by the Hasmoneans in their Jewish coinage after 100 B.C. The known weights of Seleucia Pieriae, the port of Antioch, are even more numerous than those of the main city, which hints at the great importance of the harbor for the metropolitan economy in Hellenistic and Roman times.

E. Seyrig has published a mina of 544 grams from the Chandon de Briailles collection that shows the identical anchor, three dolphins, date letters, and an official's name. The Chandon and the Fleischman

98

specimens come from the same mold, a situation rarely ever encountered. Before the casting of the heavier Chandon piece the inscription for the denomination was recut, changing HMIMNAION to MNA and the last letter of the ethnikon from N to Σ. (For one explanation of the latter change, see cat. no. 94.)

The weight of the two pieces confirms these findings. As about a good tenth of the Fleischman piece is missing, its initial weight must have been approximately 260–270 grams, or half that of the Chandon specimen.

Our early bronze elephant mina (cat. no. 94) and the two anchor pieces described above strengthen Seyrig's assumption that in Antioch and Seleucia in the third and second centuries B.C. "a mina weighed between 500 and 600 grams."

BIBLIOGRAPHY: Unpublished.

RELATED REFERENCES: E. Seyrig, *Scripta Varia*, Bibliothèque Archéologique et Historique 125 (Paris, 1985), p. 377, no. 12, pl. III.12; for the metrological problem, pp. 403–407, esp. p. 406.

—L. M.

98

Mina Weight with Elephant

Late Hellenistic, second–first century B.C.
Lead; H: 10.8 cm; W (INCLUDING PROJECTING TAB): 11.5 cm; WEIGHT: 520 g (MINA OF ANTIOCHIA SYRIAE)
Condition: Intact.

The weight has the form of a square plaque. On the obverse, framed by an astragal border, is an elephant walking toward the viewer's left. Above is inscribed, ANTIOXEIA (Antioch); between the elephant's legs, MNA (mina); beneath the ground line, ΕΠΙΖΗΝΟΦΑΝΥΟΥ (under Zenophanes). The reverse has a net pattern in low relief. On one side is a projecting tab. As the tab is attached sideways, which is quite unusual, it may have provided a place where the weight could be shaved down to calibrate its weight.

The workmanship of our lead elephant mina is coarser, the relief lower than that displayed by catalogue numbers 94 and 95 above. The reverse with the square network, however, is similar. The medium and design as well as the naming of an official and the late form of the *alpha* also indicate a later date. As the designation "agoranomos," usual on Roman weights, does not appear, a late Hellenistic date is likely. During this period the Antioch mint was a prolific producer of Seleucid silver coinage, a testimony to the economic importance of that metropolis.

BIBLIOGRAPHY: Unpublished.

RELATED REFERENCES: For weights of this period, see E. Seyrig, *Scripta Varia,* Bibliothèque Archéologique et Historique 125 (Paris, 1985), pp. 371–374. For the image of the elephant, see H. H. Scullard, *The Elephant in the Greek and Roman World* (London, 1974).

—L. M.

99

99
Weight with Theatrical Mask

Hellenistic-Early Imperial, first century B.C.–first century A.D.
Bronze; L: 1.4 cm; W: 1.4 cm; H (WITHOUT MASK): 0.8 cm; (WITH MASK): 1.6 cm; WEIGHT: 23.5 g
Condition: Intact except for wear, especially at the corners; smooth greenish brown patina.

The mask is shown in high relief on the small, thick plaque of the weight. It has a smooth youthful face with a down-turned mouth and plain center-parted hair. Mask types like this are seen on oscilla and elsewhere in the decorative arts of the Vesuvian cities, though they were already current in Hellenistic times. The high relief and good detail of the little weight might suggest a date in the first century B.C. or the first century A.D. The figure of 23.5 grams corresponds, in fact, to six full-weight silver Roman denarii in the Julio-Claudian period.

BIBLIOGRAPHY: Unpublished.

RELATED REFERENCES: For a mask-oscillum of this type, see A. Claridge and J. B. Ward Perkins, *Pompeii A.D. 79,* exh. cat. (Boston, 1978), no. 78.

—L. M.

100
Statuette of the Youthful Herakles

Hellenistic, third century B.C.
Bronze; H (WITHOUT TANG): 17.4 cm; (WITH TANG): 19.5 cm
Condition: Right arm missing from below the biceps.

Although the figure's physique has a Hermes-like nimbleness, the lion's skin around the left arm makes it certain that Herakles is intended. The piece belongs to a type known as the Hercules Bibax, as in complete versions the right hand holds out a drinking horn or a skyphos (see Boucher 1976, p. 145). Examples related in style to the piece under discussion are in Vienne (see Boucher 1971 [I owe this reference to J. R. Guy]), Paris (see Adam), Avignon (see Rolland), Verona (see Franzoni), Vienna (see *LIMC*), and the British Museum (see Walters). Although the other versions are of more modest workmanship than ours, they share its boyish face and attenuated but defined physique.

The type seems indigenous to Hellenistic Italy, but dates assigned to individual pieces have ranged from the third century B.C. down into Augustan times. A variant, in the same pose but bearded and with a more conventionally overdeveloped body, continues far into the Imperial period. Our figure's fine workmanship suggests that it belongs near the beginning of the series and makes it more informative than the other, somewhat provincial examples. The juvenile appearance of the hero has affinities with late Etruscan traditions, and the elongated but muscular physique recalls the exaggerated versions of Lysippian figures on third-century Tarantine reliefs. It seems preferable to see the style in this context rather than as a lightened and simplified late second-century interpretation of the Hellenistic baroque.

Our figure, whose face seems subtly individualized, poses a problem analogous to that of several other ambitious Italo-Hellenistic statuettes that have been considered portraits of rulers in heroic or divine guise (see Adamo Muscettola; C. C. Vermeule in *The Gods Delight*, no. 30). The delicate, short-chinned features are decidedly atypical for Herakles. However, they persist in other, much less sophisticated members of the series, suggesting that if an influential version of the image was indeed intended as a portrait, this meaning was probably lost on, or considered as secondary by, most of the craftsmen who reproduced it.

BIBLIOGRAPHY: Unpublished.

RELATED REFERENCES: S. Boucher, *Vienne: bronzes antiques* (Paris, 1971), no. 22. A. M. Adam, *Bronzes étrusques et italiques de la Bibliothèque Nationale* (Paris, 1984), no. 306. H. Rolland, *Bronzes antiques de Haute Provence* (Paris, 1965), no. 84. L. Franzoni, *Bronzetti romani del Museo Archeologico di Verona* (Venice, 1973), nos. 100, 101. *LIMC*, vol. 4, no. 773, s.v. "Herakles" (J. Boardman, O. Palagia, S. Woodford). H. B. Walters, *Catalogue of the Bronzes, Greek, Roman, and Etruscan in the Department of Greek and Roman Antiquities, British Museum* (London, 1899), nos. 1248, 1270. S. Adamo Muscettola, "Bronzetti raffiguranti dinasti ellenistici al Museo archeologico di Napoli," *Bronzes hellénistiques et romains: tradition et renouveau*, Actes du Ve Colloque international sur les bronzes antiques (Lausanne, 1979), p. 87ff.

—A. H.

100

101

Statuette of Aphrodite

Hellenistic, first half of second century B.C.
Bronze; H: 38 cm

The apple held out in the figure's left hand is an attribute of Aphrodite, but the decorous and fashionable clothing suggests that a real person, probably a queen, is depicted here in divine guise. The bold, almost swaggering posture and powerful physique, with jutting breasts and well-fleshed arms, are typical of the Hellenistic baroque style. The woman's stature is enhanced by thong sandals with high platform soles, notched between the big toe and the others, a fashion at its most extreme in the early second century B.C. Her voluminous underdress, of thick material fastened on the shoulders with disk-shaped brooches, is long enough to fan out and trail on the ground around her, creating a broad base for the figure. The folds of this heavy garment are seen through the himation, a thin outer wrap with a grid-work of press-folds, which is wound tightly around her body and drawn up to veil the back of her head. The crescent-shaped stephané, decorated with foliate scrolls and worn with a veil, resembles the one seen in a fragmentary statue of a queen, perhaps Apollonis, from Pergamon (see Smith). Our figure's delicate but fleshy oval face with long, pointed nose, puffy-lidded eyes, and a hint of double chin recalls a head, with similar hairdo and stephané, on a Type I couch fulcrum datable in the first half of the second century

101

101

102

(see Faust). The coiffure, rolled back from the face in winglike masses and drawn into a chignon behind, is a typically High Hellenistic one worn, for example, by the Juno Cesi (see Linfert) as well as by the head on the fulcrum.

The statuette, though dressed in the clinging, transparent mantle which remained fashionable in late Hellenistic art, probably belongs to the first half of the second century, before robust body types were rejected in favor of elongated, narrow-shouldered figures, and the powerful thrust and counterthrust of the composition gave way to a smooth, gliding motion. Taken together with the Type I fulcra, with which it shares a certain quality of modeling—rather soft, heavy, and generous—our statuette constitutes a rare point of reference for the character of small-scale bronze work in High Hellenistic times.

BIBLIOGRAPHY: Unpublished.

RELATED REFERENCES: R. R. R. Smith, *Hellenistic Royal Portraits* (Oxford, 1988), no. 29, pl. 23.2. S. Faust, *Fulcra: Figürlicher und ornamentaler Schmuck an antiken Betten,* RM Ergänzungsheft 30 (Mainz, 1989), pp. 36, 179, no. 170, pl. 2.1. A. Linfert, *Kunstzentren hellenistischer Zeit: Studien an weibl. Gewandfiguren* (Wiesbaden, 1976), figs. 251–253.

—A. H.

102

Roundel with Bust of Dionysos

Hellenistic, third quarter of second century B.C.
Bronze; DIAM: 22–20 cm
Condition: Large sections of border missing at top and bottom; edges sharp, as if cut rather than accidentally broken away; three holes for attachment visible, two through border and, oddly, one through central zone behind Dionysos' right shoulder.

Dionysos is seen frontally, with his head turned slightly toward his right. His chest is bare except for a mantle over his left shoulder, against which he carries a beribboned thyrsos. He is crowned with a fillet wound once low on the forehead, and again, with an ivy tendril's leaves and *corymbi* (berry clusters), above the hairline. Generous "tadpole-shaped" ends hang down. The god's hair is parted in the middle and is rolled back from the face below the double-wound fillet. Long, wavy tresses fall on the shoulders. The beardless face has placid, regular features, with small eyes, a large, broad nose, and a heavy jaw. There is a space between the rather slack lips but no real opening. The irises and pupils of the eyes are indicated in metal of a different color.

The piece has the monumentality but not the emotionalism of High Hellenistic baroque creations. Its broad, straight nose and heavy chin recall the Venus de Milo (see Pasquier, passim) or the "Inopos" bust from Delos (Pasquier, p. 84ff.), both now usually ascribed to the second half of the second century B.C. A good typological and stylistic point of reference for our Dionysos is to be found in the heads of Dionysos on the coins of Maroneia (see Head; *LIMC*) and Thasos (see Head), thought to be datable after the closing of the Macedonian mints in 148 B.C.

BIBLIOGRAPHY: Merrin Gallery, *Bronzes: Cast in the Image,* exh. cat. (New York, 1989), no. 10.

RELATED REFERENCES: A. Pasquier, *La Vénus de Milo et les Aphrodites du Louvre* (Paris, 1985); for the date, see p. 87f. B. V. Head, *Historia Nummorum: A Manual of Greek Numismatics* (1911; reprint, Chicago 1967), p. 251, fig. 158 (Maroneian coins); p. 266, fig. 164 (Thasian coins). *LIMC,* vol. 3, no. 195, s.v. "Dionysos" (C. Gasparri, A. Veneri). B. Barr-Sharrar, *The Hellenistic and Early Imperial Decorative Bust* (Mainz, 1987). R. Winkes, *Clipeata imago: Studien zu einer römischen Bildnisform* (Bonn, 1969).

—A. H.

103

Statuette of Poseidon

Hellenistic, late second century B.C .
Bronze; H (WITH TANG): 16.35 cm; (WITHOUT TANG): 15.1 cm
Condition: Three fingers of left hand and attribute held therein missing.

The god's elongated but muscular physique, his small head, and his sweeping, elastic pose, reaching for and scooping in space beyond the figure, derive from the innovations of Lysippos in the late fourth century B.C. The mannered, almost playful exaggeration of these traits in our piece, however, typifies a revival of the style in mid-to-late Hellenistic times. A very close parallel is the famous Poseidon in Munich from the Loeb collection (see Maass), considered a work of the late second century B.C. Our figure's disheveled hair and mobile,

103

restless-looking posture identify him too as the sea god Poseidon. His attributes would be a tall trident and, in the left hand, a dolphin or ship's ornament. The spiky leaves of his crown, as in the Loeb statuette, are from shore- or water-plants.

BIBLIOGRAPHY: Unpublished.

RELATED REFERENCES: M. Maass, *Griechische und römische Bronzewerke der Antikensammlungen* (Munich, 1979), p. 24f., no. 9. E. Walter-Karydi, "Poseidons Delphin: der Poseidon Loeb und die Darstellungsweisen des Meergottes im Hellenismus," *Jahrbuch des Deutschen Archäologischen Instituts* 106 (1991), p. 243ff.

—A. H.

104
Bust of a Boy with Attributes of Herakles

Late Hellenistic, late second century B.C.
Bronze; H: 17.3 cm
Condition: Pieces of lion's skin on shoulders and part of bust at proper left side missing.
Provenance: Formerly in the collection of James Coats.

104

The bust, made as an appliqué, is of a fine-featured young boy who turns his head toward his left. His silky looking, mid-length hair is drawn up into a braid along the top of his head, from which tendrils escape onto his forehead. The rest frames his face in graceful disorder. Over his head is thrown the head of Herakles' lion's skin, the front paws of which would have hung down at the sides of his neck and another piece of which drapes his left shoulder.

The subject is difficult to explain, since the always robust and curly-headed Herakles would not have been represented in this

dainty guise at any phase of his life. The wings that would assure an identification as Eros are lacking, and it is not possible to reconstruct them rather than paws of the *leontis* in the broken spots beside the neck. Herakles' mistress Omphale, who appropriated his club and lion's skin, can hardly be intended since her femininity while usurping her lover's attributes is always stressed; our bust does not have female breasts and its coiffure is that of a male child. A wingless Eros or some progeny or boy-love ascribed to Herakles seem to be the best possibilities. The conceit, in keeping with the style, may well have been a somewhat precious one.

The generous scale, the sense of fine bones but loose flesh, and the un-incised, somewhat puffy-lidded eyes speak for a late Hellenistic date. Marble parallels include the female bust from the Mahdia wreck, an ancient ship laden with works of art that sank off Tunis near the end of the second century B.C. (see Fuchs), the Aphrodite of the Delos Aphrodite and Pan group (see Pollitt, fig. 138), and especially Daphnis in the original of the much-copied Pan and Daphnis (see Pollitt, fig. 137), all probably datable in the years soon before 100 B.C.

BIBLIOGRAPHY: S. B. Matheson, "A Bronze Eros with the Attributes of Herakles," *Yale University Art Gallery Bulletin* (Winter 1982), pp. 24–31, figs. 1–3. M. L. Hadzi, *Transformations in Hellenistic Art*, exh. cat. (Mount Holyoke College, South Hadley, Mass., 1983), pp. 20–21, no. 6. J. P. Uhlenbrock, *Herakles: Passage of the Hero Through 1000 Years of Classical Art*, exh. cat. (New Rochelle, N.Y., 1986), p. 12, pl. 33. B. Barr-Sharrar, *The Hellenistic and Early Imperial Decorative Bust* (Mainz, 1987), p. 76f., pl. 51.C172.

RELATED REFERENCES: W. Fuchs, *Der Schiffsfund von Mahdia* (Tübingen, 1963), no. 54. J. J. Pollitt, *Art in the Hellenistic Age* (Cambridge, 1986).

—A. H.

105

Statuette of a Griffin with an Arimasp

Late Hellenistic, probably late second–early first century B.C.
Bronze; H: 7.9 cm
Condition: Element once attached to back between wings missing.

The griffin, with lean body, small head, and long crested neck, sits on his haunches. His wings are of volute-tipped Archaistic form. He mauls a youthful male victim, wrenching an arm upwards with his beak while pushing down on the head with his right front claw to rend his prey limb from limb. The nude body slumps, already lifeless, between the griffin's front legs.

Griffins, fabulous creatures inspired by prototypes in ancient

Near Eastern art, were thought of by the Greeks as guarding golden treasures in a land far to the north and east of the Greek world. The griffins' opponents were the Arimasps, a legendary tribe who in mythology, but not in art, were one-eyed. Our representation, with the Arimasp nude except for a Greek helmet, is atypical. Usually the Arimasps wear Phrygian caps and are shown in the trousered, long-sleeved attire of the eastern barbarians (see *Enciclopedia dell'arte antica*).

The griffin's seated, upright posture and the rubbery look of his anatomy recall the griffins that accompany Roman statues of Nemesis. The victim's pose is that of the dead warrior in the Hellenistic Pasquino Group (see Pollitt), or, closer to our piece in its graphic horror, the companion just dispatched by Polyphemos in the group of Odysseus offering the wine cup (see Andreae). The dead youth's helmet, with its high, basinlike shape, contoured cheek pieces meeting under the chin, and crest rising straight up in front but trailing behind, resembles the helmets in the census scene of the "Altar of Domitius Ahenobarbus," which is probably datable in the early first century B.C. (see Stewart).

Large, frontally seen griffins seizing Arimasps function as consoles in the trompe l'oeil architecture of a fragmentary Second Style fresco from the Villa of the Mysteries at Pompeii (see Maiuri). Although rather different in composition from our group, the fresco testifies to interest in the theme, which may have had symbolic implications, at the time the villa was decorated in the mid-first century B.C. Our piece, which has traces of an attachment at the top and seems to have been part of some larger object, is probably datable late in the Hellenistic period.

BIBLIOGRAPHY: Atlantis Antiquities, *Ars and Techne: Art and Craft of the Graeco-Roman World*, exh. cat. (New York, 1989), no. 8.

RELATED REFERENCES: *Enciclopedia dell'arte antica*, vol. 1, p. 637, s.v. "Arimaspi" (L. Vlad Borrelli). J. J. Pollitt, *Art in the Hellenistic Age* (Cambridge, 1986), fig. 119. B. Andreae, *Odysseus: Archaeologie des europäischen Menschenbildes*, 2nd ed. (Frankfurt, 1984), p. 101. A. Stewart, *Greek Sculpture* (New Haven, 1990), figs. 845, 846. A. Maiuri, *La Villa dei Misteri* (Rome, 1947), p. 192.

—A. H.

106
Statuette of a Ruler or Divinity

Late Hellenistic, late second–first century B.C.
Bronze with silver; H: 17.5 cm
Condition: First and little fingers of left hand and end of drapery missing; large hole in figure's upper back.

The piece descends from likenesses of Alexander the Great, the most influential of which were made by his court artist Lysippos. The mobile, spiraling posture with one hand set high on a spear or scepter seems to go back to Lysippos. Here, however, the image has been given a baroque interpretation. The muscles are bulky, the pose assertive; a swag of military cloak has been added to set off the heroic nude. The stylistic phase seems to be not long after that exemplified by the bronze ruler in the Terme (see Stewart), though our piece has become at once softer and blockier.

Some details have a rather naive spontaneity. The right hand, braced against the hip, is outsize and obviously made by snipping a mittenlike blank, recalling the hands of an Alexander-like late Hellenistic statuette in Naples (see Adamo Muscettola). Another figure closely related to ours and explicitly Italo-Hellenistic in style is in the Ortiz collection (see *Ortiz Collection*). It shares with our piece the odd positioning of the cloak, in a loose double loop around the raised arm. Both have affinities with the large statuette of the god Veiovis, once in Viterbo but currently on display in the Museo Nazionale Etrusco di Villa Giulia, Rome. This piece was traditionally considered late Etruscan but is now dated by some authorities in the first century A.D. (see Sichtermann). Our figure's eyes are silvered and the pupils incised, as in the Veiovis and many statuettes from Pompeii. The person represented does not have strongly individualized features. He might be intended as Alexander himself, as some later ruler, or as one of the Dioskouroi.

BIBLIOGRAPHY: Unpublished.

RELATED REFERENCES: A. Stewart, *Greek Sculpture* (New Haven, 1990), fig. 862. S. Adamo Muscettola, "Bronzetti raffiguranti dinasti ellenistici al Museo archeologico di Napoli," *Bronzes hellénistiques et romains: tradition et renouveau*, Actes du Ve Colloque international sur les bronzes antiques (Lausanne, 1979), p. 87ff., pl. 34. *The George Ortiz Collection: Antiquities from Ur to Byzantium* (Berne, 1993), no. 168. H. Sichtermann in T. Kraus, *Das römische Weltreich*, Propyläen Kunstgeschichte 2 (Berlin, 1967), no. 260.

—A. H.

106

107

Vase in the Form of a Female Head

Late Hellenistic, probably first century B.C.
Bronze with silver inlays; H: 9.1 cm
Condition: Neck and lip of vase, all but attachment plate of handle, separately made bottom, and part of inlaid decoration missing.

107

The vessel is in the form of an idealized female head. From its top sprouts a narrow vase neck which would originally have had a trefoil mouth and a back handle. The handle's attachment plate, incised with a foliate ornament, survives. The head has an oval face with rather hard but harmoniously proportioned ideal features. The eyes are shallowly hollowed to receive inlay. The hair, fine and only slightly waving, is swept back on the temples and tied up behind the head. The edge of a cap or a broad band, bordered with a row of silver beading and originally inlaid with zigzags in another metal, appears on the forehead under the turbanlike headcloth, which wraps the back of the head and is knotted at the front. The ends emerging from the knot have the form of broad, flat tabs and were inlaid with stripes. The ears are pierced for earrings.

Our head vase is related to a series whose descent from fourth-century models like the elegantly coiffed ladies of the Panajurište Treasure (see Venedikov/Gerassimov) is evident. Examples in Athens from Dodona (National Archaeological Museum 795), in the Louvre from Janina (see de Ridder), in the Kluge collection (see Eisenberg 1983), and on the New York art market from Lebanon (see Eisenberg 1985), are typologically quite similar but have puffy, Hellenistic-looking faces. The headdress, though made up of the same elements seen in our version, is smaller, decorating rather than concealing the hair. All the vessels have handles in the form of a plant stem with a growth knot bursting into leaf, a Hellenistic type that continued into the Imperial period.

A head vase in Naples (although found in Rome; see Biroli Stefanelli, no. 118, fig. 236) offers a much closer parallel. The face, like that of our piece, is a narrow oval with austere, regular features. The kerchief swathes the whole of the coiffure except at the front, where, as in ours, it is concealed by a cap or band, the lower edge of which is decorated with silver beads. Unfortunately, in both pieces the potentially revealing handle is lost.

Like the Naples vessel, the Fleischman piece differs both from the usual fleshy, obviously Hellenistic versions of its type and from Early Imperial productions like the "Sappho" head vase from Herculaneum (see Pirzio Biroli Stefanelli, no. 117, figs. 232–234), which reproduce Classical models with an academic precision. Our head vase may represent a retro-stylization of the familiar puffy-faced Hellenistic model, perhaps produced in the Greek East at the end of Hellenistic times. The rendering of the hair, straight, silky-looking, and without volume, can be a late Hellenistic feature, while the lavish use of metal polychromy recalls the "Sappho" vase. The question of date, however, must remain open.

BIBLIOGRAPHY: Unpublished.

RELATED REFERENCES: I. Venedikov and T. Gerassimov, *Thracian Art Treasures* (Sofia and London, 1975), pl. 126. A. de Ridder, *Les Bronzes antiques du Louvre*, vol. 2 (Paris, 1915), no. 2956. J. Eisenberg for the Royal-Athena Galleries, New York, *Masterworks in Miniature from the Ancient World*, 1983, leaflet, p. 1. J. Eisenberg for the Royal-Athena Galleries, New York, *Art of the Ancient World*, vol. 4, *Ancient European, Oriental, Pre-Columbian, and Tribal Works of Art*, sale cat., 1985, no. 298. L. Pirzio Biroli Stefanelli, ed., *Il bronzo dei romani* (Rome, 1990).

—A. H.

108
Statuette of a Standing Comic Actor

Hellenistic, probably second century B.C.
Bronze; H: 12.2 cm

The actor stands in a hip-shot pose, accentuating his pot belly and his prominent buttocks. The figure is solid, with strong legs and a head that is not exaggeratedly large. Over leggings and long sleeves he wears a short mantle, which is wrapped around his torso and pulled up to cover the back of his head. His right hand, clenched, is raised close to his chin in a "plotting" gesture. The mask is that of a resourceful leading slave with stiff hair brushed back in the *speira* (rolled pompadour), a snub nose, and a thick mustache over his scoop-shaped beard.

Although he wears a long artificial phallus, omitted from most later comic representations, the actor has little in common with the gnomelike terracotta figures illustrating Middle Comedy. If anything, he is closer to the more matter-of-fact South Italian *phlyax* representations (see cat. nos. 56–59). However, his sturdier and less grotesque aspect is clearly influenced by the more naturalistic style used for figures of New Comedy (see Seeberg). The broad-faced mask, confidently but rather softly modeled, is not unlike the two bearded comic masks among those which serve as the feet of bowls in the Metropolitan Museum's third-century silver treasure (see Bothmer). Full-length relatives can be found among terracotta actors from Taranto, usually thought to be early or middle Hellenistic (see De Juliis/Loiacono). However, the appearance of a not very dissimilar terracotta actor in a set from a tomb group of the second half of the first century B.C. at Myrina (see Mollard-Besques) is an indication of the tradition's wide geographical range and long continuity.

Our piece's main features seem to have third-century sources. Its soft but strong and correct modeling, not dependent on calligraphic detail or even on precisely miniaturized forms, might be at home then, or considerably later in Hellenistic times.

BIBLIOGRAPHY: Unpublished.

RELATED REFERENCES: A. Seeberg, "Bronzes Referring to New Comedy," in *Akten der 9. Tagung über antike Bronzen* (Vienna, 1988), p. 270ff. D. von Bothmer, "A Greek and Roman Treasury," *Metropolitan Museum of Art Bulletin* (Summer 1984), p. 59f., nos. 105, 106. E. De Juliis and D. Loiacono, *Taranto: il Museo Archeologico* (Taranto, 1985), no. 430. S. Mollard-Besques, *Catalogue raisonné des figurines et reliefs en terre-cuite grecs, etrusques et romains*, vol. 2, *Myrina* (Paris, 1963), pl. 1.3 (Tomb C), pl. 172 (MYR 320).

—A. H.

108

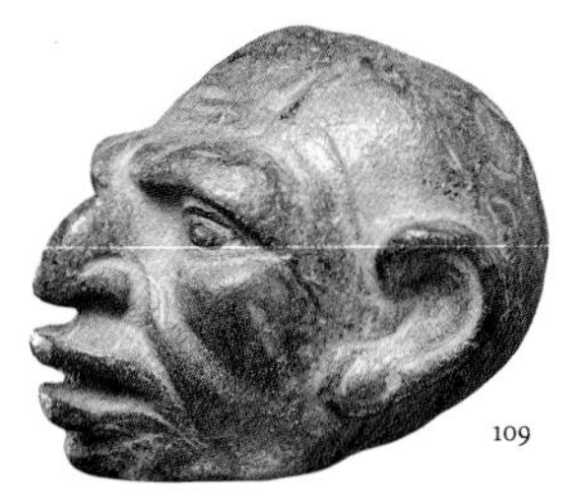

109

110

109
Weight in the Form of a Grotesque Head

Late Hellenistic-Roman, probably first century B.C.–first century A.D.
Bronze with lead; H: 3.2 cm; L: 4.6 cm
Condition: Part of cirrus missing.

The piece, which served as a steelyard weight, is the head of a brutal-looking man with thick lips, large ears, a prominent nose, and a receding forehead. His skull is shaven except for the *cirrus* (a lock of hair worn by pugilists), which served as a suspension loop for the weight. There is a square cutting on the underside of the head to receive the lead filling.

Such thick-lipped, shaven-headed grotesques are especially prevalent among Hellenistic terracottas from Egypt and may be based on local tradition. However, the taste for such representations spread to workshops in many parts of the ancient world (for a terracotta statuette made at Smyrna, see Higgins) and continued into the Imperial period (see cat. no. 146 below).

BIBLIOGRAPHY: Sotheby's, New York, *Antiquities and Islamic Art*, sale cat., June 18, 1991, lot 152A.

RELATED REFERENCES: R. Higgins, *Greek Terracottas* (London, 1967), pl. 52d–e.

—A. H.

SILVER AND GOLD

110
Appliqué in the Form of the Head of a Satyr

Hellenistic, third century B.C.
Silver with gilding; H: 3.4 cm
Condition: Reconstructed from fragments.

The frontal head is modeled in high relief and was probably shaped by hammering over a form. The young semihuman creature has a rustic handsomeness. A smile indents the corners of his mouth, emphasizing his broad nose and prominent cheekbones. The eyes and eyebrows slant impishly upward, and the ears are pointed like an animal's. In the hair is a wreath of pointed leaves. The ends of a Dionysiac pantherskin, tied around the neck, appear under the chin. The head is set on a plain background plate, approximately oval in form and edged with an incised ovolo.

The piece, shaped for attachment to a convex surface, recalls the masks serving as the feet of two early Hellenistic silver bowls in the Metropolitan Museum (see Bothmer). Our piece's flimsy construction makes it unsuitable for a position subject to much pressure or wear; perhaps it was filled with some reinforcing material. The broad-faced satyr, not yet one of the rather standardized rococo types that emerge in the second century B.C., can be dated to early Hellenistic times. There are still affinities with the Silenos emblema of a cup from Tomb II (ascribed to Philip II of Macedon but probably to be dated somewhat later, soon after 320 B.C.) at Vergina (see Andronikos).

BIBLIOGRAPHY: Unpublished.

RELATED REFERENCES: D. von Bothmer, "A Greek and Roman Treasury," *Metropolitan Museum of Art Bulletin* (Summer 1984), nos. 105, 106. M. Andronikos, *Vergina: The Royal Tombs and the Ancient City* (Athens, 1984), figs. 113, 114.

—A. H.

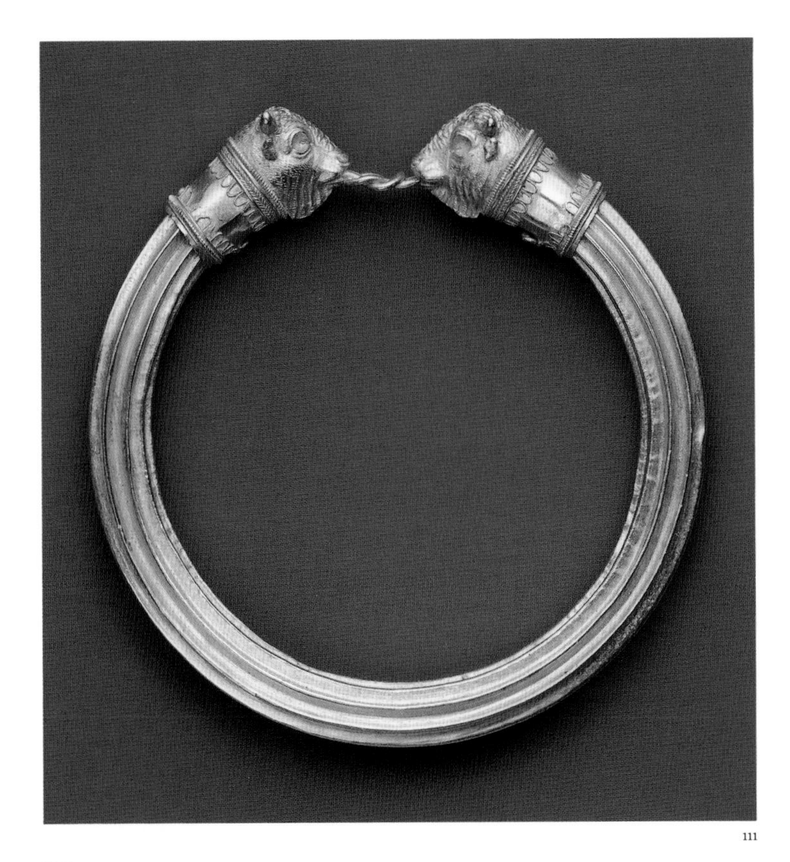

111

111

Bracelet with Bull's Head Terminals

Hellenistic, probably second half of third century B.C.
Gold; DIAM: 8.45–8.3 cm
Condition: Essentially complete except for missing enamel and inlays.

The hoop of the bracelet is made from a tube folded into nine deep flutes. It terminates at each end, by way of a cylindrical collar, in the head of a short-horned young bull. Each head, produced by hammering over a form, has forehead curls raised in relief and finely chased, a stippled coat elsewhere, and eyes hollowed for enamel. Ears and horns, separately made, were soldered in place. Simple pins hold the bulls' heads with their collars to the ends of the hoop. A small Herakles knot of gold wire links the muzzles of the bulls together.

Neither the basic fluted hoop nor the bull's head terminals belongs to a standard category, but the forms of both hint that the piece is a relatively late member of the series. Bovine heads on early penannular bracelets are calves of stylized design, inspired by Achaemenian models (see Pfrommer, p. 110ff., nos. 28–30, fig. 16), though more naturalistic bulls' heads in a Greek style are occasionally seen earlier as necklace pendants (e.g., an excavated necklace at Varna, Bulgaria; see Minchev). Well-modeled bulls' heads are found as terminals on a chain necklace in the British Museum (see Marshall); their collars, originally beads of semiprecious stone, are of a type that suggests a late third-century date (see Pfrommer, p. 322, no. TK31). Naturalistic bulls' heads with collar beads are used to attach the chain to an ornate early second-century Herakles knot in the Benaki Museum (see Pfrommer, p. 301, no. HK19, pl. 13.1).

In our piece the wide collars that join the heads to the hoop are not decorated with elaborate filigree designs, as in early examples, and the row of cut-out tongues, a structural element joining the collar to the hoop, has been suppressed. Instead, within borders made of multiple beaded wires, rows of tongues, facing inward, are outlined in filigree along the upper and lower edges of each collar. This detail recalls the filigree tongues edging the biconical, beadlike collars on some third-century hoop earrings with bulls' head terminals (see Pfrommer, pl. 25.1–6). The bulls' heads on the earrings, however, are usually rather different from ours in style.

Athough this bracelet's hoop seems to be unique, spirally fluted bracelets formed by twisting a plain fluted tube are known (see Pfrommer, p. 331f., no. TA4, pl. 10.1,3, for a pair from Mottola, with antelope heads, datable in the mid- to late third century). A fluted tube forms the base of the late third- or early second-century medallion hair-net from the Schimmel collection, now in the Metropolitan Museum of Art (1987.220: Milleker).

These parallels are only approximate, but all seem to suggest that the bracelet with the bulls is to be dated well into Hellenistic times, probably in the second half of the third century. Far from sharing the flamboyance of much High Hellenistic jewelry, our piece seems to have been designed with an almost conscious purism and restraint.

BIBLIOGRAPHY: J. Ogden, *Independent Art Research, Ltd., Report 89058,* December 6, 1989.

RELATED REFERENCES: M. Pfrommer, *Untersuchungen zur Chronologie früh- und hochhellenistischen Goldschmucks* (Tübingen, 1990). A. Minchev in *Akten des XIII. Internationalen Kongresses für Klassische Archäologie* (Berlin, 1988), p. 463f., pl. 70.2. F. H. Marshall, *Catalogue of the Jewellery, Greek, Etruscan and Roman, in the Departments of Antiquities, British Museum* (1911; reprint, London, 1969), no. 1973. E. J. Milleker, in *Metropolitan Museum of Art Bulletin* (Spring 1992), p. 50f.

—A. H.

112
Earrings with Pendant Nikai

Hellenistic, late third–early second century B.C.
Gold with glass; H: 9.7 cm
Condition: Essentially intact except for loss of some originally enameled or inset details

The earrings belong to a class, current since the second half of the fourth century B.C., with a figural pendant hanging from a decorated disk. The goddess of victory with her wings and windblown drapery is the subject of many of the most elaborate and spectacular examples. The figures of such pairs, as in our example, are usually mirror images.

The present version is a flamboyantly decorative late interpretation of the type. While the earlier Nike pendants were cast by the lost-wax process, with only the wings and the drapery behind their nude bodies added in sheet metal, our fully draped, hollow figures are assembled from front and back pieces produced by hammering into or over a form. Their billowing garments, full heart-shaped faces, and loosely knotted upswept hairdos are in the spirit of the Hellenistic baroque. Upswept sheet-gold wings, edged by spirally beaded wire, are attached; their feathers, rather than being chased as in earlier examples, are filigree-outlined compartments, originally filled in with colored enamel. Each figure carries a torch as do earlier, unpublished Nikai in Boston and a pair in the British Museum (see Marshall, nos. 1849, 1850). The torches are decorated with spirally beaded wire and end in flame-shaped settings

112

112

for colored glass or stone inlays.

Each pendant is attached to its disk by a loop at the lower end of a plain, tapering wire that is applied to the back of the disk and bent over at the top to form the ear hook. The disks represent an opulent late development of their type and have features usually found in conjunction with such late third- and second-century varieties of pendant as amphorae, figures set on pedestals, and enameled or glass-paste birds. The disks are bordered by small, outward-radiating petals, a feature very frequent in middle Hellenistic earring disks (see De Juliis, no. 85 [with glass-paste doves as pendants]). Their contour is broken at the top by a large palmette with filigree-outlined lobes, perhaps once enameled, and a pear-shaped central setting for a stone, creating a baroque, open form (see Marshall, no. 1919 [palmette-topped disk with dove-on-pedestal pendant], nos. 2332, 2333 [with amphorae on pedestals]). At the lower edge of each disk two small ivy leaf shapes partly conceal the attachment loop for the pendants. The disk proper is a concave saucer and holds an attached rosette with two layers of cut-out petals: an outer ring of twelve pointed ones, widely spaced against the plain background, and an inner ring of six close-set heart-shaped petals covered with granulation (compare De Juliis, no. 85 [bird pendant with widely spaced outer ring of lanceolate petals], no. 84 [bird pendant with inner ring of heart-shaped petals with granulation]). The center of each flower is a carnelian-colored glass bead held in place by the granulated head of a gold pin, the shaft of which is pulled through to the back of the disk and wrapped around the shank of the ear hook (for this type of pin-mounted bead, see Deppert-Lippitz, pl. XXV [earrings from Pelinna with chariots on pedestals], pl. XXVII [earring in Syracuse with granulation-covered dove pendant]).

Our Nike earrings are gorgeously colorful and effective decorative compositions that represent the end of a development from the concentrated miniature sculptures of the late fourth century. They date from middle Hellenistic times. Many of their features are paralleled in jewelry from Italo-Hellenistic contexts, and attribution to a Tarantine atelier has been suggested. However, the appearance, for example, of bird-pendant earrings stylistically identical to one another in widely scattered parts of the Greek world, from Etruria, Sicily, and Taras to the coast of the Black Sea, shows either that a type could be produced in many centers, or that such luxury objects were regularly in circulation, as trade goods, far from their place of manufacture.

BIBLIOGRAPHY: J. Ogden, *Independent Art Research, Ltd., Report 90164,* January 21, 1991.

RELATED REFERENCES: F. H. Marshall, *Catalogue of the Jewellery, Greek, Etruscan and Roman, in the Departments of Antiquities of the British Museum* (1911; reprint, 1969). E. M. De Juliis et al., *Gli ori di Taranto in eta ellenistica,* exh. cat. (Milan, 1984). B. Deppert-Lippitz, *Griechischer Goldschmuck* (Mainz, 1985).

—A. H.

113
Bowl

Hellenistic, second century B.C.
Silver; H: 5.5 cm; DIAM: 11.1 cm; WEIGHT: 260 g

The hemispherical bowl was cast, then the out-turned rim with its rounded molding was smoothed on a lathe and the decoration enhanced with additional tooling and gilding. A composite floral of acanthus and nymphaea nelumbo petals rises from a five-petaled rosette on the underside. There are four petals of nymphaea nelumbo, two with two smaller interior petals, two with one; these alternate with four acanthus leaves, windblown left and right at the top. Between the petals and acanthus leaves are floral tendrils with blossoms. The background is stippled. Above is a horizontal molding of a spirally twisted ribbon edged above and below with beading.

This format of acanthus and nymphaea is found on at least eight other silver bowls: three in Naples from Cività Castellana (see Pfrommer), two in Munich reported to be from the Fayum in Egypt (see Ahrens), one in the British Museum said to have been found in southern Bulgaria (see *British Museum*), and others in the Toledo Museum of Art (see Pfrommer) and in a New York private collection (see *Artemis*). No two bowls are quite alike in shape and decoration: the Naples bowls, for instance, are shallower than the others; more noticeably, the arrangements of the florals and the heights of the reliefs vary from bowl to bowl and on one of those in Naples some of the leaves are even undercut. On the Fleischman bowl the relief is relatively smooth.

The alternating arrangement of acanthus and nymphaea nelumbo petals is found in silver on shapes other than bowls, including jars, kantharoi, and aryballoi, and is also reproduced on four gilded glass bowls. Not surprisingly the decorative scheme recurs on moldmade, ceramic relief bowls; the closest parallels were made in Asia Minor or Alexandria. The few fragments of this design found in the Athenian Agora, for instance, were imported from elsewhere. To judge from the pottery versions, this scheme of decoration must have been introduced as early as the late third or early second century B.C.

BIBLIOGRAPHY: Unpublished.

RELATED REFERENCES: M. Pfrommer, *Studien zu alexandrinischer und grossgriechischer Toreutik frühhellenistischer Zeit* (Berlin, 1987), pp. 264–265, pls. 56–58. D. Ahrens, "Berichte der Staatlichen Kunstsammlungen, Neuerwerbungen," *Münchner Jahrbuch der bildenden Kunst* 19 (1968), pp. 232–233, figs. 5, 6. *British Museum Society Bulletin* 62 (Winter 1989), p. 47, ill. *Artemis 91–92: Consolidated Annual Report* (Luxembourg, 1993), p. 8, no. 1. For glass parallels, see pieces in the British Museum: D. Harden, "The Canosa Group of Hellenistic Glasses in the British Museum," *Journal of Glass Studies* 10 (1968), pp. 23–25, figs. 1–9; Geneva, Musée d'Art et Histoire MF 3634: J.-L. Maier, "Bols en verre à décor doré du Musée de Genève," *Genava* n.s. 25 (1977), pp. 221–225, figs. 1–8. For mold-made ceramic parallels, see S. I. Rotroff, *Hellenistic Pottery, Athenian and Imported Moldmade Bowls*, vol. 21 of *The Athenian Agora* (Princeton, 1982), p. 6, esp. n. 6; pp. 18, 88, no. 375, pls. 66, 94; H. A. Thompson, "Two Centuries of Hellenistic Pottery," *Hesperia* 3 (1934), pp. 407–409, no. E79, figs. 96a–b.

—A. O.

113

114

114
Bowl

Hellenistic, second century B.C.
Silver; H: 7.4 cm; DIAM: 15.8 cm; WEIGHT: 287 g

The bowl is conical with a rounded bottom and flaring rim and was probably cast and then raised by hammering after which it was finished on a lathe. There is a centering mark on the underside. As with all such bowls the decoration is on the interior: at the rim is a rounded convex molding with an incised wave pattern just below; lower down is the principal molding, also convex and incised to show a wreath in sixteen sections—pairs of crosshatched sections alternating with pairs of herringbone sections. Bordering it above and below are wave patterns, partially dotted. All of the moldings and patterns are gilded. The bound wreath was a standard item in the Hellenistic decorative repertory, appearing for example on a silver medallion in St. Petersburg from Aktanisovskaya Stanitza on the Taman peninsula (see Grach) and on a number of Hellenistic bronze Etruscan mirror covers (see Jucker).

Of all the conical silver bowls known, few are like the Fleischman piece, but a closely related one was recently purchased by the J. Paul Getty Museum (81.AM.84.23: Pfrommer). Three others, somewhat more elaborate, with a third zone of ornament and with separately made central floral rosettes, were acquired by the Metropolitan Museum of Art in the early 1980s (see Bothmer). Best known are the pair of bowls in the Rothschild collection found in Taranto in 1896 (see Wuilleumier). Distinctive on all is the convex molding on which the second band of ornament is drawn and the manner in which the decoration is shown in relief rather than merely incised on the smooth silver.

Metal bowls of this design clearly inspired conical bowls known in Hellenistic ceramics; noteworthy are certain non-Attic versions of West Slope pottery and the Calenian pottery of Italy, some of earlier date than the silver, in which the lower molding is sometimes done in relief (see Pagenstecher). On the ceramic bowls, as on the silver ones, the decoration is limited to the interior and the centers of the bowls are variously treated: they are sometimes plain, sometimes decorated with a starry rosette, or, in the Calenian bowls, with relief scenes.

BIBLIOGRAPHY: Unpublished.

RELATED REFERENCES: N. Grach, *Ancient Silver Metalwork* (in Russian) (Leningrad, 1985), pp. 40, 43 (ill.), no. 48, cover. I. Jucker, "Bemerkungen zu einigen etruskischen Klappspiegeln," *RM* 95 (1988), pp. 1–39, esp. pls. 2.2, 4.2. M. Pfrommer, *Metalwork from the Hellenized East: Catalogue of the Collections, J. Paul Getty Museum* (Malibu, 1993), p. 151, no. 24. D. von Bothmer, "A Greek and Roman Treasury," *Metropolitan Museum of Art Bulletin* (Summer 1984), pp. 54–55, nos. 92–94. P. Wuilleumier, *Le Trésor de Tarente* (Paris, 1930), pp. 34–40, pls. III–IV. R. Pagenstecher, *Die Calenische Reliefkeramik* (Berlin, 1909), pp. 66–67, no. 93a, pl. 12 (bowl in Berlin).

—A. O.

115
Group of Eight Drinking Vessels

Hellenistic; about 175–75 B.C. (FLUTED AND CONICAL BOWLS); about 100–50 B.C. (OVOID BOWLS AND SKYPHOI)
Silver

A.–B.
Pair of Fluted Bowls

H: 7.5 cm; DIAM: 12.5 cm; WEIGHT (OF BOWL WITH 66 FLUTES): 415 g; (OF BOWL WITH 64 FLUTES): 470 g

Fluted, or long-petaled, hemispherical bowls were a standard type in Hellenistic tableware and are known in silver, glass, and mold-made ceramic, especially after the middle of the second century B.C. On this pair of cast silver bowls, the flutes (sixty-four on one bowl, sixty-six on the other) are alternately plain and gilded with their tips slightly undercut. They radiate from a medallion on the bottom featuring a rosette of six leaves formed by a trefoil acanthus combined with a trefoil lotus; the backgound of the leaves is stippled and the floral is surrounded with a molding of a twisted cable edged with beading. The rim of each bowl has a set of relatively plain moldings, the lower part gilded. The bowls bear identical inscriptions in dotted style on the outside of the rim: ΚΤΑCΙΜΟΤΟΥ ΜΑΜΟΤΟΥ <PM ΑΡΤϵΙΜΟΥ <PM CΛΜωΤΟC (Of Ktasimotos, son of Mamotos. Year 140. Of Arteimas. Year 140. Of Samus).

Glen W. Bowersock of the Institute for Advance Study, Princeton, has kindly provided the following information on the inscriptions: The name Αρτειμου is particularly associated with Lycia (see Zgusta, p. 99, no. 108–1). The nominative is Αρτειμας. The other names are difficult, but they look Pisidian (for example, see Zgusta, p. 286, no. 855–1, 2, 3 [Μαμωτασις, Μαμοτασις]; *Tituli Asiae Minoris* III. 1.872, A VII [Μαμο[ ]). I can find no parallel for Κτασιμοτος/-ας nor for Σαμωτος/-ας (assuming the CΛ should be read as CA). The mark <PM looks like a number, perhaps a date with < for ἔτους). PM is the number 140, which—if this were the Seleucid era—would be 172 B.C. I wonder whether κτασι should be read as κτᾶσις (i.e., κτῆσις), "the possession of," but I do not like this solution for several reasons. So I prefer to read the two names in the genitive, Ktasimotos [or -as], son of Mamotos [or -as]. I cannot see why Samotos, if that is what it is, should be nominative, and I prefer therefore to see this word as a genitive (compare "Bulletin Épigraphique," *Revue des Études Grecques* [1949], no. 197 a [perhaps from a nominative Σαμυς]). Pisidian—*mot*—names and the Lycian Arteimas suggest the border regions between the two areas.

If, as Professor Bowersock suggests, the number in the inscription is a year and it refers to the Seleucid era, the date of the bowls would be 172 B.C. This, although some twenty years earlier than the date at which long-petaled ceramic bowls are thought first to have been made, is not impossibly early for a silver forerunner. In any event, on the assumption that the number is indeed a year, these bowls become the only ones known from antiquity to bear a date.

A nearly identical bowl in the Departement Orientale in the Louvre, with sixty-four flutes alternately gilded, was excavated at Susa in Iran in a late second-century B.C. context (see Byvanck-Quarles). A somewhat smaller one (DIAM: 11.6 cm) with forty-eight flutes, also alternately gilded, is in the Ashmolean Museum (see Symes), while a third, lacking or no longer possessing gilding, is in the collection of Jan Mitchell (see Sotheby & Co.). All should be assigned to the second half of the second century B.C. Earlier, fourth-century fluted bowls, probably of Persian origin, are shallower; later, first-century A.D. Roman versions embodying this decoration, in neo-Hellenistic style, have a ring foot and handles.

BIBLIOGRAPHY: Unpublished.

RELATED REFERENCES: L. Zgusta, *Kleinasiatische Personnennamen* (Prague, 1964). L. Byvanck-Quarles van Ufford, "Un bol d'argent

115A

115B

115A AND 115B, BOTTOM VIEW

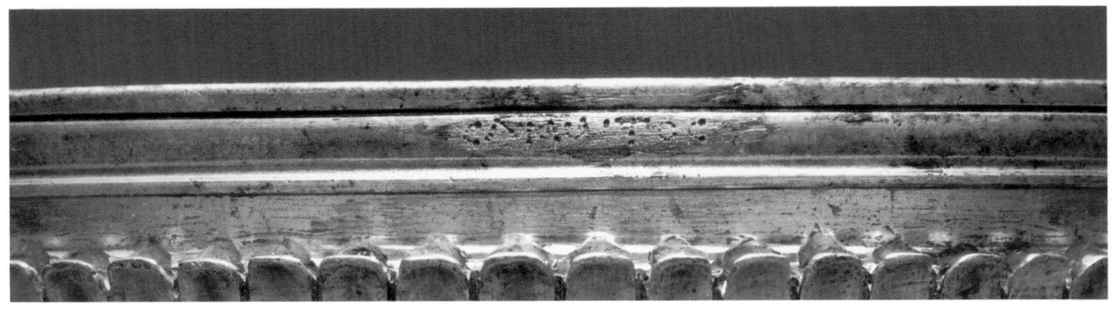

115 INSCRIPTION ON BOWL B

hellénistique en Suède," *Bulletin Antieke Beschaving* 48 (1973), p. 122, fig. 2. R. Symes, *Ancient Art* (London, 1971), p. [29], no. 31. Sotheby & Co., London, sale cat., June 12, 1967, lot 167. For glass parallels, see G. D. Weinberg, "Hellenistic Glass Vessels from the Athenian Agora," *Hesperia* 30 (1961), pp. 380–392. For mold-made ceramic parallels, see S. I. Rotroff, *Hellenistic Pottery, Athenian and Imported Moldmade Bowls,* vol. 22 of *The Athenian Agora* (Princeton, 1982), pp. 34–37. For fourth-century fluted bowls, see A. Oliver, Jr., *Silver for the Gods* (Toledo, 1977), p. 33, no. 7. For first-century Roman fluted cups, see IBM Gallery of Science and Art and IBM-Italia, *Rediscovering Pompeii,* exh. cat. (Rome and New York, 1990), p. 197, no. 102.

C.–D.
Pair of Conical Bowls

H: 13.7 cm; DIAM: 19.0 and 19.2 cm; WEIGHT (OF LARGER BOWL): 660 g; (OF SMALLER BOWL): 657 g

The bowls were raised by hammering and then finished and smoothed on a lathe. A centering mark is present on the exterior of each bowl. On the interior of the

115C

115D

rim is a rounded molding, grooved at the lower edge.

Conical bowls in silver, some with flaring rims, were widespread in the Hellenistic Mediterranean world. The shape was also standard in ceramic and glass from the mid-second to the mid-first century B.C. or a little later, particularly in the East Greek, late Hellenistic world.

There are engraved inscriptions on both bowls. On the one with the larger diameter: ΚΟΤϵΟΥϹ ΗΛΙΟΥ (Belonging to Kotês, [priest] of Hêlios); on the other: ΚΟΤϵΟΥϹ ΜΗΤΡΟϹ ΟΡϵΑϹ (Belonging to Kotês, [priest] of the mountain mother).

Glen Bowersock has kindly provided the following information on the inscriptions: On both bowls the inscriptions contain the possessive genitive of the name Κοτης (or Κοττης), particularly associated with Pisidia (see Robert 1948). That on the second of the two bowls refers to the "mountain mother" (μήτηρ ὀρεία) and Κοτης must accordingly be understood as a priest of this deity. Thus "[the cup belongs to] Kotês, [priest] of the mountain mother." In view of this interpretation we should understand ἡλίου on the first cup as a god (e.g., Apollo Hêlios), rather than as a patronymic—which is theoretically possible. Thus "[the cup belongs to] Kotês, [priest] of Hêlios." Compare the conjunction of Hêlios and a mother goddess near Motella (see "Bulletin Épigraphique," *Revue des Études Grecques* [1954], no. 233 [Ἡλίῳ Λερβηνῷ καὶ Μητρί]; Robert 1949 and 1962). But there the mother is not qualified as a mountain mother. There is, however, a mountain mother at Oenoanda (see "Bulletin Épigraphique," *Revue des Études Grecques* [1940], no. 156 [Κλοινιζόας ἀπέλυσεν τῇ Μητρὶ Ὀρείᾳ ἱεροδούλας]).

BIBLIOGRAPHY: Unpublished.

RELATED REFERENCES: L. Robert, *Hellenica*, vol. 6 (Paris, 1948), p. 12. Idem, *Hellenica*, vol. 7 (Paris, 1949), p. 57. Idem, *Villes d'Asie Mineure*, 2nd ed. (Paris, 1962), pp. 127–149. For silver parallels, see D. E. Strong, *Greek and Roman Gold and Silver Plate* (London and Ithaca, 1966), pp. 108–109. For mold-made ceramic parallels, see C. Meyer-Schlichtmann, *Die pergamenische Sigillata aus der Stadgrabung von Pergamon*, Perga-

115 INSCRIPTION ON BOWL C

115 INSCRIPTION ON BOWL D

menische Forschungen 6 (Berlin, 1988), pp. 78, 217 (type B 3), fig. 58, pl. 9. For glass parallels, see G. D. Weinberg, "Hellenistic Glass from Tel Anafa in Upper Galilee," *Journal of Glass Studies* 12 (1970), pp. 17–27; D. Grose, "The Syro-Palestinian Glass Industry in the Later Hellenistic Period," *Muse* 13 (1979), pp. 54–67.

E.–F.
Pair of Ovoid Bowls

H: 9 cm; DIAM: 9.5 cm; WEIGHT: 213 and 209 g

The ovoid bowls were raised by hammering, then finished and smoothed on a lathe. A rougher band of lathe marks is visible on the upper centimeter of the exterior at the rim, a feature matched on the pair of skyphoi (G and H below). On the inside of the rim is a rounded molding, grooved at the lower edge. There are no inscriptions. The bowls seem to be unique in silver but are paralleled in late Hellenistic pottery from western Asia Minor.

BIBLIOGRAPHY: Unpublished.

G.–H.
Pair of Skyphoi

H: 8.5 and 8.8 cm; W (WITH HANDLES): 20.2 cm; DIAM (OF BOWL): 11 cm; (OF FOOT): 9.4 cm; WEIGHT (OF TALLER SKYPHOS): 573 g; (OF SHORTER SKYPHOS): 578 g

The skyphoi were raised by hammering and then turned and finished on a lathe; the feet, soldered to the bowls, were made separately in a similar technique and are marked with lathe-turned, concentric moldings on the underside. The top centimeter of the exterior exhibits rougher lathe turning marks, as on the ovoid bowls (E and F above). The handles, designed to be held with thumb, forefinger, and middle finger, were cast in several parts and soldered to the bowls. The horizontal, flaring plate on top has tight curlicues at the rim attachment; the lower attachments have elaborate cut-out designs featuring scrollwork and an ivy leaf (largely missing on one skyphos), a decorative device found on much late Hellenistic bronze and silver tableware. The rings have small, leaflike knobs on the outer face (these match the ring elements of the handles on a pair of late Hellenistic silver stemmed cups in the Getty Museum (see Oliver). The profiles of the bowls are noteworthy in that their sides are not upright, as in Roman Imperial-period sky

115 E

115F

115G

115H

phoi, but incurving. The skyphoi lack inscriptions.

Skyphoi of this shape had a long life as tableware of the Greek and Roman world from the fifth century B.C. into the early Imperial period. Originally made in pottery, they were translated into silver as early as the fourth or early third century. In the early first century B.C., when the Fleischman silver skyphoi are likely to have been made, the shape is found in glass and is ubiquitous in ceramic. Closely related in design are late Hellenistic pottery skyphoi (belonging to a large class of Pergamene sigillata) found not only in Pergamon but also at other sites in Asia Minor to which they were exported; the Pergamene skyphoi, made in the first half of the first century B.C., share with the silver ones a similar profile and the tight curlicues ornamenting the handle. Their source and distribution help support the notion that the silver was made and used in Asia Minor.

BIBLIOGRAPHY: Unpublished.

RELATED REFERENCES: A. Oliver, Jr., "A Set of Ancient Silverware in the Getty Museum," *GettyMusJ* 8 (1980), pp. 155–158, figs. 1–7. For fourth-century silver skyphoi, see an example from Vergina: M. Andronikos, *Vergina: The Royal Tombs and the Ancient City* (Athens, 1974), p. 149, fig. 109; from Artiukhov's Barrow, Taman peninsula: M. I. Maximova, *Artichovskij Kurgan* (Leningrad, 1979), pp. 79–81, fig. 25. For Pergamene sigillata skyphoi, see C. Meyer-Schlichtmann, *Die pergamenische Sigillata aus der Stadgrabung von Pergamon*, Pergamenische Forschungen 6 (Berlin, 1988), p. 65, no. S3. For Hellenistic glass skyphoi, see A. Oliver, Jr., "Late Hellenistic Glass in the Metropolitan Museum," *Journal of Glass Studies* 9 (1967), pp. 30–33.

—A. O.

116

116
Pair of Hoop Earrings with Heads of Maenads

Hellenistic, first century B.C.
Gold; DIAM: 2.26–2.5 cm

The hoop of each earring is made of three strands of gold wire, twisted and tapering toward the back where they finally merge into a smooth point. At the front the hoop ends in a spool-shaped element, edged with beaded wire above and below. On this collar is set the head of a maenad, with soft features and upswept hair crowned with an ivy garland. The heads are hollow and were produced by hammering over a form or into a matrix. Although they are likely to have been made in halves, no seams are now visible.

The form of the spool bases under the heads shows that the earrings are datable at the end of Hellenistic times (see Ogden); they can probably be ascribed to Syria, where many late examples of the type have been found.

BIBLIOGRAPHY: J. Ogden, *Independent Art Research, Ltd., Report 90090*, June 28, 1990.

RELATED REFERENCES: For a map with geographical distribution of types, see M. Pfrommer, *Untersuchungen zur Chronologie früh- und hochhellenistischen Goldschmucks* (Tübingen, 1990), p. 181f., fig. 55; for chronological chart, pl. 30.

—A. H.

TERRACOTTA

117
Statuette of a Seated Comic Actor

Early Hellenistic, late fourth–early third century B.C.
Terracotta; H: 10.3 cm

The piece is mold-made, apparently in front and back sections, from buff clay. The rotund actor sits cross-legged, his right arm clasping the front of his body at waist level and his left raised to his chin in a "plotting" gesture. Over leggings with T-strap sandals, and a long-sleeved upper garment, he wears a short, belted costume. His

117

mask has thick hair brushed back in a *speira* (rolled pompadour) and a shovel beard that accentuates the roguishly grinning mouth. As in the bronze statuette of a standing comic actor (cat. no. 108 above), "real" lips are visible within the contour of the mask's wide and deep scoop-shaped trumpet.

Loosely comparable figures include one in the Louvre (D 3716: Mollard-Besques), tentatively ascribed to Apulia and dated at the end of the fourth century B.C.

BIBLIOGRAPHY: Christie's, London, *Fine Antiquities*, auct. cat., July 8, 1992, lot 121.

RELATED REFERENCES: S. Mollard-Besques, *Catalogue raisonné des figurines et reliefs en terre-cuite grecs, étrusques et romaines*, vol. 4, pt. 1, *Époques hellénistique et romaine, Italie méridionale-Sicile-Sardaigne* (Paris, 1986), pl. 68.

—A. H.

118

Double-spouted Hanging Lamp with the Figure of a Comic Actor

Hellenistic, perhaps late second century B.C.
Terracotta; H: 10.1 cm; W: 8.1 cm
Condition: Intact, with vent hole in middle of figure's back.
Provenance: Formerly in the Mustaki collection.

Between the lamp's two nozzles a comic actor sits cross-legged, his left arm folded in front of him and his right hand raised to support his chin. The subject is the favorite comic theme of the plotting slave. He is clad in a chiton and short mantle over the usual body-suit. His mask has a furrowed, worried-looking forehead and eyebrows which come together in a U-shaped line over small bulging eyes. His bristly hair is short and swept back in a rolled speira, with a large ring atop the head for suspension.

The fabric, of reddish clay coated with a glossy red "Roman" slip, is known, especially for lamps, in Ptolemaic Egypt (see Higgins). The pert stylization of the mask, with its small, deep mouth and beady little spherical eyes under overhanging brows, suggests the slave masks in the late Hellenistic mosaic border from the House of the Masks at Delos (see Trendall/Webster) or those in another mosaic border of similar date framing the fragmentary hunt scene in Palermo (see Wilson). The figure as a whole has affinities with the bronze version of the same subject from the late Hellenistic Mahdia wreck, an ancient ship that sank off Tunis (see Fuchs). Such comparisons suggest that our piece may belong early in the sequence of Greco-Egyptian plastic lamps, most of which date from the first century B.C. to the first century A.D. (see Bailey). The careful yet lively modeling is of a quality most unusual within the category.

BIBLIOGRAPHY: T. B. L. Webster, *Monuments Illustrating New Comedy*, 2nd ed. (London, 1969), p. 98, no. EL 1.

RELATED REFERENCES: R. Higgins, *Greek Terracottas* (London, 1967), p. 32f. A. D. Trendall and T. B. L. Webster, *Illustrations of Greek Drama*, vol. 4 (London, 1971), no. 8c. R. J. A. Wilson, *Sicily under the Roman Empire* (Warminster, 1990), fig. 22. W. Fuchs, *Der Schiffsfund von Mahdia* (Tübingen, 1963), p. 20, no. 12. D. M. Bailey, *A Catalogue of the Lamps in the British Museum*, vol. 1 (London, 1975), p. 242.

—A. H.

118

119
Statuette of a Mime

Hellenistic, probably late second–early first century B.C.
Terracotta; H: 19 cm
Condition: Fingers of left hand missing; large round vent hole in lower back.

The piece is made of buff clay covered with a light-colored base coat. It was assembled from mold-made main elements such as body and head, with lesser pieces modeled freehand. Traces of bright pink paint survive, especially on the sleeves and headdress. To judge from the remains of color on the left foot, the shoes were vermilion.

The bony figure stands with left hand on hip and right thrown out in a declamatory gesture. The chiton is pulled up over a belt, shortening it to above the knees and creating a long *kolpos* (pouch-like overfold). A mantle is drawn around both shoulders, crisscrossed at the back, and then wrapped around the waist like a sash. The hair, pulled up into a high ponytail, is held in place by a small kerchief the ends of which are drawn up and knotted over the forehead. The face has a beaklike nose and eyebrows knitted in a scowl.

The short, belted-up chiton, the mantle worn as a sash to keep it out of the way during strenuous action but handy when needed for warmth, and the maidenly headdress are all features of the huntress's costume worn by Artemis. Our ugly and querulous-looking personage is certainly not the goddess but may be a mime impersonating her, or a nymph from her

119

retinue, in some mythological burlesque. The performer, not masked, is an actor who uses gesture and facial expression to comic effect (see Beacham).

The piece, reportedly from Asia Minor, can perhaps be ascribed to Myrina (J. R. Green, verbal opinion). While exact parallels seem to be lacking, our statuette is not dissimilar in flavor to the terracotta figures of a tympanon player in Berlin and a castanet player in Athens (see Bieber, figs. 241, 242; Webster), both from Myrina and not to be dated before the end of the second century B.C. A grotesque performer in Athens (Bieber, fig. 372) and a series of finely modeled masked actors (Bieber, figs. 379, 386) are other Myrina productions, datable in late Hellenistic times, which have some affinities with our piece.

BIBLIOGRAPHY: Christie's, London, *Fine Antiquities*, sale cat., July 11, 1990, no. 239. H. Mallalieu, "Around the Salerooms," *Country Life*, August 30, 1990, pp. 114–115 (illus.).

RELATED REFERENCES: R. C. Beacham, *The Roman Theater and its Audience* (Cambridge, Mass., 1992), p. 129ff. M. Bieber, *The History of the Greek and Roman Theater*, 2nd ed. (Princeton, 1961). T. B. L. Webster, "Leading Slaves in New Comedy," *Jahrbuch des Deutschen Archäologischen Instituts* 76 (1961), p. 100ff., for the date of the Berlin figure, p. 103.

—A. H.

MARBLE

120
Statuette of Tyche

Hellenistic, second half of second century B.C.
Translucent, fairly large-crystaled Greek island marble; H: 84.5 cm
Condition: Chips from edge of mural crown, right arm from above elbow, left arm from the shoulder with a portion of the mantle, and edge of base in front of left foot missing; back of mural crown and upper back part of head and veil separately made and now lost; nose tip scuffed; hole in the right foot filled in, in antiquity, with a still-surviving marble plug.

The heavily draped female figure is identified by her turreted mural crown as Tyche, in this case surely the personification of a specific city's identity and destiny rather than the more banal goddess of good fortune in general. The left arm, made separately and attached with a metal dowel, probably held a cornucopia, though some other large attribute such as a baby or an animal is also possible. The right hand, lowered and extended to the side, is likely to have held the top of a rudder, secured below by a pin in the base beside the right foot. The ears are pierced for real earrings, and a hole through the veil to either side of the neck would have served to secure a necklace.

Our Tyche must be the miniature adaptation of a life-size, or more probably a colossal, public monument. The forerunner may have been made in early Hellenistic times, when many cities were founded by Alexander's successors and the images of their Tychai created. The basic figure type recalls the Eirene of Kephisotos or the newly found, not yet fully published peplophoros from Vergina (see Pariente). The mural crown, tall and narrow compared to the broader crowns of Roman works, has the form seen in copies of the early Hellenistic Tyche of Antioch (see cat. no. 157 below).

However, several features give the impression that, even if it is derived from an earlier model, our statuette represents a distinctively late Hellenistic interpretation of its prototype. The thong sandals with thick platform soles, indented between the big toe and the others, are typical of High Hellenistic statues and remained fashionable through much of the second century B.C. (see Morrow). The thin, crinkled chiton, gathered and sewn to a neck-band, is worn by goddesses on the Altar of Pergamon (see Kähler) and in the late second-century frieze of the Hekateion at Lagina (see Simon). The fragmentary statue of a queen from Pergamon (see Smith), like our figure, wears it as an underdress.

Over her chiton Tyche wears the peplos, a dignified, old-fashioned garment little favored by Hellenistic artists. Our figure, however, is very different from Classical peplophoroi, where the heavy dress is used to create a stable-looking grid of horizontals and verticals. The goddess stands in a langorous, undulating posture, with the left leg set far back and to the side. Her dress is long enough to trail and pool luxuriously on the ground. The fabric clings to the upper body and is dragged into deep, curving

120

folds around and between the legs. Under the peplos' long *apoptygma* (overfall) with its abbreviated, rufflelike *kolpos* (pouched overfold), rising in a steep ogee arch, the forms of the high, indented waist, narrow shoulders, and prominent breasts are clearly outlined.

The head of our figure finds a close parallel in that of a late second-century statue from the Aegean island of Kos (see Kabus-Preisshofen, no. 56). Both have the same sfumato finish over delicate, short-chinned Classicistic features, the same ovine expression, and the same center-parted coiffure, with flat, only slightly wooly-looking hair rolled back from the face in wings. Possibly our statue's most unusual element is the long veil, worn over the mural crown instead of under it, and falling from the top of the headdress to the ground. Unifying and monumentalizing the composition, the fashion, perhaps not accidentally, evokes very ancient religious images and the later cult representations derived from them (see Klemmer).

In scale and workmanship our Tyche recalls the late Hellenistic "apartment sculptures" of coastal Asia Minor, Rhodes, Kos, and Delos. At Delos, the only one of the sites to furnish a chronological point of reference, most such material is datable to the years between the sacred island's rise to a new commercial prominence in the mid-second century B.C. and the disastrous sacks of 88 and 69 B.C., which put an end to its prosperity (see Marcadé). Late Hellenistic peplophoros statuettes in the Louvre and at Bodrum, both of them headless, seem to be simplified replicas of the same type as our figure (see Gernand), while a statuette of a peplos-wearing Tyche at Kos, now thought to be late Hellenistic in date, is a variant (see Kabus-Preisshofen, p. 72f., no. 72, pl. 69). Our similar if more elegant production can also be ascribed to a workshop active, probably during the late second century B.C., in a Greek city of coastal Asia Minor or the neighboring islands. The statuette's careful workmanship and, especially, the fact that it was adorned with real jewelry suggest that the image had a devotional significance, perhaps for an association of merchants or a wealthy family who were natives of the city she personified (in this connection see the clubhouse at Delos of the Poseidoniasts of Berytos: Marcadé, p. 386ff.; see also Stewart).

There is one much later example, a statuette from Ostia (Museo Nazionale Romano 85: Parabeni). Except that the veil is looped on the right shoulder, the piece is essentially a replica of ours. It too wears a thin, crinkled chiton under the peplos, though without the Hellenistic neck- and sleeve-bands. The execution is Antonine, as attested by the profiled oval plinth. There are remains of a rudder by the right foot and seemingly of a cornucopia in the left arm. The Ostia version, so much later and geographically so far removed from the others, shows that the type had a lasting significance, especially, perhaps, in a merchant and maritime setting. Our example is by far the finest of the surviving versions and the only one whose head is preserved.

BIBLIOGRAPHY: Unpublished.

RELATED REFERENCES: A. Pariente, "Chronique des Fouilles et découvertes archéologiques en grèce en 1990," *BCH* 2 (1991), p. 899, fig. 92. K. D. Morrow, *Greek Footwear and the Dating of Sculpture* (Madison, 1985), p. 90ff. H. Kähler, *Der grosse Fries von Pergamon* (Berlin, 1948), pls. 6, 22, 44a. E. Simon, "Der Laginafries und der Hekatehymnos in Hesiods Theogonie," *Archäologischer Anzeiger* (1993), p. 277ff., fig. 3. R. R. R. Smith, *Hellenistic Royal Portraits* (Oxford, 1988), no. 29, pl. 23.2. R. Kabus-Preisshofen, *Die hellenistische Plastik der Insel Kos,* AM Beiheft 14 (1989). I. Klemmer, "Ein Weigeschenk an die delische Artemis in Paros," *AM* 77 (1962), p. 207ff., Beil. 56–63. J. Marcadé, *Au Musée de Délos,* BEFAR 215 (Paris, 1969). M. Gernand, "Hellenistischen Peplosfiguren nach klassischen Vorbildern," *AM* (1975), p. 1ff., pl. 4, fig. 2 (Bodrum statuette), fig. 3 (Louvre statuette). A. Stewart, *Greek Sculpture* (New Haven and London, 1990), fig. 831. E. Paribeni in A. Guiliano, ed., *Museo Nazionale Romano,* vol. 1, pt. 2, *Le sculture* (Rome, 1981), pp. 301–302 (IV.19).

—A. H.

121

Relief with Heroic Banquet

Hellenistic, probably second half of second century B.C.
Marble; H: 50.6 cm; W: 83.3 cm
Condition: Surface weathered; chip from upper left corner, big pieces of both lower corners, section from rim of krater, cupbearer's left forearm, middle leg of table, dog's head, and banqueter's right forearm with front edge of patera missing; damage to cupbearer's face; large cuttings (for later mounting?) under dog and table at bottom and through background at top; traces of former restoration, including dowel hole for cupbearer's left forearm.

121

The piece belongs to the category of *Totenmahl* (heroic banquet, literally "Feast of the Dead") reliefs. First made as votive offerings to various divinities, such representations could also honor a heroized deceased person or serve as grave markers. Drawing on East Greek antecedents, the type reached its canonical form in Attica during Classical times, but further development in the Hellenistic period was again centered in the eastern part of the Greek world.

The iconography of our example is drastically pared down. The scene lacks the wife who normally sits on the end of or beside the banqueter's couch, as well as the status-denoting attributes such as a horse's head or weaponry, often sketched on the background. The piece is exceptional, however, not only for its elegant workmanship but for its rich detail. The table set with ritual cakes (see van Hoorn), is the kind with three feline legs that replaced the Classical straight-legged table in Hellenistic *Totenmahl* reliefs. The couch has a fulcrum apparently belonging to Type 1 (see Faust, p. 34ff.), and it displays a Type 6 leg with stacked, disk-shaped components, the usual complement of decorated fulcra (see Faust, p. 22ff.); these forms would be at home in the second century B.C. The coverlet with its tasseled border seems to have no parallel other than the coverlet in a mosaic theatrical scene of around 100 B.C., evidently reflecting a third-century model, from Pompeii (see Pollitt). The krater on its high stand is of a type revived in neo-Attic production, while the little dog has a rococo winsomeness. The ends of the banqueter's drapery hang in stylized late Hellenistic zigzags. The sfumato idealism of his bearded head and the soft, sinuous physique of the young nude attendant reflect the return to fourth-century models in later Hellenistic times.

The piece, which belongs to an eclectic, decorative phase of its type, can probably be ascribed

122

to the second half of the second century B.C. It is likely, because of the ideal character of the figures and the absence of attributes denoting a deceased person's station in life, to have had a votive rather than a funerary purpose.

BIBLIOGRAPHY: Unpublished.

RELATED REFERENCES: G. van Hoorn, *Choes and Anthesteria* (Leiden, 1951), p. 41f. S. Faust, *Fulcra: Figürlicher und ornamentaler Schmuck an antiken Betten*, RM Ergänzungsheft 30 (Mainz, 1989). J. J. Pollitt, *Art in the Hellenistic Age* (Cambridge, 1986), fig. 242. For heroic banquets in general, see N. Thönges-Stringaris, "Das griechische Totenmahl," *AM* 80 (1965), pp. 1–99.

—A. H.

122

Fragmentary Grave Stele

Hellenistic, third quarter of second century B.C.
Marble; H: 125.4 cm; W: 53.3 cm
Condition: Lower part of stele, including main figure from hip-level down and all of accompanying figures except head of boy attendant, and upper right corner missing; cutting for a clamp at left side of top surface.

The back of the piece is roughly picked, while the sides are finished with *anathyrosis* (roughening of the contact surface). On the evidence of its form, with the sides tapering inwards as they go up, the piece was a freestanding stele, not a slab from a longer frieze. It was probably set in an architectural frame. Its main figure, carved in high relief, is a draped youth in a contrapposto pose which provides an alternation of frontal and three-quarter views. He stands with his left hand grasping the drapery on his hip and his right extended to touch a largish object, too badly damaged to be identifiable, which is presented by a boy attendant, now missing except for his frontally seen head. There may originally have been a pair of little attendants; the rounded shape set between the surviving boy's head and the large figure's hip looks as if it might have been the head of another, smaller child. The main figure wears, over a chiton, a long *chlamys* (short cloak), the so-called epaphtis, secured on the right shoulder, then drawn over the left arm to be wound across the back and around the hips, forming a roll at the waist and held in place by the left hand (for this drapery scheme, see Pfuhl/Möbius). A conspicuous ivy leaf shape on the right shoulder has a stem which is definitely stuck into the chlamys and seems to be the finial of an unusual fibula. The eagle-headed hilt of a short sword emerges at the left side. A weapon of this form is well attested, especially as an attribute, in late Hellenistic reliefs (see Firatli, p. 35, pl. LXXIII.2). The young man is beardless and his hair is sketchily rendered as a cap of short, thick curls. The facial features, though of a Classicistic regularity, still have some of the solidity and fleshiness of the Hellenistic baroque. A wreath is carved in low relief on the background toward the viewer's right.

Above the figures is a beveled, projecting ledge. A number of objects, shown in low relief and in some cases foreshortened, are set on it as if on a shelf. To our left is a plain rectangular block, possibly a closed diptych of wax writing tablets (see Firatli, p. 34, pl. LXXI.3). Next is a chest with an arched lid, which, when it appears as the attribute of a male figure, is probably thought of as containing book rolls (Firatli, p. 33 and pl. LXXI.2). Next comes a framed tablet on which is displayed a wreath, like the one on the background. This tablet is set on a low block bearing the inscription ΦΑΝΟΚΡΑΤΗ[Σ] ΦΑΝΟΚΡΑΤΟΥ (Phanokrate[s?] [son of] Phanokrates). Although one would expect the first name in the inscription to be that of the deceased, Phanokrates, in the nominative, the traces of the damaged last letter look more like a *nu* than a *sigma.*

The pose and drapery scheme of the young man belong to a standard type, at home in Asia Minor and especially in the vicinity of Smyrna (see Berger, p. 257ff.) The figure originated as part of a composition including a horse and servitors. The most impressive example is the large relief in Basel, where a heroized dead man in the same pose as the deceased in our relief is shown with a caparisoned horse, a hunting dog, athletic equipment, and several armed attendants (see Berger, pp. 253–255). The type has been identified in five other versions that include a horse and one where the draped male figure, in the same pose as ours, stands near a reclining heroized banqueter (see Berger, Beil. 27, figs. 1, 3–5; Beil. 28, figs. 1, 2). A fragment in Çannakale comes from a related composition and has a male head the style of which is very similar to ours (Berger, Beil. 29, fig. 4). The fact that our figure wears a sword is probably a reminiscence of the older composition which shows the deceased with a horse and military accessories. The invention of the scene is ascribed to the mid-second century B.C.; our example, as an excerpt of the larger composition, would be somewhat later than the first versions. On the evidence of its still robust and even swaggering style, however, it should not be too far removed from them in time.

BIBLIOGRAPHY: Unpublished.

RELATED REFERENCES: E. Pfuhl and H. Möbius, *Die ostgriechischen Grabreliefs,* vol. 1 (Mainz, 1977), p. 63. N. Firatli, *Stele funéraires de la Byzance greco-romaine* (Paris, 1964). E. Berger, *Antike Kunstwerke aus der Sammlung Ludwig,* vol. 3, *Skulpturen* (Mainz, 1990).

—A. H.

123
Portrait Head of a Man

Late Hellenistic, early first century B.C.(?), or an Early Imperial copy after an original of that date
Marble, probably Pentelic; H: 28 cm
Condition: End of nose, corner of right eye and eyebrow, and left ear missing; the back and left side of the head, behind the ear, have been battered away.

The subject is a middle-aged man with a lean, high-cheekboned face, wide mouth, and heavy-lidded eyes. His hair, worn long, is bound by a fillet. This fashion is the only clue as to his identity. The fillet is often a royal attribute, but, especially in the cordlike form seen here, can also be worn by poets. The man's long, limp hair seems to be the unkempt coiffure of an ancient intellectual, proclaiming his indifference to material things, rather than the baroque, Alexander-like mane of a Hellenistic ruler.

Our portrait's clearly structured but mature-looking oblong face, wide, firm mouth, and composed expression recall Asia-Minor influenced Athenian portraits like the fragmentary head from the Stoa of Attalos (Athens, National Archaeological Museum 3266: Stewart, fig. 27a), datable soon after the middle of the second century B.C., or the leathery-faced priest also in Athens (National Archaeological Museum 351: Stewart, fig. 26a). Other aspects of the portrait head point to a somewhat later date and have their best parallels in a series of more expressionistic portraits from Delos, made in the early first century B.C., when this island was an Athenian colony (see Stewart, figs. 19b, 20d, 22b). In the Delos portraits cartilaginous bulges, wrinkles, and loose flesh are used to convey expression and energy, rather than primarily to denote old age. The style, which influenced even our basically serene representation, is evident at Athens in such pieces as National Archaeological Museum 320 (see Stewart, fig. 25b). Our portrait's corrugated forehead and the modeling of the small eyes, set deep in their sockets but projecting as if squeezed forward between their puffy upper and lower lids, are unmistakable links with the early first-century Delian and Attic series.

123

123

Because of these similarities and because its marble seems to be Pentelic, an attribution of our portrait to Athens strongly suggests itself. It is difficult to say whether the head is an original or a copy. The carving of the face has the sensitive but forthright quality of an original. On the other hand, the rather decorative stylization of the locks of hair and the emphatic drill channel between the lips might hint at copyists' work of the Roman Imperial period. If the portrait was indeed reproduced at a time well after its creation, the fact implies that the sitter was a person of some lasting renown, not simply a prominent citizen, who might commission a funerary portrait and be the subject of one or more honorary statues during his lifetime, but whose likeness would not be meaningful to a later generation.

BIBLIOGRAPHY: Unpublished.

RELATED REFERENCES: A. Stewart, *Attika: Studies in Athenian Sculpture of the Hellenistic Age* (London, 1979).

—A. H.

124

Statuette of a Draped Female Figure

Late Hellenistic, first century B.C.
Translucent, medium-grained white marble (Parian); H: 45.1 cm
Condition: Right arm from just above elbow and separately made pieces on left foot and upper right side of head missing; left arm reattached; deposits from undersea life especially noticeable on right side and back; dark, deeply absorbed stains in many places.

The slender young woman stands in a graceful, undulating pose. She wears a gathered chiton of very thin material, belted high under the small, pointed breasts and sewn to a neck-band which has slipped off her shoulder. A voluminous mantle is secured in a not wholly logical manner (perhaps further obscured by mending of this area) on the left shoulder, then wound around her body and held at the hip by her left hand. Its ends fall in zigzag folds. The head has a sfumato finish and delicate but fleshy ideal features; the throat is marked with "Venus rings" (horizontal ridges). The straight hair is rolled back in wings over a hair band and arranged in a large chignon.

The piece, with its narrow shoulders, spreading base, and complex drapery patterns, is typically late Hellenistic in style. The technique of piecing with small added sections attached only by adhesives is attested among the small-scale Hellenistic sculptures at Delos (see Marcadé). A statuette of Tyche from the Aegean island of Kos, thought to date from Early Imperial times, is strikingly similar to our piece in its facial type and drapery style, though further advanced toward Classicistic rigidity (see Laurenzi). The proposed date for the Kos statuette seems rather late; in our piece the puffy yet delicate face and the still rococo feeling of the drapery speak against a date much after the middle of the first century. Small statues like this were produced in the artistic centers of the Aegean islands and Asia Minor, both for local use and for export.

BIBLIOGRAPHY: Unpublished.

RELATED REFERENCES: J. Marcadé, *Au Musée de Délos,* BEFAR 215 (Paris, 1969), pl. XLI.A449. L. Laurenzi, "Sculture inedite del Museo di Coo," *Annuario della Scuola Archeologica di Atene* 33/34 [n.s. 27–28] (1955–56), p. 59ff., no. 4.

—A. H.

124

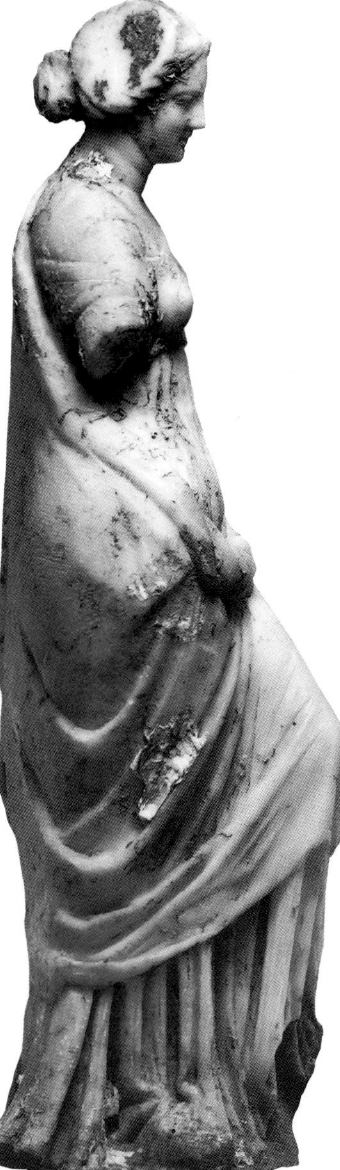

Rome and the Provinces from the Republic to the Late Antique

125

125

Fresco Fragment: Vignettes of Cityscapes

Roman, Second Style, third quarter of first century B.C.
Fresco; H: 91 cm; W: 80.5 cm; DEPTH: 4.5 cm

These two rectangular panels display the illusionism typical of Second-Style Pompeian fresco painting. The landscape scenes are bathed in what is now a light blue-green hue (the original colors have changed with the passage of time; blue pigments tend to turn green in Roman frescoes). One can imagine the room from which this fragmentary wall emerged. Based on the right-to-left orientation of the shadows on the columns, this was part of the right-hand wall upon entering the room. The yellow shelf at the bottom edge of the fresco delineates the top of the dado, signifying that it was about mid-thigh in height. The dado, or painted zone, below very likely consisted of a series of faux marble panels meeting the mosaic floor. The ceiling probably had white stucco detailing, including actual cornices above those painted in the neighboring section of the wall's upper zone (see cat. no. 126 below).

The green landscape panel at left is a sketchy evocation of a proscenium temple with a distinctive barrel-vaulted roof. A shield hangs from the side entrance to the pronaos, and a group of figures have an animated exchange on the temple platform; two or three other figures stand below the platform. Individual statues adorn the roof of the temple pronaos, which was typical of Republican temples, and a ritual column rises behind the temple at right. Other landscape features in the upper half of the painting are too faint to interpret. The right panel includes a rectangular structure at the bottom, a *tholos* (round temple) just beyond it, and another temple to the left.

The painted column at right is encircled with foliage meant to be interpreted as made of metal. The left edge of the fragmentary panel reveals that the green panels were in pairs, interrupted by a field of red background, accented with horizontal bands of faux serpentine (green) and *giallo antico* (yellow) marble. An off-white area at the top left of the fragment may be part of a larger rectangular painting.

BIBLIOGRAPHY: Unpublished.

RELATED REFERENCES: For similar paintings, see the section of wall from the villa at Portici in Naples, Archaeological Museum 8953: R. Ling, *Roman Painting* (Cambridge, 1991), fig. 39; and cubiculum M in the villa of P. Fannius Synistor at Boscoreale: M. L. Anderson, *Pompeian Frescoes in The Metropolitan Museum of Art* (New York, 1987), p. 18, fig. 23.

—M. A.

126

Fresco Fragment: Lunette with Mask of Herakles

Roman, Second Style, third quarter of first century B.C.
Fresco; H: 61 cm; W: 81 cm; DEPTH: 3 cm

The superb illusionism of Second-Style Roman wall painting is brilliantly in evidence in this fragment from the upper zone of a Pompeian wall. To judge from the scale of the fragment, the room was intimate in scale and may have been a bedroom (*cubiculum*) or dining room (*triclinium*). It is an example of Hellenistic ornamental traditions adapted to Roman domestic decoration; this adaptation is characteristic of the second of the four so-called Pompeian painting styles. Specific features of the architectural ornament, including the phallic consoles emanating from small ledges at the bottom edge of the section as preserved, link this to workshops at Oplontis and Boscoreale circa 40–30 B.C. These distinctive consoles support identical ledges, on which rest a cornice, in turn supporting a large arch, its coffers decorated with stars. A crowning filigree resembles metallic decorative elements, found on the Esquiline Hill in Rome, that decorated interiors from some years after the time of the fresco, adding weight to the assumption that such paintings depict actual ornament of the day. The painted lunette in the center shows a large mask of Herakles, a club, a cluster of poplar leaves, and a lion's skin and quiver.

The upper portion of the fresco matches precisely the upper portion

126

of a fresco section in the Shelby White and Leon Levy collection (see Bothmer) and is from the same room, as is catalogue number 125 (above).

BIBLIOGRAPHY: Unpublished.

RELATED REFERENCES: D. von Bothmer, ed., *Glories of the Past: Ancient Art from the Shelby White and Leon Levy Collection*, exh. cat. (New York, The Metropolitan Museum of Art, 1990), p. 201, no. 142. For the paintings of Boscoreale, see M. L. Anderson, *Pompeian Frescoes in The Metropolitan Museum of Art* (New York, 1987) and references therein. For the Esquiline filigree, see M. Cima and E. La Rocca, *Le Tranquille Dimore degli Dei* (Rome, 1986), pp. 113–128.

—M. A.

127

Fresco Fragment: Woman on a Balcony

Roman, Second Style, end of first century B.C.
Fresco; H: 60 cm; W: 45.2 cm; DEPTH: 3 cm

The scale of the woman standing at a balcony suggests that this fine fragment of a painted wall is of the Second Style and dates from the end of the first century B.C. Based on the right-to-left shadows cast from the bosses adorning the cruciform balustrade of the balcony, this is from the right-hand wall of a room relative to its entrance. The woman sips from a shallow drinking cup, perhaps a phiale, gracefully held in her right hand, and steadies an oinochoe on the ledge of the balcony railing. She wears a loose-fitting and sleeved tunic, cinched at the waist, along with a sakkos which matches the tunic's light green color. She looks downward to her right, and we might suppose that she is observing a street scene or those dining in the room, but she is just as likely to

253 FRESCO

be tipsy. If she is a casually attired but highborn daughter at home, the former two explanations bear consideration; if she is a maidservant avoiding her chores, as might a figure from New Comedy, the effect of her wine cup's contents may explain the languorous pose. Ariel Herrmann has made the appealing suggestion that this might allude to a work by the fourth-century Greek painter Pausias depicting *Methe,* the personification of drunkenness.

The fragment is probably from the upper zone of a dining room of the Augustan period (9 B.C.–A.D. 14); the vertical black field at left may be the top of the background of a large-format painting. It seems at first as though the artist neglected to depict the figure's legs behind the railing, but some brushstrokes in the appropriate interstices may suggest that her lower body was once depicted but has almost entirely vanished.

BIBLIOGRAPHY: Unpublished.

RELATED REFERENCES: On Pausias's painting of Methe, see Pausanias II.27.3. For similar paintings, see the panels of the upper zone of triclinium p (the Ixion Room) in the House of the Vettii: R. Ling, *Roman Painting* (Cambridge, 1991), p. 80, fig. 81.

—M. A.

128

128

BRONZE AND LEAD

128

Statuette of a Draped Female Figure

Roman, first century B.C.
Bronze; H: 25.4 cm
Condition: Essentially complete except for holes in mantle; arms and mantle reattached.

The figure, in an elegant Classicistic style, steps forward on tiptoe, as if alighting from above. She is barefoot and dressed in a peplos; a heavy mantle, draped from arm to arm across her back, blows upwards as if in a strong wind. Her lowered left hand grasps a problematical object which might be either an alabastron or the end of a rather abbreviated torch; her right hand is raised and holds up the edge of her peplos' *apoptygma* (overfall). Her fine-featured face is oblong, with hair cut in bangs at the front, bound with a headband and pulled into a chignon behind.

The head recalls those of large-scale eclectic works such as the Esquiline Venus (see Zanker, pls. 50.6, 51.2) or the Stephanos athlete (see Zanker, pls. 44, 45). The drapery aspires to Classical purity, but the treatment of the peplos betrays a certain awkwardness. Oddly, the graceful composition and the rarefied overall style are not matched by any special care in the treatment of detail, usually a strong point with high-level Roman craftsmen. It seems possible that the statuette was designed to be sheathed in another material such as silver, and that any refinements were carried out on this surface.

The figure's windblown mantle and alighting posture characterize her as an embodiment of natural, and especially celestial, forces. The attribute in her right hand, perhaps either a torch or a vessel shedding dew or dreams, hints at a nocturnal aspect. She recalls the descending female figure with down-turned torch, suggestive of nightfall and balanced by a woman with two raised torches to symbolize daybreak, shown in relief on a cuirassed statue of Flavian date in Famagusta (see Karageorghis). Our statuette has traditionally been identified as Nyx ("Night"), a personification well known in allegorical compositions but less likely to appear as an isolated figure (see *LIMC*). The divinity would, however, have been appropriate for the decoration of a lamp stand or a very large, elegant lamp. A less recherché identification would be as the moon goddess Selene. Although she lacks the decisive attribute of a lunar crescent over the brow, the pose and attire of our figure recall such images as the bronze Selene in Berlin (see *Master Bronzes*, no. 251, color pl. v) and the well-preserved marble Selene in Istanbul (see Chaisemartin/Örgen). Later, on sarcophagi with the myth of Endymion (see Sichtermann/Koch), Diana-Selene alights from her chariot, dressed in long, windblown garments; a torch-bearing Eros sometimes leads the way, and Hypnos sprinkles the recumbent shepherd boy with sleep-inducing drops. These later representations lack our figure's antiquarian sophistication and austere grace. In spirit she is closer to the Diana evoked, in her aspect as a moon goddess, by Catullus' exquisite hymn (XXXIV).

BIBLIOGRAPHY: E. Berger, ed., *Kunstwerke der Antike*, exh. cat. (Lucerne, 1963), no. B25; R. Zahn, *Sammlungen der Galerie Bachstitz*, vol. 2 (Berlin, 1921), pp. 42–43, no. 98, pl. 36.

RELATED REFERENCES: P. Zanker, *Klassizistische Statuen* (Mainz, 1974). V. Karageorghis, *Sculptures from Salamis*, vol. 1 (Nicosia, 1964), no. 48, pl. XLIII. *LIMC*, vol. 4, p. 939ff., s.v. "Nyx" (H. Papastavrou). N. de Chaisemartin and E. Örgen, *Les documents sculptés de Silatarağa*, Institut Français d'Études Anatoliennes, Editions Recherche sur les Civilisations, Mémoire 46 (Paris, 1984), p. 18ff., no. 4, pls. 12, 13. H. Sichtermann and G. Koch, *Griechische Mythen auf römischen Sarkophagen* (Tübingen, 1975), nos. 16–19.

—A. H.

129

Appliqué in the Form of a Nereid and Triton

Roman, late first century B.C.
Bronze; H: 16 cm; W: 15.2 cm
Condition: Essentially intact; cast, with traces of lead for attachment remaining in several places on back.

Although recalling Hellenistic forerunners like the creatures from the retinue of the sea god Poseidon on the "Altar of Domitius Ahenobarbus" (see Lattimore, fig. 4), our figural group is interpreted with an Augustan serenity and dignity. Rather than interacting with the Triton, the Nereid sits upright on his fishtail in a decorous sidesaddle pose. Only the windblown mantle characterizes her as a being associated with the elemental powers of sea or sky. She is a near-replica, though reversed and with clothed upper body, of the female figure riding a *ketos* (sea dragon) as a personification of the waters in the Tellus relief of the Ara Pacis Augustae (see La Rocca).

The Triton bears the heavy shield on his left arm as if it were his own fighting equipment. Tritons do carry weapons on the early Hellenistic Anape sarcophagus (see Vaulina/Waşowicz), but usually they are merely the mounts of the Nereids who bring Achilles his newly forged weapons. When armed Tritons appear in Roman art, they seem to symbolize naval victory, as on a fragmentary cuirassed statue from Oropos (see Petrakos). They are seen on the marble basin from Ospedale S. Spirito (see Guiliano; Lattimore, fig. 6), which despite its baroque appearance has been dated in Augustan times (see Kraus). Amid the sea monsters on the basin a swan, the Apollonian bird favored by Augustus, takes wing, ridden by a Cupid who may be an allusion to the Julian house and its descent from Cupid's mother Venus, goddess of love. The armed Tritons there and in our appliqué were probably meant to evoke Augustus' naval supremacy, a favorite theme after his victory over Anthony and Cleopatra at Actium (see Zanker, figs. 63, 65–67, 81, 82, 102, 231).

129

The appliqué's downward-swooping composition and the way that it is made for attachment to a surface with an extremely shallow, slightly irregular curve suggests that the piece, rather than decorating a vessel, may have adorned the cuirass of a statue. A marble cuirassed statue at Cherchell (ancient Caesarea) in Algeria, perhaps of Augustan date, has a Triton placed, as our group may have been, in a symmetrical composition on the lower abdomen (see Zanker, fig. 178). A bronze cuirassed statue, found in the sea off Cadiz (Cadiz Museum 4.584: *Los bronces romanos en España*), had part of its relief decoration cast separately, and several stray appliqués have been ascribed to such ensembles (see Vermeule). Another possibility is that the appliqué decorated a horse's chest piece, like the sea creatures on the trappings of the two gilt bronze horses from Cartoceto in Ancona (see Nicosia et al.; Bergemann).

BIBLIOGRAPHY: Unpublished.

RELATED REFERENCES: S. Lattimore, *The Marine Thiasos in Greek Sculpture* (Los Angeles, 1976). E. La Rocca, *Ara Pacis Augustae* (Rome, 1983), p. 45f. M. Vaulina and A. Waşowicz, *Bois grecs et romains de l'Ermitage* (Wroclaw, 1974), p. 87ff., no. 12. V. Petrakos, *Oropos kai to hieron tou Amphiaraou* (Athens, 1968), pls. 29, 30. A. Guiliano, ed., *Museo Nazionale Romano,* vol. 1, pt. 1, *Le sculture* (Rome, 1979), p. 255ff., no. 159. T. Kraus, Review of *Die Vorbilder der neuattischen Reliefs* by W. Fuchs, *Gnomon* 32 (1960), p. 463ff. P. Zanker, *Augustus und die Macht der Bilder* (Munich, 1987). *Los bronces romanos en España,* exh. cat. (Museo Nacional de Arte Romano de Merida, Madrid, 1990), p. 184 f., no. 41. C. C. Vermeule, "Hellenistic and Roman Cuirassed Statues," *Berytus* 13 (1959–60), p. 2ff., esp. p. 4, n. 3. F. Nicosia et al., *Bronzi dorati da Cartoceto,* exh. cat. (Florence, 1987), pls. 4, 5, 15, 31. For persuasive new dating for the equestrian statues from Cartoceto, see J. Bergemann in *RM* 95 (1988), pp. 125–128.

—A. H.

130
Lebes

Late Hellenistic (Roman ?), second half of first century B.C.
Bronze with silver; H: 58 cm
Condition: Essentially complete; some repairs and interior reinforcement to the body.

This sumptuously ornamented vessel, unique among the ancient vases known to us, is probably best characterized as a *lebes* (cauldron). Its purpose can only be inferred. The decoration alludes to Dionysos, god of wine and ecstasy. The half-length figure of a young satyr (or Pan), a member of the god's inebriated retinue, adorns the front, and a large grapevine leaf is shown in relief under the back handle. The exuberant vegetation-like forms of the handles and the intricate low-relief composition of blooms and foliage at the front seem to express the inexhaustible generative forces of nature, another Dionysiac theme. The vessel must have been connected with the service of wine, and the very functional hinged design of the lid may imply that it was made for actual use. However, the extraordinary preservation of the piece suggests that it survived in a closed chamber, probably a tomb, either as a container for the ashes of the deceased or as a funerary offering.

While the object as a whole is unprecedented, many of its separate features have revealing parallels elsewhere; Hellenistic elements predominate. The basic shape of the body, a flattened globular form with a short, wide neck and everted lip, is one familiar in decorative or ceremonial vases of late Hellenistic times (see Stuart Jones; Rasmussen/Spivey), as well as in Roman cinerary urns (see Mlasowsky). Monumental vessels of this form, usually with griffin protomes testifying to the shape's ultimate derivation from the great orientalizing cauldrons of Archaic times, figure prominently in the sanctuary scenes of Second- and Third-Style wall painting (see Riz). In these representations as in our piece the vase has a lid of flattened conical shape with concave sides, culminating in a tall, spindlelike turned finial.

The energized appearance of the vessel owes much to the fact that its body rests not on the usual profiled base but on three rollerlike spool feet. It has been noted (Bernard Holtzmann, verbally to Marion True) that similar feet support the basin with perching doves in the Capitoline mosaic, which is probably a work of the second century B.C. (see Donderer).

The fluted side handles of the lebes, bursting into leaf at the edges of their central bead, have palmette attachment plates. They belong to a Hellenistic type seen, for example, on a bronze hydria in the Metropolitan Museum of Art (66.11.12: Sotheby's; MMA 66.119.1 is a stray handle of the same type); variants, including an example on the New York art market (Fortuna Gallery, 1993), continue into the Roman period.

The magnificent foliate back handle has a close parallel in two bronze attachments from the Mahdia wreck (an ancient ship, laden with works of art, that sank off Tunis at the end of the second century B.C.) (see Fuchs). They appear to have a slightly different curvature but are strikingly similar to the lebes handle in style and basic composition. On the lebes, the toothed acanthus of the handle proper continues beyond the hinge, breaking like a wave onto the lid, where the foliage is modeled in low relief and has frothing, ruffled edges like the acanthus on Italo-Hellenistic capitals.

Hellenistic rococo inspiration is obvious in the bust of a satyr decorating the front of the vase. The handsome, high-cheekboned young rustic snaps his fingers and bares his teeth in a wild, impudent grin; his eyes and teeth are silvered. His facial features and his gesture recall the Kroupezion Satyr and the Young Centaur, large-scale Hellenistic works usually dated in the second half of the second century B.C. (see Pollitt). The finely modeled bust seems to reproduce a prototype of this time with almost academic precision. One detail, the wart near the right nostril, has a realism almost unknown outside Roman portraiture. (There are at least two other bust appliqués with the same composition, but they are in completely different styles from our piece and have both been dated, perhaps on insufficient evidence, relatively early in Hellenistic times [see Bothmer; Barr-Sharrar]).

130

130

Other elements carry us further into Roman Republican or Early Imperial times. The vessel's lip is decorated front and back with flat plates of cut-out scrollwork. This cut and curled ornament is a Roman fashion, at its height in the metalware of the first century B.C. (see Strong, pl. 34). The miniature cup held by the young satyr is decorated around the lip with scrollwork in the same style and is fitted with thumb-plates, loop handles, and finger-rests like those familiar from silver drinking cups of the first centuries B.C. and A.D. (see Strong, pls. 33B, 35B). The cut-out ivy-leaf-shaped plates connecting the spool feet with the body of our piece are also very Roman. The large repoussé vine leaf under the back handle seems a spontaneous creation, loosely inspired by the appliqué vine-leaf attachment plates of oinochoe handles.

The delicate foliate ornament worked in relief below the bust recalls another forerunner for this part of our vessel's decoration: the relief designs under the handle attachments of Hellenistic situlas. It springs, via a vestigial leaf chalice, from an inverted "flaming" palmette. The palmette, with thin, undulating lobes around a much longer lanceolate central lobe, is an elaboration of an early Hellenistic type, further refined by the introduction of tiny silver rosettes, each on an incised stem, between the lobes. This kind of palmette is revived in Types II, III, IVb, and VI of the standardized decorative schemes for the bases of Roman marble candelabra (see Cain).

130

The overall designs of these reliefs, combining foliate and abstract ornament, have telling affinities with the composition adorning the lebes—their creation is ascribed to Early Imperial times.

On the lebes, the foliage springing up from the inverted palmette has the treelike arrangement, perfectly symmetrical for all its apparent lushness, that appears in the reliefs of the Ara Pacis Augustae, constructed between 13 and 9 B.C. (see Kraus; Börker). As on the Ara Pacis, the design is composed around a candelabrum-like vertical axis of stacked vegetal forms. Many details, including some whole flowers, evenly distributed for a starry effect and always symmetrically disposed, are picked out in silver. Regularly spaced acanthus scrolls, each encircling a flattened, frontally seen blossom, are the basic units of the construction. The jagged, crisp-looking foliage on the lebes, still related to late Republican decoration, recalls the early Augustan frieze from the Temple of the Divus Julius (see Zagdoun). The acanthus scrolls are flattened as if pressed between the plain, solid background and an imaginary front plane. The continuity and linear energy of the stems, rather than the plasticity of the leaves and blooms, is emphasized (cf. the decoration on the seat of the Augustan goddess from Cumae: Zanker). Each growth knot has paired wrapper-leaves, one short and one elongated, whose tips turn back, sometimes folding over to appear in "perspective," or curling around a neighboring stem. Abstract, ribbonlike helices, their tips rolled around silver buttons, mingle with the foliage; these are a feature of Early Imperial ornament. As in the Divus Julius frieze, slender, wandering shoots intertwine with the more regular elements to create a lacy unity. A few naturalistic touches are admitted, such as the little, ragged-looking bunches of immature leaves and the pairs of soft bud capsules. Two lightly sketched birds, symmetrically placed, are gestures to a vogue for inhabited foliage.

The additive, inorganic combination of decorative features, many of them derived from earlier models, and the sudden jumps in scale and degrees of relief recall vessels from the cities destroyed by Vesuvius in A.D. 79. Details picked out in silver and lacy low-relief ornament, even on bulky vessels, are well known on the bronze vases from Pompeii, Herculaneum, and the villas of the surrounding countryside (see Pirzio Biroli Stefanelli, p. 212f., figs. 190, 191 [silver details on a situla]; p. 250, fig. 237 [relief ornament on a krater]). The vase's bulging belly and tall conical lid, as well as the manner of the four-sided composition's placement over three supports, obviously suggest the famous Pompeian "samovar" (see Pirzio Biroli Stefanelli, p. 225, fig. 207; Pernice).

The pieces from the Vesuvian sites, however, furnish only a terminus ante quem, since some of them were probably not new when they were buried. The Hellenistic components in the decoration of our lebes seem to represent survivals of Mahdia-like forms down past the middle of the first century B.C. Elements more specific to the decorative vocabulary of late Republican Rome and still attested in early Augustan times, but disappearing thereafter, show that the piece is unlikely to be much later. Its sophistication, vitality, and grace suggest that it was produced by one of the mainstream ateliers whose work inspired the Pompeian pieces. The lebes is to those vessels as the deliciously inventive decoration of the Farnesina House by the Tiber is to routine Third-Style Roman wall painting.

The inside of the vessel is rather summarily finished. The body is hammered and the vine leaf under the back handle raised in repoussé. The relief foliate ornament at the front, however, was evidently worked by another technique, since it seems to have left no "ghost" on the interior (unless this is disguised by reinforcing material). No interior seam is evident along the ridge at the juncture of body and shoulder, though one would expect these elements to have been made separately. The cone of the lid is hammered and turned, and a lining piece has been soldered to its interior. The turned finial, as well as the cast handles, bust, scrollwork plates, and feet, were separately made.

BIBLIOGRAPHY: Unpublished.

RELATED REFERENCES: H. Stuart Jones, *A Catalogue of the Ancient Sculptures Preserved in the Municipal Collections of Rome: The Sculptures of the Palazzo dei Conservatori* (Oxford, 1926), p. 175f., no. 10. T. Rasmussen and N. Spivey, eds., *Looking at Greek Vases* (Cambridge and New York, 1991), fig. 82. A. Mlasowsky, "Die Aschenurne des Germanicus," *RM* 98 (1991), p. 223ff., esp. p. 225, n. 8. A. E. Riz, *Bronzegefässe in der römisch-pompejanischen Wandmalerei* (Mainz, 1990), p. 96ff., pls. 55–57. M. Donderer, "Das kapitolinische Taubenmosaik: Original des Sosos?," *RM* 98 (1991), p. 189ff., pl. 49. Sotheby's, London, *Antiquities,* auct, cat., July 25, 1966, lot 267. W. Fuchs, *Der Schiffsfund von Mahdia* (Tübingen, 1963), no. 34. J. J. Pollitt, *Art in the Hellenistic Age* (Cambridge, 1986), figs. 139, 145. D. von Bothmer in *Glories of the Past: Ancient Art from the Shelby White and Leon Levy Collection,* exh. cat. (Metropolitan Museum of Art, New York, 1990), p. 186, no. 133. B. Barr-Sharrar, *The Hellenistic and Early Imperial Decorative Bust* (Mainz, 1987), p. 57, no. C95 ter. D. E. Strong, *Greek and Roman Gold and Silver Plate* (London, 1966), pl. 34. H.-U. Cain, *Römische Marmorkandelaber* (Mainz, 1985), p. 71f., Beil. 3–4. T. Kraus, *Die Ranken der Ara Pacis* (Berlin, 1953). C. Börker, "Neuattisches und Pergamenisches an den Ara Pacis-Ranken," *Jahrbuch des Deutschen Archäologischen Instituts* 88 (1973), p. 283ff. M.-A. Zagdoun, *La Sculpture archaïsante dans l'art hellénistique et dans l'art romain du haut-empire,* BEFAR 269 (Rome, 1989), fig. 159, no. 378. P. Zanker, *Augustus und die Macht der Bilder* (Munich, 1987), fig. 245. L. Pirzio Biroli Stefanelli, ed., *Il bronzo dei romani* (Rome, 1990). E. Pernice, *Die hellenistische Kunst in Pompeii,* vol. 4, *Gefässe und Geräte aus Bronze* (Berlin and Leipzig, 1925), p. 9, pl. 11.

—A. H.

131

Handle with Satyr Heads

Roman, late first century B.C.–early first century A.D.
Bronze with silver; H: 14.1 cm; W: 9.9 cm
Condition: Small losses at lower edge of attachment plates and to central petal over right-hand head.
Provenance: Formerly in the collection of Martin Stansfeld.

The center of the U-shaped handle is marked by a rowel-like ornament, derived from much earlier metal and ceramic models (see Pernice, p. 15ff.). Here the element is characterized as a wool fillet, tied off at regular intervals to create a beaded effect. This is flanked by narrow bands of wave pattern and then by areas that are smooth except for, in each, a wrapper-leaf shown in low relief. Below these smooth zones are zones of fluting, which splay at the bottom into blossomlike forms over the attachment plates. Each attachment plate has the frontal head of a young Pan, or satyr, against the background of a cloven-hoofed fawn- or goat-skin with the legs knotted below the chin. The whites of the satyrs' eyes are silvered, as are their horns and the goatlike glands under their chins.

The handle is an especially elaborate and careful variant of a well-known type, usually used for calyx kraters, which can have various kinds of heads, often those of bearded silenoi but sometimes of tritons, long-haired youths, or even that of Hermes (see Pernice, p. 41, figs. 50, 52). The refined modeling of our satyr heads and the use of

131

silver details find a close comparison in the "marine Medusa" faces on a krater from Pompeii (see Pernice, p. 25, pl. XIV; Pirzio Biroli Stefanelli).

BIBLIOGRAPHY: Sotheby's, New York, *Antiquities,* auct. cat., December 2, 1988, lot 57.

RELATED REFERENCES: E. Pernice, *Die hellenistische Kunst in Pompeii,* vol. 4, *Gefässe und Geräte aus Bronze* (Berlin and Leipzig, 1925), p. 15ff. L. Pirzio Biroli Stefanelli, ed., *Il bronzo dei romani* (Rome, 1990), figs. 238, 240.

—A. H.

132
Statuette of an Old Woman

Late Hellenistic (Roman?), first century B.C.–first century A.D.
Bronze; H: 12.6 cm
Condition: Left thumb, left foot, and objects figure once held missing.

The frail old woman stands in a posture which may be that of a spinner, holding up the distaff with unworked wool in her left hand while twisting the thread away from her body with the lowered right (see Hughes/Forrest). Her head is tilted downwards and her glance seems to follow the progress of the thread. Her withered face is sensitive and delicate. She wears a long-sleeved chiton, slipping off her right shoulder, with a mantle wound around the hips and over the left shoulder. In the center, at waist level, a loop of the mantle is pulled up to shorten and secure the garment. The woman's hair is parted in the middle and drawn back under a kerchief.

A bronze statuette in Vienna (see Gschwantler et al.) is essentially a replica of ours, though the execution is less refined, and, since the hands are missing, their action is uncertain. Other figures with the same stance and drapery scheme are in marble and larger, though still under life-size (see Laubscher; de Lachenal; Wrede). Lists differ according to each observer's definition of the type. Most examples are fragmentary, but enough remains of them to make clear that they originally had different attributes and that the same basic figure could play a variety of roles. A version in the Terme carried a large jar (Wrede, pl. 45.2), and one formerly on the art market a clutch of chicks (pl. 44.1), while another, once in the Woodyat collection, held up a sheep or goat (pl. 44.3); an example in the Vatican adapted the design for an elderly male figure clasping a statuette of Horus (pl. 45.1). Our piece, then, is not necessarily a key to the meaning of the original, but may represent only one of several subjects for which the type could be pressed into service.

Two features of the figure's costume, the small headcloth and the mantle looped up at the center front, are thought to have a ritual meaning (see Wrede, passim). They are worn in scenes of Isiac, Dionysiac, or Priapic worship, and elsewhere by figures of country people whose activities or attributes can be construed as having Dionysiac overtones. Exactly the same attire, however, is worn by the elderly nurses and serving women so frequently seen on mythological sarcophagi. The garb, which became widespread in Hellenistic and Roman art, was a humble one, practical for women engaged in domestic or agricultural tasks. In religious contexts it may have served to characterize the female votaries as being, or assuming the roles of, rustic devotees. Our figure wears a chiton with long, tight sleeves, the same garment worn by servant girls on Classical and Hellenistic grave stelai, not the short-sleeved chiton and separate arm coverings, sometimes seen in conjunction with the headcloth and knotted mantle, which are convincingly identified by H. Wrede as cult attire.

If our figure's activity can indeed be identified as spinning, the undemanding but productive task characterizes her as a good, industrious old woman. She might be the priestess of a rural cult or a kindly peasant or wise nurse from mythology. The slipped shoulder of her chiton, common to all the examples, need not imply dissolute habits as it does in the famous image of the Drunken Old Woman (see Zanker). Seen in many other representations of female old age, this motif proclaims the liberation of the elderly from sexual considerations and the restrictions they impose on women of childbearing years.

The prototype of our figure with its frontal pose and layered, fluttering drapery can perhaps be

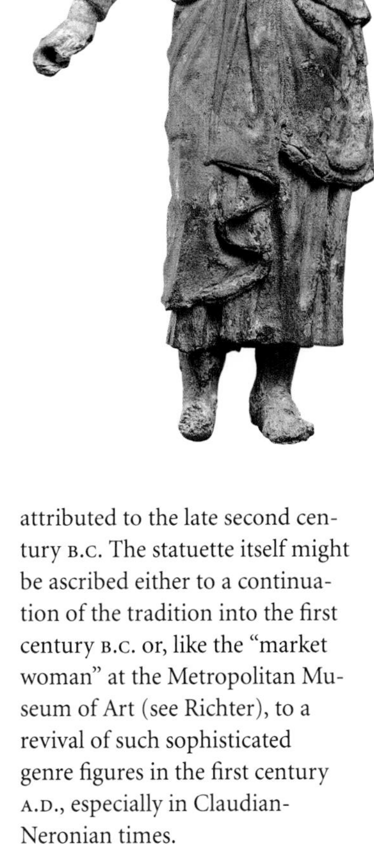

132

attributed to the late second century B.C. The statuette itself might be ascribed either to a continuation of the tradition into the first century B.C. or, like the "market woman" at the Metropolitan Museum of Art (see Richter), to a revival of such sophisticated genre figures in the first century A.D., especially in Claudian-Neronian times.

BIBLIOGRAPHY: Unpublished.

RELATED REFERENCES: M. Hughes and M. Forrest, eds., *How the Greeks and Romans Made Cloth,* Book II of *Cambridge School Classics Project,* Classical Studies 13–16 (Cambridge, 1984), p. 71. K. Gschwantler et al., *Guss+Form,* exh. cat. (Vienna, 1986), no. 182. H. P. Laubscher, *Fischer und Landleute: Studien zur hellenistischen Genreplastik* (Mainz, 1982), p. 121ff., sect. A4–A5. L. de Lachenal in A. Giuliano, ed., *Museo Nazionale Romano,* vol. 1, pt. 2, *Le sculture* (Rome, 1981), pp. 298–299, no. IV.17. H. Wrede, "Matronen im Kult des Dionysos," *RM* 98 (1991), p. 163ff. P. Zanker, *Die Trunkene Alte* (Frankfurt, 1989). G. M. A. Richter, *Catalogue of Greek Sculptures* (Cambridge, Mass., 1954), no. 221. S. Pfisterer-Haas, *Darstellungen alter Frauen in der griechischen Kunst* (Frankfurt, 1989).

—A. H.

133
Statuette of Athena Promachos

Roman, first century B.C.–first century A.D.
Bronze; H: 20.7 cm
Condition: Front parts of both feet, tail and left wing of griffin, helmet crest, spear, shield, and twenty silver snakes from border of aegis missing; crack through right wrist; right wing of griffin bent.

Athena strides forward as if to do battle, her right arm held high to brandish a now-missing spear and her left arm advanced to support a shield. The aegis forms a semicircular breastplate in front and hangs down, following the forms of the body, to below the buttocks in back. It is set with a small silver gorgoneion over the middle of the chest and was once edged with silver snakes. Under it Athena wears a sleeved chiton and a long Archaic diagonal mantle with overfold, fastened on the right shoulder so that its edges form a cascade of zigzags down the right side. Her austere-looking face has silvered eyes. Her hair is arranged in a pair of long curls to each side at the front, and a wavy mass, ending in points, behind. Her Attic helmet is topped by a crouching griffin and has a stephané-like front piece, picked out with silver ornaments in the center and at the sides.

The figure is related to an Archaistic Promachos type known from a headless marble version in Dresden, as well as from statuettes in Vienna and London (see Fullerton, p. 50ff., figs. 18–20). The creation has been considered either the neo-Attic adaptation of an early fifth-century original (see Willers), or an invention of the late second–first century B.C. (see Zagdoun; Fullerton, p. 52f.). The Dresden-Vienna-London type has an aegis hanging far down in back like that of our figure, but two of the snakes are knotted across the front to form a belt. Our piece's sucked-in waist seems to imply knowledge of the belted type; the missing snakes at waist level might even have been knotted across the front, but the edge of the aegis from which they emerge is not stretched forward as in the Dresden figure. Another Archaistic Promachos type, represented by a headless, much restored marble version in Palermo and a fragment in the Vatican (see Fullerton, p. 53ff., figs. 21, 22), wears similar attire, unbelted, but does not have the indented waist. The overfall, in our piece, hangs in bunches of stepped folds below waist level, as in the Dresden type, rather than rising in the more authentically Archaic high inverted V seen in the Palermo statue.

Our statuette's head is of the same basic design as the well-preserved head of the London statuette (see Fullerton, fig. 20) except for having two rather than three long locks to each side and being surmounted by a sphinx rather than a griffin. The tight sleeves recall the London bronze but have bizarre cufflike turn-backs at the edges, suggesting that our piece may be at several removes from its prototype.

The workmanship of our bronze statuette seems not inconsistent with the finds from the Vesuvian cities; the emphasis on ribbonlike locks of hair, inspired by Archaic forerunners, can be a Claudian mannerism. Despite some awkward features, our ambitious and well-preserved miniature version provides new hints as to the character of the original. The stiff, elongated figure is similar in its proportions to those on the Caesarean Civita Castellana altar (see Strong). The oblong face with its thin, straight nose and stern expression is related to Classicistic creations like the Stephanos athlete rather than being derived from genuine Archaic works. The type, not the quiet Athena as protectress of intellectual life but a martial Promachos, lending itself to political imagery, seems to express the naively intense imitative fervor of late Republican times.

BIBLIOGRAPHY: Unpublished.

RELATED REFERENCES: M. D. Fullerton, *The Archaistic Style in Roman Statuary,* Mnemosyne Suppl. 110 (Leiden and New York, 1990). D. Willers, *Zu den Anfängen der archaistischen Plastik in Griechenland,* AM Beiheft 4 (1975), p. 59f. M.-A. Zagdoun, *La sculpture archaïsante dans l'art hellénistique et dans l'art romain du haut-empire,* BEFAR 269 (Paris and Athens, 1989), p. 60. D. E. Strong, *Roman Imperial Sculpture* (London, 1961), fig. 19.

—A. H.

133

134

134
Statuette of a Comic Male Figure

Roman, probably first century B.C.–first century A.D.
Bronze; H: 11.4 cm
Condition: Essentially intact.

The pudgy figure, completely nude, lunges forward. His torso twists as he flings both arms outwards in a declamatory gesture. He wears a pointed beard; tufts of hair at the forehead and nape set off a bald pate. The large projecting ears are detailed with harsh, concentric semicircular grooves.

Although its movement and character suggest a comic actor, the figure seems not to be masked and is definitely nude rather than dressed in the theatrical costume that, however indecorous, always clothes an actor. The personage is silenos-like but lacks the tail or animal ears of Dionysos' companions. His rather infantile plumpness and small genitals set him apart from typical Hellenistic grotesques, who tend to be emaciated but overendowed. The blurring of what in Greek art were clearly separated iconographic categories points to a date in Roman times. The large, shapeless, and vaguely characterized body has some affinity with that of the lyre-playing silenos in the Dionysiac frieze of the Villa of the Mysteries in Pompeii (see Andreae), but the bold, even coarse workmanship may speak for a somewhat later date.

BIBLIOGRAPHY: Merrin Gallery, *Bronzes: Cast in the Image*, exh. cat. (New York, 1989), no. 6.

RELATED REFERENCES: B. Andreae, *The Art of Rome* (New York, 1977), pl. 29.

—A. H.

135

135
Handle with the Figure of a Seated Comic Actor

Roman, first century B.C.–first century A.D.
Bronze; H: 8.8 cm
Condition: Figure's right foot and blade of implement missing.

The handle has the form of a foliate chalice the leaves of which are rather lushly though summarily modeled and have incised veins. It is topped by the figure of a seated comic actor. He perches, legs crossed, on a folding stool and rests his head on his right hand in a "plotting" gesture. His masked, bearded face is turned to his left and upward, to face the viewer.

This piece belongs to a class of small figural handles rising from leaf-chalice bases (see cat. no. 136 below). Their subjects are usually risqué, humorous, or related to the world of theatrical and amphitheatrical entertainment (see Perdrizet, nos. 88, 90, 91 [grotesques], 108 [actor], 112 [gladiator], 116 [figure in Egyptian dress], 133 [hound]). The subjects seem to imply a certain frivolity, not necessarily feminine. Although their compositions are always Hellenistic in inspiration the production of these objects seems to have continued well into Roman times, on the evidence of such pieces as the Fouquet gladiator.

BIBLIOGRAPHY: Unpublished.

RELATED REFERENCES: P. Perdrizet, *Bronzes grecs d'Égypte de la collection Fouquet* (Paris, 1911).

—A. H.

136
Handle with the Figure of a Seated Baboon in Human Attire

Late Hellentistic (Roman ?), first century B.C.–first century A.D.
Bronze; H: 6 cm
Condition: Essentially complete except for the blade of the implement.

The piece belongs to the category of small handles with figures set on leaf chalices. A monkey, dressed in a short tunic belted with several cords wound tightly around the chest and with a hood hanging on the shoulders, sits wearily on the ground with a lantern beside him. The creature is definitely identifiable by his long, doglike muzzle as a member of the genus *papio*, the baboons, all five species of which are native to Africa and Arabia. He looks more like *papio porcarius*, a lean creature with short, wiry hair, than the more familar ruffed *papio hamadryas*. A splendid statuette in the Borowski collection (see Leipen), has the head, evidently a mask, of the same kind of monkey. To judge from his tail and the construction of his arms, legs, and buttocks our figure is a real baboon, not a costumed actor.

Domesticated monkeys were a fairly familiar sight in the late Hellenistic and Roman world, probably more often as performers, especially in the case of rather large simians like baboons, than as pets. They could, ancient authors report, learn to distinguish letters, to dance, to play musical instruments, and even to drive chariots (see Toynbee).

Our creature's posture and the lantern at his side might characterize him as a parody of a figure often represented in small-scale Roman sculpture—the little slave, dressed in a hooded garment and equipped with a lantern, who waits sleepily for his master during a nocturnal revel (see Menzel, no. 65, for a typical example). However, Michael Padgett has noted (verbally to Lawrence Fleischman) that the chest bands resemble the dress of a charioteer; in addition, as Lawrence Fleischman has himself pointed out, there is a relief of an unclothed monkey charioteer, driving a pair of camels, in the reserves of the Museo Nazionale Romano (see Zschietzschmann). Perhaps we do indeed have here a monkey

136

charioteer, though this would not explain the lantern and the creature's somnolent attitude. Another possible reason for the wearing of chest bands is simply that, as in the charioteer's costume, they were used to hold loose garments in place during strenuous activity. It has always been tempting to dress tame monkeys in human clothes, but ancient drapery could hardly have been worn by the unpredictable creatures without some special provision for securing it.

The object belongs to the same class of small handles as catalogue number 135. Although examples are found all over the Greco-Roman world, the type is especially well represented among finds from Egypt (see Perdrizet). In our piece, the accurate representation of an African animal, the baboon, might hint at an Alexandrian origin, and the subtlety of conception and modeling suggest a date relatively early in the series.

BIBLIOGRAPHY: Unpublished.

RELATED REFERENCES: N. Leipen in *Glimpses of Excellence*, exh. cat. (Toronto, 1984), no. 40. J. M. C. Toynbee, *Animals in Roman Life and Art* (Ithaca, N.Y., 1973), p. 55ff. H. Menzel, *Bildkataloge des Kestner-Museums, Hannover*, vol. 6, *Römische Bronzen* (Hannover, 1964). W. Zschietzschmann, *Hellas and Rome* (London and Tübingen, 1959), pl. 74a. P. Perdrizet, *Bronzes grecs d'Égypte de la collection Fouquet* (Paris, 1911). For observations on monkeys in human guise, see P. Blome, "Affen im Antikenmuseum," in M. Schmidt, ed., *Kanon: Festschrift E. Berger*, Antike Kunst Beiheft 15 (1988), p. 205ff.

—A. H.

137
Clipeus (Roundel) with Bust of Nike

Roman, late first century B.C.–early first century A.D.
Bronze; DIAM: 20.2 cm; H (OF NIKE BUST): 11 cm

The roundel is carefully finished behind as well as in front. On it is mounted the separately cast bust of Nike, personification of victory. Wings lowered in repose, she wears a peplos over a button-sleeved chiton. Both hands are outstretched, probably with the victor's palm

137

137

once held in the left and a wreath in the right. Her hair, rolled back over the ears, falls in long tresses on her shoulders. Straying tendrils on her forehead, like the wing feathers, are incised. The irises of the eyes are hollowed to hold inlay.

In its overall form as well as in the workmanship of the disk, the object recalls similar roundels from Pompeii with busts, rostra, or animal protomes (see Barr-Sharrar, no. C146), though none of the images compare in quality with our Nike. Later decorative roundels survive but are rather different (see Barr-Sharrar, no. C169). Like all of these *imagines clipeatae* (literally, "images on shields"), our piece descends from the votive shields dedicated in Greek temples. Their appearance is most strikingly conveyed by the Second Style frescoes of the Oplontis Villa (see de Franciscis). Sculptured roundels echoing these votive shields adorned such late Hellenistic buildings as the Mithridates Monument on Delos or the Heroon at Kalydon. Our decorative version, in a serene and precise style inspired by Classical models, should be datable in or close to Augustan times.

BIBLIOGRAPHY: Unpublished.

RELATED REFERENCES: B. Barr-Sharrar, *The Hellenistic and Early Imperial Decorative Bust* (Mainz, 1987). A. de Franciscis, *Die pompejanischen Wandmalereien in der Villa von Oplontis* (Recklinghausen, 1975), figs. 11, 12. *Enciclopedia dell'arte antica,* vol. 2 (Rome, 1959), p. 718ff., s.v. "Clipeate, immagini" (G. Becatti). R. Winkes, *Clipeata imago: Studien zu einer römischen Bildnisform* (Bonn, 1969). For a full-length statuette of a peplos-clad Nike in Naples (MN 5010): W. Wohlmayr, *Studien zur Idealplastik der Vesuvstädte* (Buchloe, Austria, 1991), pl. 73.

—A. H.

138

138

Handle with a Tritoness over a Triton Mask

Roman, late first century B.C.–early first century A.D.
Bronze with silver inlays; H: 13 cm
Condition: Intact except for small losses along edges of mask; much corrosion and pitting.

The piece served as a handle of a vessel, perhaps a small, ornate jug. The attachment plate is a relief mask of a triton or "marine Medusa" with long, windblown locks, fins on the face, and crab claws in the hair. Above the mask is the figure of a tritoness in the round, with a single fishtail springing from foliage at the waist and curving back under her left forearm. Her upper part has the form of a slender, austerely lovely young woman with long hair streaming down her back and fastened by a low barette. In her right hand is a ship's stern ornament (*aphlaston*). Silver inlays pick out details such as the triton's face-fins, the eyes, and the disk of the aphlaston.

The finely modeled, silver-inlaid masks decorating a bronze krater from Pompeii (see Pirzio Biroli Stefanelli) are comparable in style and iconography to the mask on our piece (see also the masks on cat. no. 131 above). The mermaidlike tritoness is less easy to parallel; a tritoness attachment from Smyrna in the Petit Palais (see Petit) also has a single tail but is quite different in style. Although disfigured by corrosion, the workmanship of our handle was originally crisp and careful. The combination of Classical with Hellenistic elements and the emphasis on linear detail speak for a date in Early Imperial times.

BIBLIOGRAPHY: Unpublished.

RELATED REFERENCES: L. Pirzio Biroli Stefanelli, *Il bronzo dei romani* (Rome, 1990), fig. 240. J. Petit, *Bronzes antiques de la collection Dutuit* (Paris, 1980), no. 25.

—A. H.

139

Statuette of Dionysos with Base

A. Archaistic Draped Dionysos

Roman, second half of first century B.C.–first half of first century A.D.
Bronze with silver and traces of gilding; H: 40 cm
Condition: Essentially intact except for numerous cracks and holes; hollow cast, very uneven in thickness and with many flaws; carefully patched and finished in antiquity.

To judge from the disk-shaped attachment on the head, the figure had a caryatid-like function as part of an ornate piece of furniture or equipment (see Zanker, fig. 214). Details added in other metals enhanced its decorative quality. The lips, now missing, would have been of copper; the whites of the eyes are silver. The now-missing sandal straps and fillet were separately applied, perhaps again in silver. Some traces of gold leaf on the back of the mantle suggest that this garment was gilded.

The Archaic flavor of the piece comes chiefly from its stance and coiffure. The figure's attire, though of a mannered complexity, falls in free, natural-looking folds, the artist having made no attempt to copy rigid Archaic drapery patterns. Dionysos stands in the stiff pose of a kouros, with feet placed close together and weight almost evenly distributed between them. His left hand, at his side, plucks at a fold of his garment, while his right, extended, held out a now missing attribute, perhaps a kantharos. He wears a chiton of thin, crinkled material, fastened where it is visible on the left arm and shoulder with eight buttons and, diagonally over this, a Bacchic fawn- or goat-skin, belted at the waist and fastened on the left shoulder. A clinging but voluminous mantle is wrapped around the hips and thrown over the left shoulder, falling in zigzag folds at the back. The hair, looped back over the ears, falls on the shoulders in a smooth mass with large horizontal waves underlying the delicate vertical incisions on the surface. This treatment, like the spade-shaped, neatly trimmed beard with an upper section set off from the main mass, is a borrowing from Archaic art frequently seen in neo-Attic creations.

Our piece's realistic rendering of thin, layered drapery sets it apart from more literal Roman imitations of Archaic statues, but in its careful, compact, sober quality it is also unlike the mannered and usually rather frivolous rococo-Archaistic "Roman works in Rhodian style" (see Fullerton). Its typology recalls the "Sabazios" caryatids on Dionysiac sarcophagi (see Koch/Sichtermann), which do not make their appearance until Antonine-Severan times but are

139A

clearly inspired by a late Hellenistic prototype created in the Greek East. Our Dionysos, comparable in workmanship and stylistic feeling to the sphinxes on the Augustan tripod from the Isis temple at Pompeii (see Zanker, fig. 213), must represent an earlier influence of such models on craftsmen producing luxury objects in Italy.

BIBLIOGRAPHY: Unpublished.

RELATED REFERENCES: P. Zanker, *Augustus und die Macht der Bilder* (Munich, 1987). M. D. Fullerton, *The Archaistic Style in Roman Statuary*, Mnemosyne Suppl. 110 (Leiden and New York, 1990), p. 147f. G. Koch and H. Sichtermann, *Römische Sarkophage* (Munich, 1982), fig. 230. M.-A. Zagdoun, *La Sculpture archaïsante dans l'art hellénistique et dans l'art romain du haut-empire*, BEFAR 269 (Paris and Athens, 1989).

—A. H.

B. Tripod Base

Roman, perhaps first century B.C.
Bronze; H: 77 cm; W: 32.3 cm; DEPTH: 28 cm

Each time a new form emerges from scientific excavation, it is heralded as a discovery; when it first appears in a private collection, it is assumed to be a forgery. A great deal more variety in the typology and style of late Hellenistic and Early Imperial bronzes exists than one might suppose. The imagination of late Greek and early Roman artists working in bronze was no less fluid than the medium itself. In addition, the unrestrained and often uninspired experimentation of Archaistic style—a first-century revival of Archaic style from the late sixth century B.C.—often resulted in uncanonical figural types and combinations.

Aspects of the bronze tripod base in the Fleischman collection do not fit neatly into any well-known typology. The size and proportions of the object itself are in keeping with what we would expect of lamp stands from the late Republic or early Empire, yet the combinations of figures are difficult to correlate with known prototypes. The upside-down chalice-flowers, which have puzzled some scholars, are found in Roman painting of the period from which the tripod purports to date, as in the House of Augustus on the Palatine; there is much room for variation in first-century decorative arts. The proportions of the base, along with the various vegetal elements which decorate its legs, are in keeping with what one could expect of an eclectic and Archaistic bronze.

Among the concerns voiced about this work are technical ones, including the fact that the patina is not ancient. These doubts were largely dispelled by detailed scientific investigation undertaken in 1973 by W. J. Young of the Museum of Fine Arts, Boston, and by a technical examination by John D. Cooney of the Cleveland Museum of Art. Young's examination consisted of energy-dispersive x-ray analysis to establish the metallurgy, ultraviolet surface examination, microscopic analysis, and thermoluminescence testing. The results of each method gave credence to an ancient date for the bronze casting, which was subsequently repatinated; Young concluded his report by stating that "the figure (no. 2a) and the base (no. 2b) were originally together and are unquestionably of ancient origin." Cooney concurred: "The relief panels retain very extensive areas of cuprite, a red substance which appears only as the result of long burial. The ancient origin of this tripod is beyond all doubt. Its date must fall in the first century B.C. or early in the following century."

Leaving aside the favorable conclusions of two highly respected specialists, there are several concerns that have been expressed by art historians. These have focused on the iconography of the three figural panels, which deserve individual consideration.

Panel A shows a scene of sacrifice with a woman to the left of an Archaistic statuette of Athena, and a man holding a kid to its right; two goats advance from the right. The scene seems in keeping with what we might expect of a rustic sacrifice; less easily explicable is scene B, with a statuette of a youthful male on a pillar and a beardless male to the left, perhaps Apollo, holding a deer by the horns. To the right is a beardless man holding a thyrsos, an attribute of the wine-god Dionysos. The apparent standoff of Apollo and Dionysos finds no exact parallel, and is somewhat artlessly conceived, with a stiffly executed Apollo and the wine-god looking away. Side C shows a scene with a club-bearing Herakles walking to the right, grasping the right arm of a woman walking behind him. Her left hand seems to rest placidly on his arm, while her right

139B, PANEL C

139B, PANELS A AND B

hitches up her drapery as she advances coquettishly. The scene recalls images of Herakles and Auge or, perhaps, the ill-fated Alcestis, being led back from Hades.

Many of the awkwardnesses in the figural scenes may be compared with the eclecticism of two-figured "Campana" reliefs of the Augustan period, which include face-offs that are successful in varying degrees and often involve a central vertical monument, such as a betylus, candelabrum, krater, tripod, or other elements that function like the statue bases in panels A and B. The odd features of this bronze are thus either the consequence of a free-associating artist of the first century or one of the nineteenth or twentieth. The burden of scientific evidence argues on behalf of its being an eclectic antiquity.

BIBLIOGRAPHY: Unpublished.

RELATED REFERENCES: Technical reports filed on August 9, 1974, for the Museum of Fine Arts, Boston, by W. J. Young (Examination no. 73.329) and on June 5, 1973, by John D. Cooney, for the Cleveland Museum of Art. For comparison, see a bronze base in the British Museum: *Catalogue of Bronzes* (London, 1899), no. 284. For the upside-down chalice-flowers, see the upper cubiculum (no. 15) of the House of Augustus: G. Carettoni, *Das Haus des Augustus auf dem Palatin* (Mainz am Rhein, 1983), fig. 17.

—M. A.

140

Stamp with Panther Pattern

Roman, circa first century B.C.
Bronze; H: 2.7 cm; W: 1.9 cm; WEIGHT: 40.8 g
Condition: Intact; solid green patina.

The tall, narrow object is square in plan and flares toward the top. On its upper surface is a panther protome in high relief, seen in profile, truncated at the base of the front legs and behind the shoulders.

The weight of the object, 40.8 grams, corresponds to a twentieth of the heavy Roman mina current in the Levant from the first century B.C. to the second century A.D. Although it is related to the mina, comparison with analogous objects suggests that this piece may

140

141

have been a jeweler's punch. Such punches, held in the fingers and hit with a mallet from above, were used to stamp designs into thin sheet gold that had been placed on a yielding surface such as pitch. The relief of the panther protome on the Fleischman piece is high enough that it could create an impression sufficiently deep to be cut out and used as one half of a full-round protome of the type that served as terminals of hoop earrings of middle to late Hellenistic date.

BIBLIOGRAPHY: Unpublished.

RELATED REFERENCES: R. Higgins, "Jewellery," in D. Strong and D. Brown, eds., *Roman Crafts* (London, 1976), p. 53ff, fig. 54. For the form and construction of hoop earrings with protomes, see H. Hoffmann and P. Davidson, *Greek Gold* (Mainz, 1965), nos. 29, 30.

—K. H./L. M.

141

Weight with Head of a Lion

Late Hellenistic–Early Roman Imperial (from the Levant), first century B.C.–first century A.D.
Bronze with lead fill; H: 5.8 cm; W: 5.5 cm; DEPTH (AVERAGE, WITHOUT LION HEAD): 2.2 cm; WEIGHT: 455.8 g
Condition: Intact; even brownish patina.

The weight, a square, thickish block, has a lion's head shown frontally in high relief. Details of the mane are incised. The mouth opening has been hollowed from side to side as if prepared for a ring handle. Beginning at the bottom left and running around the edge is the inscription: ΑΠΟΛΛΩΝΟΣ (of Apollo) in large, dotted Greek letters. The sides are bordered with a narrow strip molding above and below. The back is plain and has a large rectangular hollow which has been filled with lead, with signs of scraping to adjust the weight.

The inscription implies that the weight was conserved in a temple of Apollo. It was a common practice to place official weights in temples for safekeeping. The feline most usually associated with Apollo was the lynx, but the animal here has a lion's rounded ears, though somewhat understated, rather than the pointed ones of the lynx. Lions could also be associated with the god, as is the case with the Archaic lion statues along the sacred way to the sanctuary of Apollo at Didyma or on the terrace of the Naxians at Delos. Many animals have been associated with Apollo: the wolf, lynx, and lion, but also the roe deer and birds such as the raven, hawk, and swan, not to forget the mouse and the grasshopper. Only lately has the importance of the lion been stressed.

The weight is clearly a mina of the light standard common in late Hellenistic and Early Imperial times in the region of the Eastern Mediterranean. An origin in the Levant is likely.

BIBLIOGRAPHY: Unpublished.

RELATED REFERENCES: For a bronze lion inscribed "I belong to Apollo," see H. A. Cahn, "Die Löwen des Apollon," *Museum*

Helveticum 7 (1950), pp. 185–199 = *Kleine Schriften zur Münzkunde und Archäologie* (Mainz and Basel, 1975), pp. 17–32, fig. 2. For a description of Apollo with a crouching lion on the black-figured amphora in London (British Museum 49): J. Boardman, "Leaina," *Enthousiasmos: Festschrift Hemelrijk* (Amsterdam, 1986), pp. 93–96, ill. p. 95. For a black-figured cup (Karlsruhe 69/61) with Apollo between two lions: H. A. Shapiro, *Art and Cult under the Tyrants in Athens* (Mainz, 1989), p. 59, pl. 29.

—L. M.

142

142
Weight with Relief of Scales

Roman, circa second century A.D.
Lead; H: 6.6 cm; W: 5.6 cm; DEPTH (AVERAGE): 1.1 cm; WEIGHT: 213.8 g
Condition: Whitish, crusty surface.
Provenance: Said to be from Asia Minor.

The rectangular plaque has plain frames and low relief designs on both sides. Side A has a balanced scale on a footed pedestal. Side B has a footed, two-handled cup, with a pair of bell- or dome-shaped objects above.

The importance of weights and measures for commerce is evidenced by this specimen. The distinctly shaped balance scale on the front is self-explanatory. The same goes for the cup on the back, which would be suitable for measuring grains and liquids. One could propose the same purpose for the two dome-shaped objects above, but it is difficult to imagine how they would function. It is possible that these enigmatic objects represent the helmets of the Dioskouroi. On coins and other bronze objects from Tripolis Phoeniciae the two helmets of the Dioskouroi are often depicted with a star above each helmet. Although here the stars are lacking, perhaps these objects are the two helmets seen from the side. The protruding parallel projections at the bottom of the domes could then be understood as cheek pieces or chinstraps.

The weight of 213.8 grams corresponds to the quarter mina of the heavy Roman standard used in the second century A.D.

BIBLIOGRAPHY: Unpublished.

RELATED REFERENCES: For a weight with two helmets attributed to Tripolis Phoeniciae, see S. Qedar, in Münzzentrum, Cologne, *Gewichte aus drei Jahrtausenden*, vol. 4, sale 49, 1983, no. 5092.

—L. M.

143
Head of the Young Dionysos

Roman, first half of first century A.D.
Bronze with silver; H: 21.6 cm
Condition: Head broken off through upper part of neck; inlaid irises missing; rather thick, even cast with a large square patch at front left side of neck.

The god is shown as a beardless adolescent with an oval, regular-featured face. His longish wavy hair, parted in the center, is encircled by an ivy wreath twisted from two strands, with leaves and bunches of berries. Free-falling locks on the neck were cast separately. The silvered whites of the figure's eyes have deeply cut circular cavities with pinholes at the back for fixing inlaid stone or glass irises. The eyebrows are incised with feathery strokes.

The head belongs to a series of sculptures that are all variants of the same basic type, alike in their

143

143

facial features and the disposition of the locks of hair. In two cases, the entire figure is preserved. Especially close to our piece, though flimsier and more mannered, is the head of a bronze statue from Volubilis in Morocco (see Zanker, pls. 33.1; 35.2, 4; 36.3, 6). This figure, a decorative Roman creation, functioned as a lamp-holder. The Volubilis boy, like ours, wears the Dionysiac ivy wreath over mid-length locks parted in the middle. The more luxuriant hair of the Volubilis boy, however, covers his ears, which are visible on the Fleischman head.

A bronze head in Munich (see Zanker, pls. 35.1, 3; 36.2, 5) is a very different permutation of the image, with a sober expression and shorter, more compact hair. However, the oval face, the center parting of the hair, the fillet around the head, and the arrangement of locks at the back of the head leave no doubt that all three works go back to a common prototype.

There are further affinities with a marble statue at Aphrodisias (see Erim), which has been considered a version of the Polykleitan Diskophoros (see Berger). The head of this statue resembles the Munich bronze; though the hair is even shorter, the disposition of the locks and the rather unusual part over the forehead are repeated. The diadem is in this case the emblem of an athletic victor. The body has Polykleitan muscular definition, but the pose is enough like that of the Volubilis statue to suggest that the ultimate source for the two is the same.

The series is revealing as to how the same basic model can be adapted to different iconographic requirements and different stylistic preferences. The body, in its attenuated form, is also used with other heads, for example in the bronze boy lamp-bearer of Antequera (see *Los bronces romanos*), the Idolino (see Zanker, pl. 33.2, 3) and two bronze lamp-bearing statues from Pompeii (see Pozzi et al.).

Our head, probably once part of a full-length figure, is comparable in style to many of the elegant eclectic works from the cities destroyed by Vesuvius in A.D. 79. It may come from a purely decorative image or from one that served as a lamp-holder, a favorite use for Classicistic figures of handsome youths. The underlying sensitivity of the workmanship and its constrast of smooth surface areas with crisp linear detail might suggest a Tiberian date.

BIBLIOGRAPHY: Unpublished.

RELATED REFERENCES: P. Zanker, *Klassizistische Statuen* (Mainz, 1974). K. T. Erim, *Aphrodisias: City of Venus Aphrodite* (New York and Oxford, 1986), p. 86. E. Berger, *Antike Kunstwerke aus der Sammlung Ludwig,* vol. 3 (Mainz, 1990), p. 117ff., replica list, p. 120ff., no. 13. *Los bronces romanos en España,* exh. cat. (Museo Nacional de Arte Romano de Merida, Madrid, 1990), no. 174. E. Pozzi et al., *Le Collezioni del Museo Nazionale di Napoli,* vol. 1, pt. 2 (Rome and Milan, 1989), p. 146f., nos. 251, 252.

—A. H.

144
Statue of the Infant Dionysos (?)

Roman, probably first half of first century A.D.
Bronze with silver and copper; H (WITHOUT TANGS, WHICH ARE SET INTO THE BASE): 64 cm

The chubby toddler stands in a swaybacked pose. His body is treated with a rather appropriate lack of detail except in the pudgy-toed little feet with their finely modeled toenails. The modeling of the head, tipsily or dreamily tilted, is more studied. The large eyes, silvered and with incised irises, roll toward the left in the plump face. The full lips are slightly parted over the small, fat chin, giving the expression an abandon at once babyish and suggestive of Bacchic trance. The medium-short hair lies in flat, beautifully modeled and chased curls, with elegant tendrils framing the face. The flesh was originally coated with tin to give it a silvery gloss which would have contrasted with the golden bronze-colored hair.

The child wears a foliate wreath with a broad fillet twisted around it, leaving the long, flowing ends, made of sheet copper, to hang down on the shoulders. This crown is the only real clue as to the identity of the supernatural baby, and as such it is somewhat ambiguous. Vegetation in ancient sculpture is not always botanically accurate, but the foliage here does seem to be the leaves and immature clusters of a grapevine. The toddler's almost inebriated expression, too, suggests an identification as Dionysos.

144

grape clusters of our baby's fillet-wrapped crown do seem to prove that the big, jagged leaves are those from Dionysos' grapevine rather than Herakles' poplar. In its composition the figure seems to be a charming improvisation, unless one prefers to think of it as a playful adaptation of the adult type known as the Hope Herakles (see Waywell). Although the subject has religious overtones, the image may have had a primarily decorative function, like much of the ideal sculpture found in luxurious dwellings at Pompeii and Herculaneum. In style the little statue is close to works from these cities.

BIBLIOGRAPHY: Unpublished.

RELATED REFERENCES: *LIMC*, vol. 4, p. 728ff., nos. 363, 365, 369, 655, 986, 1143, 1199, 1201, 1202, 1253, 1643, s.v. "Herakles" (J. Boardman, O. Palagia, S. Woodford). G. B. Waywell, *The Lever and Hope Sculptures* (Berlin, 1986), Hope no. 4, pl. 48.2.

—A. H.

144

However, the *corona tortilis* (fillet-wrapped crown), combined with foliage, is characteristic of Herakles (see *LIMC*) and of athletic victors. Dionysos' usual wreath, on the other hand, is a simple if sometimes luxuriant circlet of grapevine or ivy tendrils, often worn with his other typical adornment, the *mitra* (a flat band tied low on the forehead).

The question, as for the elegant bronze figure of a divine baby in St. Louis (see *Master Bronzes*, no. 128), comparable in some respects to our piece and probably contemporary with it, is not easily decided. Both Herakles and Dionysos have important infancy myths, and both are frequently represented as small children. Herakles, given to drunken revels, definitely has a Dionysian aspect and often, for example, joins Dionysos' triumphal rout on sarcophagi. Dionysos, on the other hand, never has anything Herculean about him. The nascent

145
Statuette of Herakles (Farnese type)

Roman, A.D. 40–70
Bronze with silver-inlaid eyes; H: 15.2 cm
Condition: Fingers of left hand and club, which must have been made separately, missing; green, heavily encrusted surface.

The muscular hero rests from his labors. Behind his back in his right hand, he holds the apples from the garden of the Hesperides. He leans on his (now-missing) club, which is cushioned by the skin of the Nemean lion.

The composition is based on the weary Hercules created by Lysippos in the late Classical period

145

and named for a replica formerly in the Farnese collection and now in the Museo Nazionale in Naples. Here, however, Herakles is less aged and exhausted than usual. He is more upright and self-supporting, and unlike the usual shaggy Herakles, he does not look down wearily but glowers out at an adversary with hostility. His short hair gives him the brisk, menacing look of a professional athlete. He even seems to be balding; his plump curls, which are rendered with incision, are limited to the top and back of his head. His elongated proportions—especially his small head—and the crisp shifting of axes make him more elegant and sinuous than other examples of the type (see *LIMC*).

The combination of (relatively) slender elegance and emphatic plasticity seems especially close to works that have been dated to the Claudian period, such as the Basalt Doryphoros in Florence or Polykleitan bronze statuettes from Weissenburg, Germany, and Maladers, Switzerland (see Maderna-Lauter/Leibundgut-Maye). The fluid shifts of axis and soft modeling of surface recall some of the later decorative marble sculpture buried by Vesuvius, such as the satyr with flute from Pompeii (Naples, Museo Nazionale 6343: Wohlmayr), and raise the possibility of a Neronian date.

BIBLIOGRAPHY: Unpublished.

RELATED REFERENCES: *LIMC*, vol. 4, pp. 762–765, s.v. "Herakles" (O. Palagia). C. Maderna-Lauter and A. Leibundgut-Maye in *Polyklet: Der Bildhauer der griechischen Klassik*, exh. cat. (Liebieghaus, Frankfurt am Main, 1990), p. 333, fig. 202, nos. 199, 201. W. Wohlmayr, *Studien zur Idealplastik der Vesuvstädte* (Buchloe, 1991), pp. 67, 117, fig. 51, no. 60.

—J. H.

146

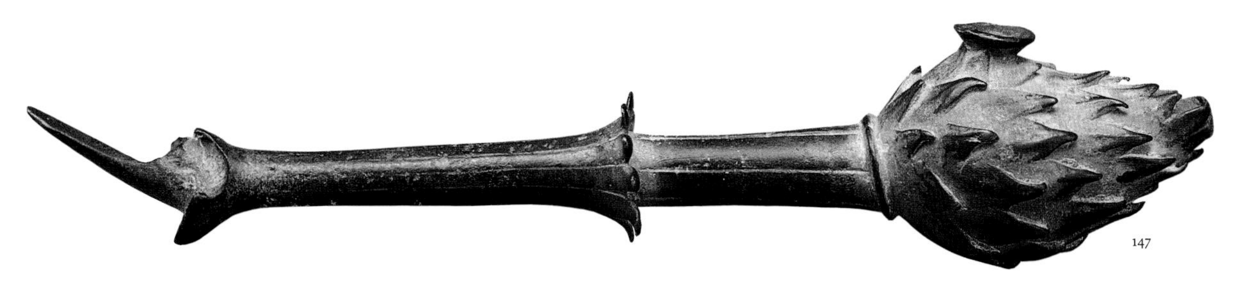
147

146
Weight in the Form of a Pugilist

Roman, probably first century A.D.
Bronze with silver and copper, lead fill;
H: 10.8 cm

The subject of this bust is an uncouth-looking man with a stubble haircut which forms an emphatic widow's peak on his forehead. At the back of his head he wears a *cirrus* (the long lock of hair that was the professional emblem of a boxer or wrestler). His brow is furrowed, his eyes bulging. The powerful bare chest dwindles to small, thin arms, indicating that the object was conceived as an excerpt rather than a complete figure. A loop in the top of the head makes clear its function as a steelyard weight.

Such images of pugilists were popular throughout the Roman period. A full-length figure in Cleveland is an unusually sophisticated Classicistic example (see M. True in *The Gods Delight,* p. 147ff.), but versions of the theme continued down to the end of antiquity. The workmanship of our piece, comparable to that seen in bronzes from the Vesuvian cities, suggests a date relatively early in Imperial times.

BIBLIOGRAPHY: Unpublished.

—A. H.

147
Lamp in the Form of a Thyrsos or Torch

Roman, first century A.D.
Bronze; L: 30.5 cm
Provenance: Said to be from the Talleyrand collection.

A fluted stalk ends in a lamp shaped like the tip of a thyrsos or, perhaps, the burst of flame at the end of a torch. At the other end is a spike for mounting the stalk to a wall attachment or possibly to a large, composite lighting fixture of some sort. A similar torch-shaped lamp, designed to project horizontally and with its tip in the form of a flame, comes from Pompeii (see Valenza Mele). Two versions of the same basic type, now in Vienna, have tips of definitely floral form. They come from a set of bronze objects discovered at Montorio near Verona in 1830 (see Gschwantler et al.; Beschi). The group has been ascribed to the second half of the second century A.D., but the lamps may, on the evidence of the Pompeian version, be earlier than other pieces in the find.

BIBLIOGRAPHY: Unpublished.

RELATED REFERENCES: N. Valenza Mele, *Museo Nazionale Archeologico di Napoli: Catalogo delle lucerne in bronzo* (Rome, 1981), no. 358 (with reference to an unpublished example in the Aquileia Museum). K. Gschwantler et al., *Guss+Form,* exh. cat. (Vienna, 1986), no. 208. L. Beschi, *I bronzetti romani di Montorio Veronese* (Venice, 1962), p. 102ff.

—A. H.

148
Statuette of a Striding Comic Actor

Roman, first century A.D.
Bronze; H: 7.1 cm

The little actor wears a bearded mask, close-fitting undergarments which are not grotesquely padded (as is usual in such figures, see cat. nos. 56–58, 59, 62, 64), and a himation wound round the hips and thrown over the left shoulder. He strides forward, looking boldly outward, while gripping his drapery with his left hand and making a sweeping rhetorical gesture with his right. It has been suggested (verbally to the author by J. R. Green) that he may be the comically irate father who is a stock character in many comedies. This vibrant miniature, simply conceived but full of energy, survives in other examples. One is in Lyon (see Boucher) and another was atop a lamp in the Passavant-Gontard collection (see Reinach). Parallels for the form of the Lyon

148

149

figure's base and of the Passavant-Gontard lamp can be found at the Vesuvian sites, suggesting an approximate date for the actor statuettes.

BIBLIOGRAPHY: Unpublished.

RELATED REFERENCES: S. Boucher, *Bronzes romains figurés du Musée des Beaux-Arts de Lyon* (Lyon, 1973), no. 183. S. Reinach, *Répertoire de la statuaire grecque et romaine,* vol. 6 (Paris, 1930), p. 182, no. 2 = R. von Passavant, *Sammlung R. von Passavant-Gontard* (Frankfurt, 1929), no. 50.

—A. H.

149
Statuette of a Gladiator

Roman, mid-first century A.D.
Bronze; H: 5.9 cm
Condition: Both feet missing; rather glossy brown patina.

The gladiator strides to the attack, with the cheek pieces or visor of his helmet closed to cover his face completely. His left arm is protected by a curved rectangular shield (*scutum*). His right arm is wrapped in straps (*manica*), and he holds a short sword. His waist is protected by a metal belt, under which is his loincloth (*subligaculum*). His thighs are protected with straps and his shins with greaves (*ocreae*).

Small bronze or terracotta figures of gladiators along these lines are a well-known category. They bear a variety of weapons that undoubtedly correspond to the different types of gladiators that battled against one another in the arenas of the Roman empire. Identification of these types is often difficult, as it is here. With his short, straight sword, his rectangular shield, and his helmet without crest, this figure could be a *myrmillo* (a type of gladiator of Gallic origin who had a fish as emblem); a similar bronze figure in Berlin has been so identified (see Pflug). On the other hand, he wears two shinguards, while the myrmillo customarily has one only, on his forward (left) leg. It is possible that he is a myrmillo whose armament has been modified for combat with a net fighter (*retiarius*); the second greave would protect him from the retiarius' long trident, and by omitting the helmet crest, a projection was eliminated that could catch in the retiarius' net. It is worth noting that the shield-fighter (*hoplomachus*) also wore two greaves. In an inscription from Brescia, a gladiator is designated as both a myrmillo and a hoplomachus (see Patroni; Garcia y Bellido; Coarelli). This figure, who combines the weaponry of both, could have had such a double designation.

The brim of the figure's helmet is still relatively level, and he looks out through eye-holes rather than through a network of openings. Judging by the most advanced gladiatorial helmets from Pompeii, both features would have been out of date by A.D. 79.

BIBLIOGRAPHY: Unpublished.

RELATED REFERENCES: H. Pflug in *Antike Helme* (Mainz, 1988), p. 369, fig. 4. G. Patroni, *Notizie degli scavi* (1907), pp. 720–721. A. Garcia y Bellido, "Lapidas funerarias de gladiatores de Hispania," *Archivo español de arqueologia* 33 (1960), pp. 123ff; F. Coarelli, "Il rilievo con scene gladiatorie, "*Studi miscellanei* 10 (1966), p. 91, n. 23. For figures of gladiators, see J. Ward Perkins et al., *Pompeii A.D. 79* (Boston, 1978), nos. 288–291; J. Petit, *Bronzes antiques de la collection Dutuit* (Paris, 1980), no. 77; C. Vermeule and M. Comstock, *Sculpture in Stone and Bronze* (Boston, 1988), nos. 122, 129.

—J. H.

150
Aryballos (Unguentarium)

Gallo-Roman, perhaps made at Anthée near Namur (Belgium), A.D. 70–100
Bronze with red and blue champlevé enamel;
H (WITHOUT HANDLE): 10.5 cm
Condition: Stopper and trunks of elephants on handle attachments missing; many losses in the enamel—particularly on handle; metal elements composing vase apparently reattached; ball made in four parts—bottom pentagon, two middle rows of pentagons, and top pentagon; neck, mouth, and handle attachments cast separately.

The vessel is shaped like an Archaic Greek aryballos, but the handle has a typically Roman Imperial arrangement; two animal protomes (here elephant heads) curve from the rim toward the body and on top of them are mounted a pair of rings, to which the handle is attached. Loops of wire hold the handle proper. The vase has no base—a virtually obligatory feature of Roman bronze aryballoi. This heavy

150

150, BOTTOM

vase could never have been carried by the weakly attached neck, mouth, or handle. These elements must have been detachable and merely for show; the entire vase may have been unfunctional.

The handle has a pattern of enameled squares flanked by triangles. The top of the rim is embellished with a row of alternately red and blue triangles and its side with a stylized laurel wreath. The body is covered with a network of twelve pentagonal panels, each of which has an inner border with a scroll, and a red central pentagon enclosing a roundel. The roundels contain either a bird, a rosette, or (in one case) a radiate crown of triangles. The birds are surrounded by stylized leaves, and the rosettes have four, five, or six petals. The edges of the pentagonal fields are serrated to increase the adherence between enamel and bronze. The vase's opening occupies the central roundel in the top pentagon; the neck covers some of the pentagon's enameling when in position.

This aryballos is clearly connected with a number of other bronze vessels with champlevé enameling that have been convincingly ascribed to northern Gaul. The use of pentagonal panels links it particularly to a locally excavated casserole in Bad Pyrmont in eastern Nordrhein-Westfalen (see Stupperich), a bowl from Rochefort (Jura, France) in the Metropolitan Museum of Art, and two bowls from the cemetery of La Plante in the Archaeological Museum at Namur, Belgium (see Forsyth). The bowls have been ascribed to a workshop that may have been at Anthée, also near Namur. They have been dated to the second century, but such a chronology may well be too late. A closely related enameled bronze bowl found in Braughing, Hertfordshire, has recently been dated to the first century on the strength of its connection with designs on contemporary South Gaulish *terra sigillata* (fine red-gloss pottery) (see Potter). The pentagonal patterns alone suggest that the casserole, the bowls, and the aryballos are relatively early; pentagonal networks were most common in the Hellenistic period, on hemispherical bowls and spherical inkwells, both silver and ceramic. Pentagons also decorate the body of a glass aryballos from Thebes dated to the first century A.D. (Athens, National Museum 2701: Weinberg). Further comparisons indicate a later first-century date for the Fleischman aryballos. A bronze pitcher embellished with a comparable network of geometric pattern—incised hexagons and diamonds filled with stylized flowers—has been found at Pompeii, buried in A.D. 79 (Naples, Museo Nazionale 118295: Spinazzola; Chiurazzi). Two pavements at Aquileia of about A.D. 66–100 have hexagons with very similar decoration—a band of wave pattern and a central roundel with a six-petaled rosette (see Donderer). The cold regularity of the mosaics has, however, been varied and enlivened in this colorful aryballos, which is, in any case, the most Classical and Mediterranean of the Gaulish enameled bronzes.

BIBLIOGRAPHY: Unpublished.

RELATED REFERENCES: R. Stupperich, *Römische Funde in Westfalen und Nordwest-Niedersachsen*, Boreas Beiheft 1 (Münster, 1980), pp. 42–43, fig. 2. W. Forsyth, "Provincial Roman Enamels Recently Acquired by the Metropolitan Museum of Art," *Art Bulletin* 32 (1950), pp. 302–303, figs. 5, 6. T. Potter, *Roman Britain* (Cambridge, Mass., 1983), p. 54, fig. 63. G. Davidson Weinberg, *Glass Vessels in Ancient Greece* (Athens, 1992), pp. 72, 125, 126, frontis., no. 97. V. Spinazzola, *Le arti decorative in Pompei e nel Museo Nazionale di Napoli* (Milan and Rome), fig. 276; for the provenance and number, see S. Chiurazzi, *Chiurazzi, riproduzioni di opere classiche in bronzo e marmo* (Naples, n.d.), no. 266. M. Donderer, *Die Chronologie der römischen Mosaiken in Venetien und Istrien bis zur Zeit der Antonine* (Berlin, 1986), Aquileia 93, 95. For Roman aryballoi in general, see F. Brommer, "Aryballoi aus Bronze," *Opus Nobile: Festschrift zum 60. Geburtstag von Ulf Jantzen* (Wiesbaden, 1969), pp. 17–23, pl. 4; J. Hayes, *Greek, Roman, and Related Metalware in the Royal Ontario Museum* (Toronto, 1984), pp. 92–93, no. 147; M. Lista in E. Pozzi et al., *Le Collezioni del Museo Nazionale di Napoli: I Mosaici, le pitture, gli oggetti di uso quotidiano, gli argenti, le terrecotte invetriate, i vetri, i cristalli, gli avori* (Rome, 1986), pp. 178–179, nos. 45, 48. For pentagonal forms on silver and ceramic bowls and inkwells, see D. von Bothmer, "A Greek and Roman Treasury," *Metropolitan Museum of Art Bulletin* 42 (Summer 1984), pp. 56–57, no. 96; H. Thompson, "Two Centuries of Hellenistic Pottery," *Hesperia* 3 (1934), pp. 381–383, 398, figs. 69, 86, nos. D 38, E 58; J. Schäfer, *Hellenistische Keramik aus Pergamon*, Pergamenische Forschungen 2 (Berlin, 1968), p. 117, pl. 51, G3; S. Rotroff, *Hellenistic Pottery: Athenian and Imported Moldmade Bowls*, vol. 22 of *The Athenian Agora* (Princeton, 1982), no. 403.

—J. H.

151
Situla with a Frieze of Athletic Contests

Gallo-Roman, late Flavian, late first century A.D.
Bronze; H: 9.5 cm; DIAM: 14 cm
Condition: Intact, though probably missing lid and handle; square hole later cut in bottom, as if for attachment of a foot or base; crusty brown and green patina.
Provenance: Formerly in the collection of Charles Patin.

The bowl is cast, with a frieze of nine figures in high relief, some parts standing completely clear of the background. A flattened row of stylized leaves (a *lesbian kymation*) ornaments the zone between the neck and the figured zone. Two rounded zones of damage at diametrically opposed positions on the kymation were in all probability made by the attachments for a handle. The inner edge of the rim has two rectangular notches, which undoubtedly served to secure a lid of some sort. A vase of this type dredged up in the Saône was fitted with a ladle that also served as a cover (see Bonnamour).

The reliefs show contests between nude athletes. A draped judge carries palm branches and a wreath in his right hand, while pushing a boxer into action with his left. The bearded boxer to the right is kicking as well as hitting. To their right is a large prize kantharos holding palm branches on a stand. To the right again, a beardless wrestler throws an adversary backward by pulling his right ankle and arm. To their right, a table with three panther-headed legs is heaped with wreaths. Beyond them is a pair of wrestlers—the one at the left wearing a protective cap. To their right is a basin with handles on a stand. Finally, a fourth pair of wrestlers are locked together, with one holding the other's thigh.

The piece is one of the most elaborate of a group of similar cast bronze containers covered with a variety of figured scenes in high relief. Many have a bucket handle, which probably also existed here. Some have been excavated in datable contexts; an example from Köshing must be later than A.D. 80, and others from Herstal and Tongres around the middle of the second century (see Faider-Feytmans 1979, pp. 25–26, nos. 347, 367, 374; Petit). The majority come from Gaul or the northern provinces, and Gaul has been proposed as the place of manufacture for at least some of them. A Gallic origin seems quite likely since their shape recalls some low, sub-spherical vessels of fine red-gloss pottery (*terra sigillata*) of Déchelette Form 67 produced in south Gaul during the late first century (see de la Bédoyère). The form also appears in northern provincial glassware of the first to early third centuries (see Isings). A particularly significant parallel is an example from Barlow Hills, Essex; not only is it decorated with the distinctively Gaulish technique of enameling, but the vine and laurel patterns on its body seem to be those of the first century (see British Museum).

In spite of a certain clumsiness in some details, the handsome profiles and the good basic anatomical structure of the figures suggest that this bucket/bowl is an early example of the type. The elaborate kymation on the shoulder is an unusual Classicistic touch; it has palmettes between the main leaves, reviving a form of the fourth century B.C. (see Yalouris et al.). The kymation still appears occasionally in Pompeii (see De Carolis). In spite of the rather shallow relief and confused treatment of detail, it is unlikely that such a kymation is later than the first century A.D.

BIBLIOGRAPHY: C. Patin, *Familiae Romanae in Antiquis Numismatibus* (Paris, 1663), detail of frontispiece [pointed out by Robert Guy]; B. de Monfaucon, *L'antiquité expliquée* (Paris, 1719–24), Suppl. 3, p. 161, pl. 68; Korban Gallery, London, advertisement, *Apollo,* February 1985, pp. 52–53.

RELATED REFERENCES: L. Bonnamour, "Vases en bronze d'époque romaine trouvés dans la Saône," *Actes du IVe Colloque International sur les bronze antiques* (Lyon, 1976), pp. 24, 28, fig. 3. J. Petit, *Bronzes antiques de la collection Dutuit* (Paris, 1980), nos. 79, 80. G. de la Bédoyère, *Samian Ware* (Aylesbury, 1988), pp. 14, 24, figs. 5, 17. C. Isings, *Roman Glass from Dated Finds* (Groningen, 1957), pp. 88, 111, types 67c, 94. British Museum, *Guide to the Antiquities of Roman Britain* (London, 1956), p. 56, no. c 1, pl. 21. N. Yalouris et al., *The Search for Alexander* (Boston, 1980), nos. 120, 127, 164. E. De Carolis in IBM, Gallery of Science and Art, and IBM-Italia, *Rediscovering Pompeii* (New York and Rome, 1990), no. 59.

—J. H.

151

151

152

153

152
Statuette of a Dwarf Pugilist

Roman, first–second century A.D.
Bronze; H: 5.4 cm

The robustly built little figure strides forward and puts up both fists. Bald-headed and bearded, he is nude except for *caesti* (boxing gloves) (see Poliakoff, p. 75). He has rather short legs and oversize genitals but is otherwise well proportioned. The earthy but not sensationalistic mood of the piece and the accurate, unpretentious small-scale modeling might be datable somewhere between Flavian and Antonine times. The first-century A.D. boxer with caesti in Baltimore (see Poliakoff, fig. 80) is Classicistic compared to our piece. A kicking pankratiast in the Louvre from Autun (1067: Poliakoff, fig. 60) represents a more comparable stylistic phase, though its brutality is less appealing than the jaunty energy of our game-looking figure.

BIBLIOGRAPHY: Unpublished.

RELATED REFERENCES: M. B. Poliakoff, *Combat Sports in the Ancient World* (New Haven, 1987).

—A. H.

153
Statuette of a Comic Actor

Roman, probably first–second century A.D.
Bronze; H: 6 cm

The chubby but dynamic-looking figure squats with buttocks projecting. With a derisive gesture, he sticks two fingers of his right hand into the corner of his mouth, probably to help produce a flatulent noise, while his left hand is thrust back to emphasize the appropriate action of his rear end. He wears a comic mask and a tight-fitting "body-suit" with a round, sweater-like neckline, covered with a fine crosshatched pattern. This costume has an anus indicated on the tights and is completed by a large, limp penis which dangles to the ground between his legs. A narrow scarf, decorated with incised stripes, is worn over his left shoulder.

Spontaneous though it may seem, the posture is a traditional one, on the evidence of a Ptolemaic faience dwarf in the same position (see Reeder), as well as a comparable figure in terracotta (see Ugaglia). A bronze lamp in the form of a similarly posed but even more outrageous figure, which, if ancient, is evidently a Roman exaggeration of the type, was once in the Sambon collection (see Reinach). A number of Ptolemaic terracotta grotesques tug at one or both corners of their mouths, usually thrusting their tongues out, in a rude facial contortion which can indeed produce an eructive noise (see Perdrizet, pl. CX, especially the three fragments in the top row).

Our figure, unlike these forerunners, is characterized as an actor wearing a mask. His attire also seems to have Egyptian connections. Close-fitting network costumes are worn by at least two Ptolemaic terracotta actors (see Perdrizet, pl. LXXXVII, top right, pl. LXXXVIII, bottom right). The

153

dress appears in late antiquity as the characteristic garb of the comic muse Thalia on sarcophagi (see Schöndorf). Clearly, however, it had been in actual use for a long time before becoming so canonical as to be ascribed to the Muse. The scarf over the shoulder is seen in a series of Ptolemaic statuettes that have been connected with the Lagynophoria, a bibulous Dionysiac festival (see Schürmann).

Representations of actors from the third century A.D. sometimes have a hatched or "quilted" pattern on their tights and sleeves (see Webster). The late antique parallels, which include the Thalia figures, have led some observers to suggest a date in the late second or the third century A.D. for our statuette. The crisply defined modeling does indeed seem to be of Roman rather than Hellenistic times. However, the shameless impropriety of attire and gesture would seem to have been out of place in the increasingly prudish atmosphere of late paganism. A date in the first or early second century A.D. seems preferable, especially now that the network body-suit can be shown to have Hellenistic origins, perhaps based on older Egyptian craft traditions (see *Egypt's Golden Age*).

BIBLIOGRAPHY: Unpublished.

RELATED REFERENCES: E. Reeder, *Hellenistic Art in the Walters Art Gallery* (Baltimore and Princeton, 1988), p. 200, no. 102. E. Ugaglia, *Art grec de la terre a l'image* (Toulouse, 1990), p. 94f., no. 97. S. Reinach, *Répertoire de la statuaire grecque et romaine,* vol. 5 (Paris, 1924), p. 302, no. 5. P. Perdrizet, *Les terres cuites grecques d'Égypte de la collection Fouquet* (Paris and Nancy, 1921). H. Schöndorf, "Zum Kostum der Thaleia auf den Musensarkophagen des 3. Jahrhunderts," *Archäologischer Anzeiger* (1980), p. 136ff. W. Schürmann, "Zur Deutung der Fransentücher im hellenistisch-römischen Ägypten," *Festschrift für Nikolaus Himmelmann* (Mainz, 1989), p. 297ff. T. B. L. Webster, *Monuments Illustrating New Comedy,* 2nd ed. (London, 1969), p. 62f, no. AT 31. *Egypt's Golden Age,* exh. cat. (Boston, 1982), no. 199 (a loincloth and other pierced leather garments of the New Kingdom [D. Nord]); no. 209 (net clothing made of linen threads [L. Salmon]).

—A. H.

154

154

Lamp in the Form of a Comic Mask

Roman, late first–early second century A.D.
Bronze; H: 6.9 cm; L: 12.5 cm
Condition: Handle detached and the part that joined the body missing.

The lamp, the body of which has the form of a comic mask set on a low turned foot, is cast and of substantial weight. The nozzle emerging below the beard of the mask has a tip of almost triangular form when seen from above and is surrounded by a flat raised border; the space on the top of the nozzle between the lamp body and the small, round mouth is decorated with a pointed leaf. The mask type is that of the "Leading Slave," a sly and resourceful stock character in numerous comedies. His open, upturned mouth, fringed with a finely incised semicircular beard, serves as the filling hole. Most of the hair is covered with a kerchief, from which short tassels of corkscrew curls emerge at the sides. The headdress is crowned with a wreath of ivy leaves and berries shown in high relief, which continues below to adorn the zone where the handle was attached. A fragment of the upswept loop handle, set low on the body and of stylized vine-form with a turned-back leaf to serve as a thumb rest, survives.

Our piece's cast high-relief foliate decoration, like its low-set vine-loop handle, finds loose parallels in objects from the Vesuvian cities (see Valenza Mele, nos. 58, 61, 121–124); the nozzle shape is most like that of Loeschke's Type XX, well represented at Pompeii and Herculaneum (see Valenza Mele, p. 91ff.). However, the analogies are not truly close, and the forms seem to have had a long life. The combination of precise detail with powerful, even flamboyant modeling suggests that our lamp was made in the period of maximum Roman artistic attainment and confidence, probably in Flavian through Hadrianic times. In style the mask seems to represent a development beyond those on mid-first-century reliefs from the Vesuvian cities (see Spinazzola), in the direction of the colossal marble masks from Hadrian's Villa (see Giuliano).

BIBLIOGRAPHY: Unpublished.

RELATED REFERENCES: N. Valenza Mele, *Museo Nazionale Archeologico di Napoli: Catalogo delle lucerne in bronzo* (Rome, 1981). V. Spinazzola, *Le arti decorative in Pompeii e nel Museo Nazionale di Napoli* (Milan and Rome, 1928), pls. 52, 54–57. A. Giuliano, ed., *Museo Nazionale Romano,* vol. 1, pt. 2, *Le sculture* (Rome, 1981), p. 189f., nos. 4, 44.

—A. H.

155

Statuette of Mars Ultor

Roman, late first–second century A.D.
Bronze; H: 10.3 cm
Condition: Right foot missing.
Provenance: Formerly in the collection of James Coats.

The statuette is a miniature version of the colossal statue of Mars from the Forum of Nerva, which is preserved today in the Capitoline Museum. Small reproductions of this statue are very common and come from all over the Roman world (see Reinach). Many may have served as private devotional images or as votive offerings to the god of war. Most are of a scale appropriate for display in the *lararium* (household shrine). The number would be even greater if one were to include the simplified examples that omit the military cloak, and the many adaptations and variants. The huge Flavian sculpture probably inspired the ancient popularity of the type, even if it in turn reflected a less flamboyant cult image set up by Augustus. Most of the reproductions must have been made when the sensation caused by the statue was still fresh. Our small-scale example, like several others, has its legs fully preserved and shows the detail of the greaves with their thunderbolt reliefs, which is of interest because the lower legs of the Capitoline statue are restorations.

BIBLIOGRAPHY: Unpublished.

RELATED REFERENCES: S. Reinach, *Répertoire de la statuaire grecque et romaine,* vol. 2 (Paris, 1898), p. 189, nos. 5, 6, 8; p. 190, nos. 2, 9; p. 793, nos. 3–6; vol. 3 (1904), p. 244, no. 6;

155

vol. 4 (1913), p. 108, nos. 1, 2, 7; vol. 5 (1924), p. 267, no. 1; p. 510, no. 4 (for miniature versions). For the Capitoline statue and a relief reflecting its more restrained Augustan forerunner, see H. G. Martin in *Kaiser Augustus und die verlorene Republik*, exh. cat. (Berlin, 1988), p. 256f., figs. 151, 152.

—A. H.

156
Statuettes of Asklepios and Hygeia

Roman, probably first half of second century A.D., after Hellenistic models
Bronze; H (EACH): 16.5 cm

The bearded Asklepios, god of healing, wears an ample himation over his bare torso. This garment is wound around his waist, then tucked up under the right arm before being drawn across the back to be wrapped forward around the left shoulder and arm. It drapes with generous elegance in back and forms a cascade of zigzag folds at the left side.

Our piece belongs to a series known as the Albani type after an example from the Albani collection (see *LIMC*, vol. 2). At first glance it resembles the most famous Classical Asklepios image, the Giustini type (see *LIMC*, vol. 2), which has an identical pose and garment but telling differences in the arrangement of the drapery. The himation of the Giustini type, with its top folded over horizontally to form a sort of shelf, is pulled across the torso and directly onto the left arm, unifying and immobilizing the figure. In our piece and others of the Albani series, the himation is wound lower around the body, its deeper, diagonal fold-over form-

156

156

ing an apronlike triangle, and the drapery over the left shoulder, which comes from behind, is not a continuation of that wrapping the waist. The arrangement frees the arm from the torso and allows a hint of twisting movement, like that of half-draped Hellenistic male statues. The way the himation of our figure is tucked up into the right armpit creates a visually powerful, if in real life inadequately secured, festoon across the back. Ornate openwork shoes of the type favored for imposing figures in Pergamenian sculpture add to the baroque feeling. A statue in the Louvre (*LIMC*, vol. 2, no. 257) is an almost exact large-scale parallel to our piece, repeating the way the chiton is tucked into the armpit and even the position of the serpent; other versions are looser adaptations of the same type.

The statuette of Asklepios' daughter Hygeia (Health) is of the same scale and workmanship; the two were clearly made as a pair. The dignified young goddess wears heavy, many-layered drapery. Around her arms is twined one of Asklepios' healing serpents, whose head she holds confidingly close to her face. The image belongs to a series known as the Belvedere type (see *LIMC*, vol. 4).

High Hellenistic influence might be seen in the complex layering of the drapery and the way the himation is twisted into a bulky roll at the top. The full but refined features of the heart-shaped face and the piled-up, bow-knotted hair with artfully straying wisps also hint at a late third- or early second-century model. A marble version of the figure in Berlin retains its original head, which reflects, in a cold, classicizing interpretation, the same model as the bronze statuette.

The Asklepios and Hygeia are examples of widespread types. Although traditional in conception, they represent distinctively Hellenistic variations on Classical themes. It is tempting to see them as ultimately derived from sculptures made for Pergamon or Kos, where famous sanctuaries to Asklepios flourished in the Hellenistic period. Small images of the healing gods must frequently have been votive offerings or objects of household devotion for people acknowledging or hoping for cures from illness as well as for physicians. Although our beautifully modeled statuettes are of far higher quality than the typical somewhat banal small figures of Asklepios and Hygeia, they may have served a similar purpose. Unusually literal in their reproduction of large-scale statues and modeled with academic precision, they are perhaps to be ascribed to the time of Hadrian.

BIBLIOGRAPHY: Unpublished.

RELATED REFERENCES: *LIMC*, vol. 2, nos. 257–260 (Albani type), p. 894f., nos. 154–233 (Guistini type), s.v. "Asklepios" (B. Holtzmann). *LIMC*, vol. 5, nos. 191–194 (Belvedere type), s.v. "Hygeia" (F. Croissant). For a related standing Asklepios type perhaps to be connected with the Asklepieion of Pergamon, see P. Kranz in *Festschrift für Nikolaus Himmelman* (Mainz, 1989), p. 289ff. For an example of the Belvedere Hygeia and discussion of the type, see A. Linfert, *Die antiken Skulpturen des Musée Municipal von Château-Gontier* (Mainz, 1992), no. 8.

—A. H.

156, ASKLEPIOS

156, HYGEIA

157

157
Statuette of the Tyche of Antioch

Roman, second century A.D., after an early Hellenistic model
Bronze; H: 12 cm
Condition: Right hand and accessory figure of the River Orontes missing.

The draped goddess sits, cross-legged and leaning forward, on a high outcropping, her left hand braced on the rock behind her and her right hand resting on her knee. Over a finely pleated chiton, the upper half of her body is wrapped in a himation with visible press-folds, drawn up to veil the back of her head. On her feet are sandals with thick double soles.

The piece is a reduced copy of the colossal statue made by Lysippos' follower Eutychides for Antioch on the Orontes soon after the city's founding in 300 B.C. Of the significant features known from other versions, only the little swimming figure of the river Orontes, emerging under Antiocheia's rocky seat, is missing. The statuette is among the better of at least fourteen small bronze versions. Most appear to have been made in the same workshop, or rather, since they do vary as to quality, the same ambient. The production seems to be datable no earlier than the second century A.D. (see Balty in *LIMC*, p. 843). Among the bronze statuettes very similar to ours are examples in a German private collection, the Metropolitan Museum of Art, the Bibliothèque Nationale, Paris, and Turin (see *LIMC*, nos. 7, 10, 11, 13). The statue is also reflected in a late silver version, on gems, coins, and lamps, and even on mold-blown glass bottles. Sometimes it appears with secondary figures or under a baldicchino.

Like other large public monuments, the Antiocheia evidently did not lend itself to full-size reproduction. The statue is, however, reflected in a few small stone versions, of which the headless marble example in Budapest (see *LIMC*, no. 19) is the most informative. It agrees with the bronzes in having a forward-leaning posture and a voluminous, pleated underdress. A marble statuette in the Vatican (*LIMC*, no. 27) with more upright posture and smooth, rubbery-looking drapery, long the best-known version of the type, is no longer considered an accurate reflection of the original.

Small bronzes like the present carefully modeled example are valuable evidence as to the appearance of the head, missing in the larger copies but also reflected in intaglios (see *LIMC*, nos. 124–126). The hair, parted in the middle, is rolled up and back in wings from the sides of the rather fleshy face, to be gathered in a large, low chignon in back; crescent-shaped curls are brushed forward on the cheeks. The mural crown is narrow-based and tall, with projecting towers. Like the drapery style and the design of the shoes, the head has an almost High Hellenistic flavor which seems to reflect neither the early third century B.C., when the original is known to have been created, nor the Roman period, when our statuette was made.

BIBLIOGRAPHY: Unpublished.

RELATED REFERENCES: *LIMC*, vol. 1, p. 840ff., s.v. "Antiocheia" (J. C. Balty). T. Dorhn, *Die Tyche von Antiocheia* (Berlin, 1960).

—A. H.

158
Statuette of Jupiter

Roman, second century A.D.
Bronze; H: 30.5 cm
Condition: Left hand and the fingers, except for third, of right hand missing; nose tip and penis scuffed; deep scrapes and dents in several places, especially across the abdomen, left thigh, shoulder blades, and buttocks; deep hollow in top of left foot; ancient rectangular patches used to repair numerous casting flaws; bottoms of the feet without tangs; glossy blue-green patina; brassy gleams in some places.
Provenance: Found in Beauvais.

The god's pose, with its upward-spiraling rhythm and its implied support outside the body, is based on innovations of the fourth century B.C. Many reflections of the same Jupiter type, which was especially favored for large, ambitious statuettes, survive. Complete versions carry a thunderbolt in the right hand. The model for the type has been identified as the Jupiter Tonans (Thunderer), or Zeus Brontaios, by the fourth-century sculptor Leochares, which had been brought to Rome and stood in the temple of Jupiter on the Capitoline (see Donnay). The type and its variants are well represented among the Roman bronzes found in Gaul (see Boucher 1976, p. 67ff.).

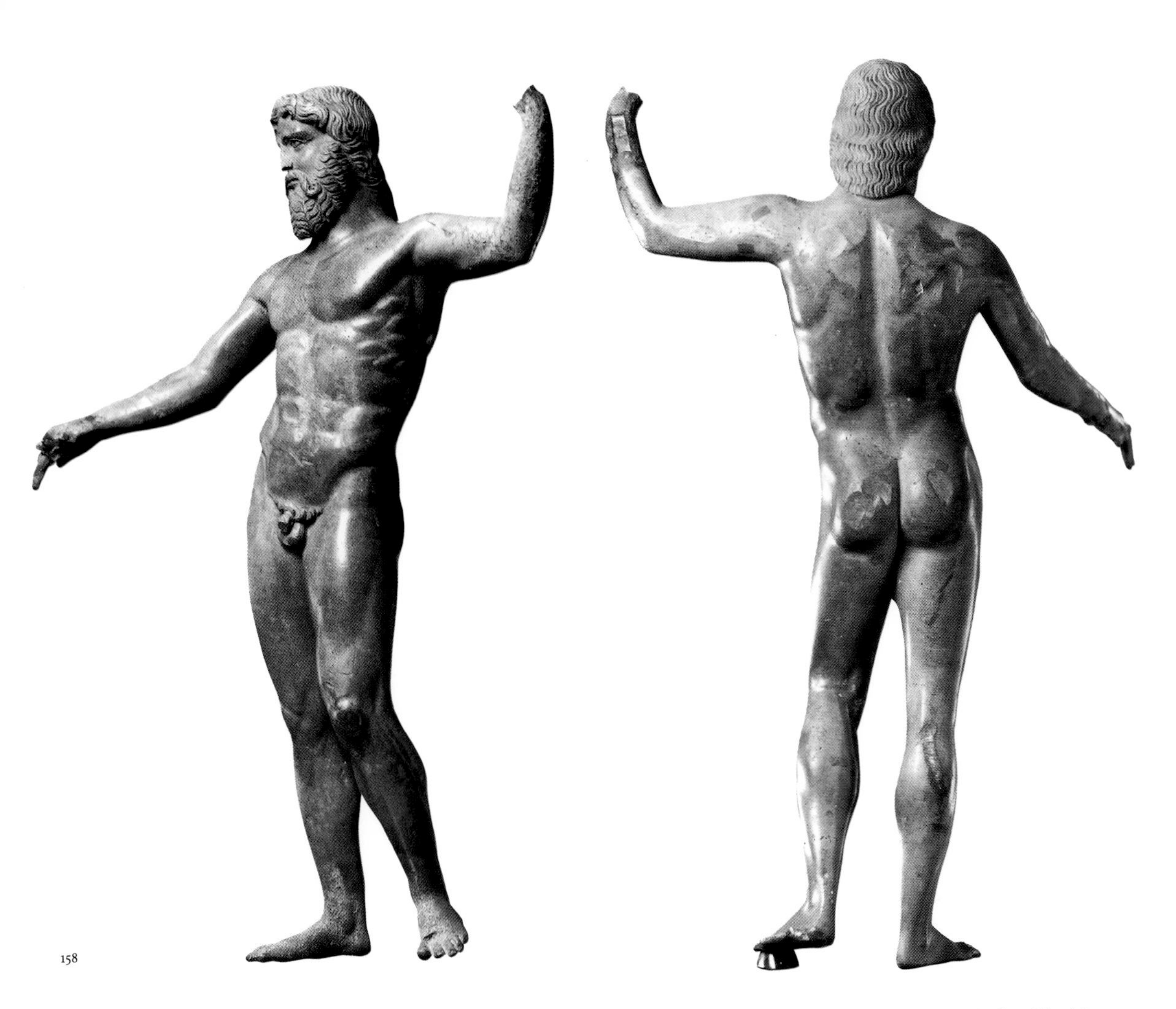

158

Large, elegant statuettes like ours, however, were often produced by sculptors less concerned with accurate reproduction of a famous prototype than with the creation of an impressive devotional image, sometimes combining features from a variety of sources (see Menzel).

The present piece has both a more baroque swing to its pose and more explicit retro-stylization of detail than typical members of the series. The slender, hard anatomy, with sharp divisions of the abdomen, long thighs, sinewy calves, and emphatic kneecaps, shows a surprisingly direct awareness of works in the style of the Riace Warrior A. Other features go beyond early Classical style to evoke Hellenistic conventions.

The head of our piece is unlike those, vaguely fourth century in inspiration, seen on most statuettes of the series and is ultimately derived from the fifth-century Dresden Zeus type (see Vierneisel-Schlörb). This is clear from the frame of center-parted locks on the forehead and the way the short whiskers at the sides of the beard stand out as a kind of ruff to meet it. The knitted brows and the horizontal "Michelangelo bar" of the forehead, however, have later sources. The stylization of the beard in corkscrew curls is seen in the Jupiter of Brée and another elegant

159

eclectic Jupiter in Florence (see Boucher 1976, fig. 148), two statuettes quite different from ours in compositional type. The long mustache with rolled-up ends is an Archaistic feature, as is the mass of hair with soft horizontal waves underlying fine vertical striations (see cat. nos. 139a, 179). Our piece makes use of borrowings from various sources, some of them quite rarefied, with amazing confidence and panache. The hard, defined physique and linear rendering of detail might speak for a Hadrianic or slightly later date.

BIBLIOGRAPHY: Pescheteau-Badin-Ferrin, Commissaires-Priseurs associés, Paris (J. Roudillon, expert), *Antiques*, Drouot-Richelieu, auct. cat., May 9, 1990, lot 23.

RELATED REFERENCES: G. Donnay, "Le Iuppiter Tonans du Capitole romain et ses imitations dans les bronzes figurés," *Toreutik und figürliche Bronzen römischer Zeit: Akten der 6. Tagung über antike Bronzen, 13–17 Mai, 1980* (Berlin, 1984), p. 107ff. H. Menzel, "Die Jupiterstatuetten von Brée, Evreux, und Dalheim und verwandte Bronzen," *Toreutik und figürliche Bronzen*, op. cit., p. 186ff. B. Vierneisel-Schlörb, *Glyptothek München: Katalog der Skulpturen*, vol. 2, *Klassische Skulpturen des 5. und 4. Jahrhunderts v. Chr.* (Munich, 1979), p. 147ff.

—A. H.

159
Rotary Key with Horse-Head Handle

Roman, circa second century A.D.
Bronze handle with iron shaft and bit; L: 15.6 cm; H: 5.5 cm; DEPTH: 2.7 cm
Provenance: Said to be from Germany.

The horse's head extends straight from the neck. Its ears are flattened against the poll, its forelock and pulled mane swept back as on a steed in racing gallop or a stallion about to strike. Three strands of forelock are incised on the forehead. The horse's neck emerges from a four-petaled calyx bound at the bottom by a torus molding which clamps onto a hollow iron shaft terminating in an elaborate, eight-slotted bit. It probably opened a lever lock, this type being according to W. H. Manning, "the most advanced form . . . used in the Roman period and in its principle . . . basically the same as the modern lever-lock."

Bronze horse heads appear fairly frequently as key handles, but not as often as lions, panthers, and bears among the published northern European collections. This key may be the only complete example of its specific type. The iron shafts on other examples cited below are heavily corroded or missing. The Fleischman key is remarkable in every aspect: the high quality modeling of the horse head, the intricacy of the key bit, and the excellent preservation of the details.

This type of key was manufactured by enclosing the end of the pre-forged iron shaft in a wax model of the proposed bronze handle, encasing the wax in a clay matrix, and then casting the bronze into the mold and thus around the end of the iron shaft, forming an excellent bond between the two metals.

Of nine parallels known to this author, the Metropolitan Museum of Art's horse-head key handle (Rogers Fund 1906, 06.176.24: unpublished) is among the closest in style and quality. Four similar objects (identified as knife handles) in relatively worn condition are in Bonn (see Menzel), one is in Lyon (H.1819: Boucher), another in Belgium at the Musée de Mariemont

(3695: Faider-Feytmans 1957), and two have been recorded in Belgium but are now lost (see Faider-Feytmans 1957; Faider-Feytmans 1979). A similar bronze indeed used as a knife handle was found in Crutched Friars, London (London Museum A 11872: *London in Roman Times*).

BIBLIOGRAPHY: Unpublished.

RELATED REFERENCES: W. H. Manning, *Catalogue of the Romano-British Iron Tools, Fittings and Weapons in the British Museum* (London, 1985), p. 94. H. Menzel, *Die Römischen Bronzen aus Deutschland,* vol. 3, *Bonn* (Mainz, 1986), pp. 113–115, nos. 264, 266, 269, 272, pls. 117–118 (inv. nos. 13718 [from Bonn]; 7132 [from Neuss]; U 1268 [prov. unknown]; 11007 [from Bonn]). S. Boucher, *Bronzes romains figurés du Musée des Beaux-Arts de Lyon* (Lyon, 1973), p. 150, no. 242. G. Faider-Feytmans, *Recueil des bronzes de Bavai,* Gallia Suppl. 8 (Paris, 1957), p. 109, no. 258, pl. XLII (Mariemont); p. 109, no. 259, pl. XLII (found at Bavai in 1818, now lost). G. Faider-Feytmans, *Les Bronzes romains de Belgique* (Mainz, 1979), no. 272, pl. 104 (found in environs of Tongeren before 1836, now lost). *London in Roman Times,* London Museum Catalogues, no. 3 (London, 1930), p. 78, fig. 19.3.

—A. K.

160
Disk-shaped Mina Weight

Roman Imperial (from the Levant), second century A.D.
Lead; DIAM: 12 cm; DEPTH: 0.8–0.9 cm; WEIGHT: 885 g (HEAVY MINA)
Condition: Intact; glossy dark patina.

The piece has a beaded edge and, on side A, a wide border inscribed ZHNOBIOY (of Zenobias) in evenly spaced relief letters. Within this, to the left, is a front-facing bearded herm, probably of Herakles; in the center, crossed cornucopias; to the right, an ear of wheat.

160

Above, in relief letters: ΕΤΟΥΣΔ (year 4) and ΔΗΜΟΣΙΑ (public), and across the bottom, MNA (mina). Side B is plain except for a raised circle with three concentric rings.

The previous attribution of the weight to Palmyra rests on a misinterpretation of the name in the border as referring to Zenobia, Queen of Palmyra. Instead it is the male name Zenobios and must refer to an official responsible for weights and measures. Such a reference, which is usually combined with the title *agoranomos*, points to a period after the end of the Seleucid empire, roughly the first century B.C. to the second century A.D., when economic conditions in the Levant called for an organized, controlled course of business.

Our piece has an excellent pedigree, but its weight was erroneously indicated in the sale catalogue as 695.6 grams. The correct weight, in fact, is 885 grams, that is, the heavy mina, current in the region and period. That this was the weight of the heavy mina is confirmed by a half mina of 418 grams, labeled HMIMNAION that names a certain Diogenes as agoranomos (see Qedar).

The round form, the large raised border displaying the official's name, and the carefully conceived design and script combine to make this an outstanding specimen.

BIBLIOGRAPHY: J. Hirsch, Munich, sale 24, collection of E. F. Weber, 1909, no. 3444 (listed as weighing 695.6 grams and "Aus Kleinasien"); Hesperia Arts, *Auction-Antiquities*, part 1 (New York, 1990), no. 41; R. Göbl, *Antike Numismatik*, vol. 2 (Munich, 1978), p. 138, no. 225, pl. 21; P. Gatier, "Poids inscrits de la Syrie hellénistique et romaine (1)," *Syria* 68 (1991), p. 435, pl. 1a, b (attributes our weight and three similar ones to Seleucia Pieriae).

RELATED REFERENCES: For the Diogenes-half mina, see S. Qedar in Münzzentrum, Cologne, *Gewichte aus drei Jahrtausenden*, vol. 4, no. 5103.

—L. M.

161
Statuette of Mercury

Gallo-Roman, A.D. 120–140
Bronze with silver and copper, H: 15 cm
Condition: Wing at right side of head, caduceus missing; smooth golden patina.

The god of wealth reaches forward to proffer a moneybag placed on the palm of his right hand. His lowered left hand would have held a caduceus, his herald's staff. Wings in his hair mark him as the speedy messenger of the gods. His hair is bound tightly on his small cranium by a fillet, but his face is surrounded by a mane of dense curls. His eyes are inlaid with silver and his nipples with copper.

A nude, curly-haired Mercury shown in exactly this pose and with just these attributes is a very popular Gaulish type—one which hardly ever reached the shores of the Mediterranean (see Boucher 1976, pp. 89–91, map 13, pls. 159–168). The pose itself is a highly eclectic compilation of Classical and Hellenistic Greek elements, but in this figure the torso is given a particularly strongly Polykleitan flavor. This reference does not necessarily reflect any Polykleitan elements in some "archetype," but instead is a re-styling in keeping with earlier Imperial classicizing fashion.

The elaborate anatomical definition and the swagger of this statuette recall Flavian figures like the Herakles from Herculaneum in the Louvre (see Wohlmayr). Details like the rippling surfaces of the back and rib cage also have a Flavian richness. On the other hand, the forward surfaces of the torso are relatively broad and simplified and thus seem closer to a Trajanic or Hadrianic approach. The rich treatment of the curly hair evokes late Hadrianic or even Antonine portraiture. Similar corkscrewing curls are to be seen in the over life-size bronze bust of Hadrian found in the Thames (see Wegner) and a miniature bronze bust of an Antonine general in Wels, Austria (see Jucker; Fleischer). A very similar Mercury in Paris has been dated by Annelis Leibundgut to the second century (see *Polyklet*). Such statuettes of Mercury seem to have been used primarily in household shrines (*lararia*) (see Ward Perkins et al.).

BIBLIOGRAPHY: Unpublished.

RELATED REFERENCES: W. Wohlmayr, *Studien zur Idealplastik der Vesuvstädte* (Buchloe, 1991), pp. 67, 117, fig. 77, no. 60. M. Wegner, *Hadrian, Plotina, Marciana, Matidia, Sabina* (Berlin, 1956), p. 101, pl. 30c. H. Jucker, *Das Bildnis im Blätterkelch* (Lausanne and Freiburg I. Br., 1961), pp. 58–59, pl. 17. R. Fleischer, *Die römische Bronzen aus Österreich* (Mainz am Rhein, 1967), no. 223. *Polyklet: Der Bildhauer der griechischen Klassik*, exh. cat. (Liebieghaus, Frankfurt am Main, 1990), no. 197. J. B. Ward Perkins et al., *Pompeii A.D. 79*, exh. cat. (Boston, 1978), pp. 63–64, 81, 191, no. 216.

—J. H.

161

162

Statuette of Mars/ Cobannus

Gallo-Roman, about A.D. 125–175
Bronze; H (WITH BASE): 76 cm; H (OF FIGURE): 65 cm
Condition: Weapons, part of visor, separately cast elements of the coiffure and helmet around the face missing; many holes and casting flaws; iron strut runs through inside of head from side to side at ear level; smooth patina, brown with golden undertones.

The sculpture as preserved is assembled from seven pieces (head, helmet, body, hands with parts of forearms, legs). Beneath the cloak, the left side of the figure's torso is missing; it may have been filled in with another piece of bronze or with a material such as wax or pitch. There is a flat zone with lightly incised hair between the curls over the forehead and the front edge of the helmet. Holes in the helmet rim over the ears were probably for the attachment of the now-missing cheek pieces. The area where the ears should be is blank, and on the left side there is a pin hole; the cheek pieces probably covered this area.

The base bears the dedicatory inscription: AVG[vsto] SACR[vm] DEO COBANNO / L[vcivs] MACCIVS AETERNVS / IIVIR EX VOTO (Sacred to the venerable god Cobannus, Lucius Maccius Aeternus, duumvir [dedicated this] in accordance with a vow).

The Maccii are known epigraphically from a few inscriptions in Gaul. This individual must have belonged to the highest levels of provincial society since a duumvir was one of the two chief magistrates of a Roman colony. His proper Roman three-part name may be an aspect of his status.

The figure is remarkable in wearing a contemporary Roman helmet rather than the classicizing Greek types customary in Roman works of art. The helmet corresponds to the Niederbieber type; that is, it has a smooth, deep back culminating in an angled or horizontal neck-guard, protective flanges around the ears, and a "visor" plate above the forehead. The front edge of the helmet, moreover, is tilted up. The Niederbieber type has recently been dated generically to the second and first half of the third centuries (see Waurick, pp. 341, 357, 359, figs. 6, 7). In favor of a relatively early date for this example are the low reinforcing bands crossing on top of the helmet, which are indicated only by incised lines. The ornamentalized rivet crowning the helmet, furthermore, has two angular steps and an upper hemisphere, as does the helmet from Niedermörmter, which belongs to the earlier Weisenau type but must be later than A.D. 119 (see Waurick, pp. 336–337, 357, fig. 5.4).

The warrior is warmly dressed in a long cloak fastened with a round brooch that has the stepped form of the helmet's crowning rivet. Under the cloak, the front corner of which is weighted to insure a graceful flow of material, the warrior wears a long-sleeved tunic with fringed lower edge and close-fitting, footed leggings, over which he wears T-strap sandals. A standing spear was originally supported with the right hand and a shield steadied with the left. The warrior has youthful, idealized features and long wavy hair that springs up above his forehead in an *anastole*—a fashion inspired by Alexander the Great.

The figure undoubtedly represents Cobannus—a Gaulish god obviously identified with Mars—rather than either a generic soldier or the duumvir who dedicated the statuette. This is the first known inscription to Cobannus, but such identifications of obscure local Celtic divinities with Mars are, in fact, common and are an important manifestation of the cult of Mars in the northern provinces (see *LIMC*, pp. 567–580).

The composition is not based on a standard type of Mars. The pose with grounded shield and spear is typical of Mars Ultor, but that divinity is bearded and wears Classical armor (see *LIMC*, pp. 515–517, 554–555, pls. 384–387, nos. 24–49). This Cobannus seems to be a Gallo-Roman creation. In addition to arming him with a contemporary legionary's helmet, he is given northern dress—specifically, the long-sleeved tunic and long leggings (*braccae*). This costume was employed for other Gallic divinities. The most conspicuous application was for Succellus or Dispater, the Celtic Zeus (see Boucher 1976, pp. 164–170, figs. 299, 300, 306–309). The costume is also used for the Enarabus/Mars in Bastogne (Luxembourg), and there

162

162

are several other analogies with this Cobannus as well; Enarabus is a war-god, he has long wavy hair and his pose is the same (see Faider-Feytmans 1979, no. 52). It should be noted that long sleeves and long trousers are worn by the Dacian enemies of Rome on the Column of Trajan, while Roman soldiers wore short sleeves and leggings that stopped just below the knees (cf. Webster). Thus both Enarabus and this Cobannus are distinctly barbaric versions of Mars.

Cobannus, on the other hand, is far from barbaric in his body language; his expression is mild and benign—even languid—and his pose is unusually fluid and buoyant. This impression of lightness and mobility was probably intended to express the god's dynamism, inspiration, and superiority, following a Greek tradition going back to the fourth century B.C.

The workmanship of the figure is simple, large-scale yet conscientious. The elongated oval face, the slender lower jaw, and the long undulating locks of hair recall the Castor from the late Julio-Claudian (?) hoard of bronzes found at Paramythia in Epirus (see Swaddling; Herrmann). The sculptor of this Cobannus must have emerged from the same Classicistic Mediterranean environment, but his cultivated talents were applied in progressive, fully Roman directions. The boldness and simplicity of the workmanship may reflect the Hadrianic or Antonine date suggested by the helmet.

This entry was written with the benefit of important observations by David Cahn, Ariel Herrmann, and R. R. R. Smith.

BIBLIOGRAPHY: Unpublished.

RELATED REFERENCES: G. Waurick, "Römische Helme," in *Antike Helme* (Mainz, 1988), pp. 341, 357, 359, figs. 6, 7. *LIMC,* vol. 2, pp. 517ff., s.v. "Ares/Mars" (E. Simon). G. Webster, *The Roman Imperial Army* (London, 1969), p. 123, pl. 12. J. Swaddling, "The British Museum Bronze Hoard from Paramythia, North Western Greece: Classical Trends Revived in the 2nd and 18th Centuries A.D.," in S. Boucher, ed., *Bronzes hellénistiques et romains: tradition et renouveau,* Cahiers d'Archéologie Romande 17 (1979), pp. 103–105, pl. 50.3. J. Herrmann in *The Gods Delight,* pp. 348–349.

—J. H./A.v.d.H.

162

163

Offering Box

Gallo-Roman, second century A.D., perhaps 130–180
Bronze; H: 51 cm
Condition: Complete except for missing figure on top and some sections of cable molding that borders panels and masks junctures.

The hexagonal structure has a cast base resting on three feline paws. The main body is assembled on the base from flat panels outlined by a half-round molding with incised cable pattern, which masks the joins. At the bottom of one side is a low door, whose key is still in place. Above the door is an ancient repair with a reinforcing plate riveted to the inside. Above a cast molding is a roof with six concave sides and incised imbrications. Incised cable pattern again separates the sides, and spheres ornament the corners. There is a small round hole in the middle of the side of the roof over the door. At the top, there is an elaborately profiled platform with a swelling S-curve molding (*kymation*) and a kind of angularized cavetto. The upper corners of the platform are again embellished with spheres. On the platform are the footmarks of a now-missing statuette, which stood at right angles to the side with the door. Between the footmarks is a coin slot.

The container is said to have been found with the statuette of Cobannus in the Fleischman collection (cat. no. 162), and it could have been intended for the deposit of written vows to the god, as has been suggested by Dyfri Williams (note in collection files). Alternatively, it might have been an offer-

163

TOP 163

ing box in a shrine to the god, and this explanation seems more likely since the damage above the door suggests an ancient attempt to steal valuables.

The form of the container mimics large-scale architecture—particularly tomb buildings. Roofs with concave sides—often imbricated and usually culminating in a capital—seem to have been especially popular throughout the Empire during Augustan and earlier Julio-Claudian times. In the northern provinces, such roofs appear as early as about A.D. 40 at the mausoleum of L. Poblicius in Cologne, but they still can be found in the mausoleum of the Secundini at Igel of about A.D. 245 (see *Enciclopedia classica;* Horn et al.). The hexagonal form of the bronze container is, on the other hand, non-architectural; it is hard to find hexagonal rooms or mausolea before the late third century. Hexagonal shapes are, on the other hand, popular in mosaic pavements throughout the Imperial period and in bronze working. Numerous bronze bottles and pyxides have this shape in the second and third century (see cat. no. 165 below), and pedestals for statuettes occasionally are octagonal—perhaps as early as the first century A.D. (see Faider-Feytmans 1979, no. 52 [Enarabus]; Boucher). Hexagonal pedestals are apparently not found at Pompeii, and thus this container is probably later than A.D. 79. The bottles and pyxides, it might be noted, lack the forceful moldings of this container. The profiled pedestal topping the roof seems like a highly articulated version of the pedestals topping the legs of tripods, as in the one in the Fleischman collection (see cat. no. 167 below; Manfrini-Aragno).

If there is a connection with the figure of Cobannus, the container may be more or less contemporary with it. The decorative spheres on the container, however, have a less logical quality than the profiled rivets on Cobannus' helmet, and the two were probably produced in different workshops.

BIBLIOGRAPHY: Unpublished.

RELATED REFERENCES: *Enciclopedia classica* 3, vol. 12, pt. 1 (Turin, 1959), pp. 256, 564, figs. 287–289, 748, s.v. "architettura romana" (L. Crema). H. G. Horn et al., *Die Römer in Nordrhein-Westfalen* (Stuttgart, 1987), p. 245, fig. 181. S. Boucher, *Les Bronzes figurés antiques: Musée Denon, Chalon-sur-Saône* (Lyon, 1983), no. 141. I. Manfrini-Aragno, *Bacchus dans les bronzes hellénistiques et romains* (Lausanne, 1987), figs. 182–210.

—J. H.

164
Lar/Genius of Aurelius Valerius

Roman, third century A.D.
Bronze; H (OVERALL): 30.8 cm; H (OF FIGURE): 23 cm
Condition: Copper or silver inlays in eyes and tunic missing; figure reattached to base with modern bolt; modern support added under right foot; crusty black and green patina.

The bushy-haired youth dances forward on tiptoe. He decants wine from a *rhyton* (drinking horn) with a goat protome into a phiale. He is dressed in a short tunic decorated with rows of triangles, which origi-

164

164

nally would have been inlaid. A long sash or mantle is knotted around his waist and looped over his arms. He is shod with tall, open-toed boots with animal-skin liners.

The inscription on the base reads: ΓΕΝΙω·ΑΥΡ/ΦΑΛЄΡΟΥCΤΡΑ/ΤΙωΤΗCΠΡΑΙ/ΤωΡΙΑΝΟC (To the Genius of Aur[elius] Valerius, praetorian soldier).

The figure is identified by the inscription, which is a mixture of Latin and Greek written in Greek characters, as the genius of a specific soldier—a genius being the protective spirit of a male person in the Latin-speaking world. The image, however, is one normally used to represent the Roman Lar, a protective spirit of a specific place or, by extension, of an organization—for example, a family. In the Greek world, the Lares were identified as "heroes" (ἥρωες), and it is a considerable novelty to see this image labeled as a genius. The association is perhaps a natural one, however, since Augustus had already linked the cults of the Lares and the Genius of the Emperor (see *LIMC*, pp. 205, 212). A similar association appears in a Latin inscription dedicated to the *genio larum horrei Pupiani* (the genius of the lares of the granary of Pupianus) (see *Corpus Inscriptionum Latinarum*, vol. 11, no. 357).

The name of the dedicant offers a chronological reference point. The name Aurelius attained vast popularity after A.D. 212, when Caracalla granted citizenship to all free inhabitants of the Empire (see Woodhead). The combination Aurelius Valerius does not appear to occur in the first or second centuries and seems, in fact, especially common in the later third century and after. The presence of a Greek-speaking soldier in a praetorian unit is itself an indication of a relatively late date. Small numbers of Easterners began to be admitted to the praetorian guard from the time of Marcus Aurelius onward, but access to the unit seems to have become much easier under Septimius Severus (see Durry). Greek inscriptions to praetorian soldiers are, in any case, rare; a few are known from Bulgaria and from Rome (see R. Cagnat, *Inscriptiones Graecae ad res Romanas pertinentes*, vol. 1, nos. 58, 186, 266, 739, 1499; vol. 4, no. 537).

In spite of the late date of the inscription, the figure is surprisingly faithful to Early Imperial models; at first glance, a Lar with a goat-protome rhyton from Pompeii seems virtually identical (see Sodo). In the Genius, however, the forms are bulkier and more tranquil, and certain details—such as the toes and boots—are greatly simplified. A similar treatment appears in the (lower quality) Lar from Felmingham Hall, Norfolk, which has been dated by T. Tam Tinh to the third century (see *LIMC*, p. 208, pl. 100).

BIBLIOGRAPHY: Unpublished.

RELATED REFERENCES: *LIMC*, vol. 6, pp. 205, 212, s.v. "Lar, Lares" (T. Tam Tinh). A. G. Woodhead, *The Study of Greek Inscriptions* (Norman, Oklahoma, and London, 1992), p. 60. M. Durry, *Les Cohortes prétoriennes* (Paris, 1938), pp. 246–249. A. M. Sodo in IBM, Gallery of Science and Art, and IBM-Italia, *Rediscovering Pompeii* (New York and Rome, 1990), no. 6.

—J. H./A.v.d.H.

165

Bottle (Unguentarium)

Gallo-Roman, possibly made at Anthée, Belgium, about A.D. 200–250
Bronze with millefiori enamel; H (WITHOUT HANDLE): 10.7 cm
Condition: Essentially complete; damage to some enamel panels.

165

This unusual footed vessel has its handle attached to two plates that are secured with braces to the bottle's lip. A knobbed lid swings from a hinge mounted on the lip. The sides of the vessel are decorated with recessed panels filled with millefiori glass. In the horizontal middle panels on each side, the glass is decayed and discolored. Above and below, however, are blue panels set with tiny white four-petaled rosettes. Around the neck are two rings of millefiori glass—one is again white rosettes on blue and the other severely deteriorated. The handle plates have rectangular panels of blue enamel above and below a diamond-shaped panel of vermillion. The lid is decorated with another wide band of vermillion.

In its construction and decoration, this bottle/unguentarium is related to a group of hexagonal pyxides with millefiori enamel decoration ascribed to Gaul or the Rhineland. Since remains of a workshop making use of the millefiori enameling technique were found at Anthée near Namur, it is quite possible that this bottle was made there. The Frankish invasions of the mid-third century seem to mark the end of this kind of millefiori enameling. The lively and elegant form of this bottle can best be paralleled in hexagonal bottles from Lyon (see Boucher/Tassinari) and from Anape (see *Aus den Schatzkammern Eurasiens*), on the northeast coast of the Black Sea.

BIBLIOGRAPHY: Unpublished.

RELATED REFERENCES: S. Boucher and S. Tassinari, *Musée de la Civilisation Galloromaine à Lyon*, vol. 1, *Inscriptions, statuaire, vaisselle* (Lyon, 1976), no. 192. *Aus den Schatzkammern Eurasiens*, exh. cat. (Zurich, 1993), no. 97. For millefiori enameling in general, see W. Forsyth, "Provincial Roman Enamels Recently Acquired by the Metropolitan Museum of Art," *Art Bulletin* 32 (1950), pp. 296–300, figs. 1–3; N. Thierry, "A propos d'une nouvelle pyxide d'époque romaine a décor d'émail 'millefiori'," *Antike Kunst* 5 (1962), pp. 65–68, pl. 24; J. Cody in *Wealth of the Ancient World* (Fort Worth, 1983), no. 53.

—J. H.

166

166
Hexagonal Litra Weight

Roman Imperial, year 7 of the reign of Gordian III Pius (beginning 244 A.D.)
Lead; H: 10.5 cm; W: 11.1 cm; DEPTH: 1.4 cm; WEIGHT: 506.6 g (litra)
Condition: Essentially intact; crusty white patina, blue discoloration on side A.

The weight is a hexagonal plaque (weights in the shape of hexagons are rare). On each side, a molded border, itself stamped with inscriptions and designs, frames a long inscription in relief letters.

Side A. ΕΤΟΥC Ζ ΑΥ / ΤΟΚΡΑΤΟΡΟ / C ΚΑΙCΑΡΟC Μ / ΑΝΤΩΝΙΟΥ ΓΟ / ΡΔΙΑΝΟΥ ΕΥC / ΕΒΟΥC ΕΥΤ / ΥΧΟΥC C / ΕΒΑCΤΟΥ (Anno 7 of Imperator Caesar M[arcus] Antonius Gordianus Pius Felix Augustus). On the rim: CΕΚΟΥΝΔΙΝΟΥ ΓΑΥΡΟΥ (Of Secundinus Gaurus)

Side B. ΥΠΑΤΕΥΟΝΤΟC / ΤΗC ΕΠΑΡΧΕΙΑC / ΤΙΒΕ ΚΛΑΥΔΙΟΥ / ΑΤΤΑΛΟΥ ΠΑΤΕΡΚΛΙ / ΑΝΟΥ ΚΑΙ ΛΟΓΙCΤΕΥΟΝ / ΤΟC ΚΟΙΝΤΟΥ ΤΙΝΗΙΟΥ / CΕΥΗΡΟΥ ΠΕΤΡΩΝΙΑΝΟΥ / ΑΓΟΡΑΝΟΜΟΥ[Ν] ΤΟΕ Ι / ΟΥΒΕΝΤΙΟΥ CΕΚΟΥΝ / ΔΕΙΝΟΥ ΓΑΥΡΟΥ (Under the Consular of the province, Tibe[rius] Claudius Attalus Paterclianus, and the chief accountant, Quintus Tineius Severus Petronianus and the agoranomos Iuventius Secundinus Gaurus). On the rim: ΛΕΙΤΡΑ (litra).

The agoranomos reviewed the weight and certified it as correct by attaching his name again on the rim, a procedure which might have also served as protection against abuse. The several inscriptions on this weight attest to the strict order of authority descending from the emperor to the governor of the consular province (Bithynia) and the chief financial officer down to the market overseer of weights and measures.

The weight of 506.6 grams is the heavy Roman litra, the *litra agoraios* of the second and third centuries A.D.

BIBLIOGRAPHY: P. Weiss in *Festschrift H. Chantraine* (forthcoming).

RELATED REFERENCES: For comparanda in hexagonal shape, see two weights from Laodicea in the Beyrouth Museum: E. Seyrig, *Scripta Varia,* Bibliothèque Archéologique et Historique 125 (Paris, 1985), p. 56, fig. 35 (dated 178/179 A.D.); J. F. Rochesnard, *Album des poids antiques,* vol. 2 (n. d.), p. 73 (dated 3/4 A.D.). A. de Ridder, *Les Bronzes antiques du Louvre,* vol. 2 (Paris, 1915), no. 3342. S. Sahin, *Epigraphica Anatolica* 16 (1990), pp. 139–146. J. Nollé, *Jahrbuch für Numismatik und Geldgeschichte* 37/38 (1987/88), pp. 93–97. W. Eck, *Zeitschrift für Papyrologie und Epigraphik* 90 (1992), pp. 199–206.

—L. M.

167

Folding Tripod

Late Roman, A.D. 250–300
Bronze; H: 106 cm
Condition: End of tail of horse drinking from kantharos missing; crusty black patina.

The tripod has legs terminating in feline paws topped by acanthus leaves. One of the legs has a handle for carrying in the form of a panther protome. Atop the three legs are compositions showing horses in different phases of their life cycle. A mare suckling a foal presumably opens the series. It continues with a bridled stallion apparently in the prime of life rearing beside a small tree. Under his chest and legs are various connecting struts, which are pour-channels left over from the casting process. The last of the series is a quietly standing, perhaps aged stallion. He wears a bell around his neck and lowers his head to drink from a kantharos, which may represent a prize for victory and/or a Bacchic paradise at life's end. Each group stands on a flat base plate, from the scalloped inner edge of which projects a robust L-shaped hook to support a (missing) basin. The X-shaped braces are fixed to hinges above and to square sliding bands below.

In Roman Imperial times, tripods were used for sacrifices to the gods. Offerings were burned in the basin carried on the three legs, and libations were poured into them; these rituals are represented on coins, triumphal reliefs, and sarcophagi (see Kunckel; Reinsberg). Tripods were also strongly con-

167

nected with oracles and forecasting the future.

In this tripod, the kantharos from which the "old" horse drinks has strikingly ornate S-shaped handles; they have closed loops at either end and are treated as vines with foliate branchings. The unfunctional effect recalls the symbolic vases in so many Late Antique mosaics. A mosaic of the first half of the fourth century in Carthage offers a number of tantalizing parallels; it shows not only kantharoi with similar vegetalized handles but also horses. The allegorical language has much in common; in the mosaic, vines grow from the kantharoi, and race horses eat seasonal plants from three victory cylinders—linking horses, victory, the cycle of time, and allusions to a Bacchic paradise (see Parrish). The use of horses to convey this message of good fortune in both cases probably reflects a favorite interest of the owner.

In spite of the similar imagery, the tripod may well antedate the mosaic. Depictions of kantharoi with S-shaped handles had already appeared in second- and third-century mosaics (see Becatti), and most panther-handle tripods have been attributed to the third century (see Hill; Kaufmann-Heinimann). To be sure, some figures from tripods, such as the examples from Assoros, Sicily, Tigava, Algeria, and Devon, England, seem rough and/or stylized enough to belong in the Tetrarchic period or later (see Orsi; Lantier; *LIMC;* Royal Albert Museum). The silver tripod from Polgárdi in the Hungarian National Museum seems virtually certain to be well into the fourth century (see *LIMC,* vol. 6, p. 797, no. 153a; Thomas).

Nonetheless, the fluency of modeling and the realism of this tripod do not link it closely with the latest pieces. Its figures have much in common with other bronzes that have not received such late dates. A similar though rougher horse appears on a hunting group from a cart in the Museo Arqueologico Nacional, Madrid (see *Los bronces romanos,* no. 302). A shaft-holder from the theater of Merida is embellished with animals and "trees" just like those on this

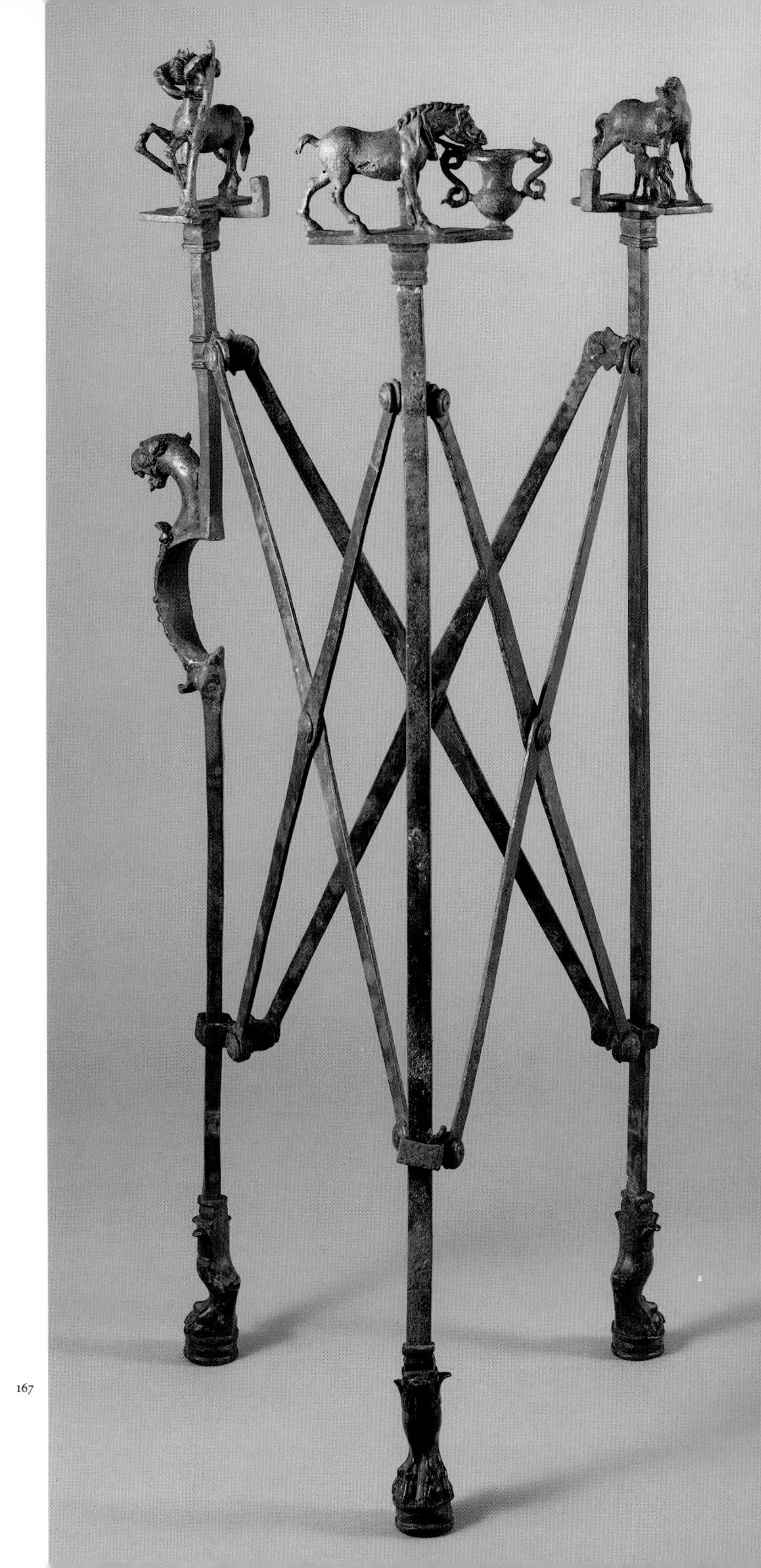

167

tripod (see *Los bronces romanos,* no. 66). All these pieces have the conspicuous connectors seen here, and the two Spanish bronzes have also been dated to the third century. A small bronze Bacchus in Parma also holds a kantharos with S-shaped handles (see Manfrini-Aragno). This figure has a doughy, irregular modeling comparable to that of the horses on the tripod although much less robust and commanding.

BIBLIOGRAPHY: Unpublished.

RELATED REFERENCES: H. Kunckel, *Der römische Genius,* RM Ergänzungsheft 20 (1974), M IV 7, M V Aquileia 1, S 5, S 11, S 16, pls. 4, 5, 22, 26, 28. C. Reinsberg, "Das Hochzeitsopfer—eine Fiktion," *Jahrbuch des Deutschen Archäologischen Instituts* 99 (1984), pp. 291–317, figs. 1, 6, 8b, 9. D. Parrish, *Seasons Mosaics of Roman North Africa* (Rome, 1984), pp. 52, 108–110, no. 8. G. Becatti, *Scavi di Ostia,* vol. 4, *Mosaici e pavimenti marmorei* (Rome, 1961), nos. 17, 79, pls. 192, 193. D. Hill, "Roman Panther Tripods," *American Journal of Archaeology* 55 (1951), p. 346. A. Kaufmann-Heinimann, *Die römischen Bronzen der Schweiz 1, Augst* (Mainz, 1977), no. 189. P. Orsi, "Nuovi acquisti per il Museo Archeologico di Siracusa," *Bollettino d'arte* 1, 3 (1907), p. 10, fig. 7. R. Lantier, "Les grands champs de fouilles de l'Afrique du Nord (1915–1930)," *Jahrbuch des Deutschen Archäologischen Instituts* 46 (1931), pp. 468–470, fig. 3. *LIMC,* vol. 6, p. 797, s.v. "Nereides" (A.-V. Szabados). Royal Albert Museum, Exeter, *Romans and Barbarians,* exh. cat. (Boston, 1976), no. 94. E. B. Thomas, "Spätantike und frühbyzantinische Silbergegenstände im mittleren Donaugebiet, innerhalb und ausserhalb der Grenzen des Römerreiches," in N. Duval and F. Baratte, *Argenterie romaine et byzantine* (Paris, 1988), pp. 138–140, pl. 4. *Los bronces romanos en España,* exh. cat. (Museo Nacional de Arte Romano de Merida, Madrid, 1990). I. Manfrini-Aragno, *Bacchus dans les bronzes hellénistiques et romains* (Lausanne, 1987), p. 69, fig. 68.

—J. H.

167

168

Key with Ring Handle

Roman, third–fourth century A.D.
Bronze; H: 9 cm; W: 3.3 cm; DIAM: 3.2 cm

The ring, cast in one piece, is formed by an inward-curving, rectangular-section volute resting on a profiled entablature flanked by two outward scrolls. The key's bit is L-shaped with six slots in the main plate and three slots, one straight and two oblique, cold-sawed into the angled plate. Assuming that the plate is functional and not merely decorative, the complexity of this key would have allowed only the quarter turn necessary to open a lift lock.

There are several published comparisons for this key: one each at the Cooper-Hewitt Museum in New York (1909.2.278: Spilker), the Musée Carnavalet in Paris (AM 825(3): Formi et al.), and the Kestner Museum in Hannover (1932,394: Menzel), two in Vienne, France (see Boucher), and seven in the Bibliothèque Nationale, Paris (1888–1894: Babelon/Blanchet). Another, present location unkown, was recorded in the nineteenth century. Some of these have three or four prongs projecting at a right angle from the main plate instead of the Fleischman key's slotted angle.

BIBLIOGRAPHY: Unpublished.

RELATED REFERENCES: B. Spilker, *Keys and Locks in the Collection of the Cooper-Hewitt Museum* (Smithsonian Institution, Washington, D. C., 1987), pp. 14–15, no. 18. P. Forni et al., *Les Bronzes antiques de Paris: Musée Carnavalet* (Paris, 1989), no. 198, p. 213. S. Boucher, *Inventaire des collections pu-*

168

bliques françaises, vol. 17, *Vienne, bronzes antiques* (Paris, 1971), pp. 191–193, no. 528 (L. 8.1), no. 542 (L. 6.0). H. Menzel, *Bildkataloge des Kestner-Museums: Römische Bronzen* (Hannover, 1964), p. 62, no. 162. E. Babelon and J.-A. Blanchet, *Catalogue des bronzes antiques de la Bibliothèque Nationale* (Paris, 1895), p. 641, nos. 1888–1894 (only 1888 and 1889 illus.). E. Guhl and W. Koner, *Everyday Life of the Greeks and Romans* (New York, 1989), p. 465, fig. 464 (after G. de la Vincelle, "Arts et Metiers" (1819), pl. XXXVI).

—A. K.

169
Situla

Late Roman, probably made in Asia Minor, about A.D. 350–420
Bronze with tin plating; H: 33.5 cm; DIAM (OF RIM): 27.2 cm; DIAM (OF BASE): 17 cm
Condition: Handle and bottom missing; repairs, especially in area above chariot of Dionysos.

The slender vessel flares out slightly above. The shape is like that of a Greek kalathos. Traces of attachments for a swinging handle show that the piece originally functioned as a bucket or situla. The horizontal bands of decoration are outlined with dots and reserved against a tinned background. Plain tinned stripes alternate with rows of other ornament: (from the top) disks, a grapevine, Dionysos and his retinue, laurel, and dolphins swimming amid stylized waves.

In the main frieze, a Dionysiac procession moves toward the viewer's right. The god reclines on a chariot drawn by panthers, and in his left hand he holds a thyrsos and in his right a vessel shaped like this one bedecked with grape clusters and a ribbon. Eros holding up bunches of grapes escorts the team, and a nude young satyr holding a shepherd's staff and an animal skin runs on tiptoe before them. In front of the satyr is a large grapevine, topped by an ovoid object with a shadowed or hollow center. Next is a dancing maenad holding a thyrsos and a grape-filled kalathos. She is preceded by a goat-legged, lyre-playing Pan and a fat old silenos, also playing the lyre, next to another exuberant plant. In the dolphin frieze is a horn—probably a Bacchic drinking horn rather than a horn-shaped snail. Outlines of figures are created with short chisel strokes. Inanimate objects may be outlined either by this method or by rows of dots made with a stylus.

Based on the kalathos, a tall, flaring basket, the shape of this vessel has had a long history. Such baskets could have many uses, but they are especially associated with wool-working and—more relevant for our piece—the harvest, especially the grape harvest. Grape harvesters or seasonal genii, for example, often carry a kalathos full of produce on sarcophagi, and on late sarcophagi these baskets often have handles (as this bronze example once did) (see Koch/Sichtermann). Ever since Hellenistic times, smaller versions of the shape had been used for drinking cups (see cat. no. 175 below; Hochuli-Gysel).

The style of drawing is highly stylized in a lively, cartoonlike sense. The figures' anatomies are lobed and sausagelike and often greatly elongated and bent. With their deeply indented contours and undulating terminations, the forms have a gaily ragged quality. The insubstantial effect is augmented by the use of tinning to represent bands of color or shadows on inanimate objects or on drapery. Lively stylized effects had long been a kind of subculture within Roman art; they can be traced in media such as ivory carving or marble intarsia as early as the first century (see Ward Perkins et al.).

169

169

The whimsical extremes reached here, however, are certainly a product of Late Antiquity. A good chronological (and perhaps even geographical) reference point for this kind of spirited drawing is offered by the silver dish with the triumphant Constantinius II (A.D. 337–361) found in the Crimea and presumably made in Constantinople (see Kent/Painter). In the dish, the figures are more solid, and the drawing elaborates anatomies with considerably more internal detail. Drapery folds are built up with delicate, sketchy fans of line. This sketchy technique of rendering drapery reappears in a tinned bronze plaque in the British Museum depicting Achilles and Briseis (see Walters; Carandini, p. 17, fig. 10; *Age of Spirituality,* no. 195). The figures on the plaque, however, are less detailed than those on the silver bowl, and they have lobed, sausagelike anatomies, much like those on the Fleischman situla. The situla and the London plaque could be roughly contemporary works somewhat later than the Constantinius bowl. The situla and the bowl have a special link through the use of dotted lines.

In the London plaque, the figures are plated with tin and the background is left bronze; the technique parallels the silver, copper, and niello inlay so popular in Late Antiquity (see *Age of Spirituality,* nos. 76, 77, 94, 137). Ornamental patterns of tinning can, however, be found as early as gladiator's armor of around A.D. 100 (see Pflug). In the Fleischman situla, the tinning technique comes into its own; the background is tinned, leaving the figures with their delicate incision largely unobscured, except, as noted above, for some shadows and ornamental bands.

A group of Late Antique bronze buckets seems decidedly later than the Fleischman piece. Closest to it is a situla in Madrid found in Bueña (see Carandini, pp. 26–28, figs. B, 33–38; *Los bronces romanos*). There are numerous similarities; the background (rather than the figures) is tinned and is spotted with reserved vegetation. A row of reserved disks forms an upper border, as on the Fleischman situla, and the figures have something of the same ragged liveliness. On the other hand, the differences are also significant. Texturing is not achieved with dots but with rings, arcs, and dents created with hollow-tipped or solid round punches. The Bueña bucket is also drier and more angular than the Fleischman piece—in the figures as well as the conical shape of the situla itself. Two other late conical copper and bronze buckets lack tinning; one is in Rome, the Doria bucket, and another from Zerzevan is in the Istanbul Museum. The Bueña and Doria buckets have been dated by A. Carandini to the second half of the fifth century and ascribed to Alexandria, and Carandini puts the Zerzevan piece, which is decorated with crosses, in the sixth century (see Carandini, pp. 28–29, figs. 13–31, 39–40; M. Bell in *Age of Spirituality,* no. 196).

BIBLIOGRAPHY: Unpublished.

RELATED REFERENCES: G. Koch and H. Sichtermann, *Römische Sarkophage* (Munich, 1982), figs. 236, 242. A. Hochuli-Gysel, *Kleinasiatische glasierte Reliefkeramik* (Bern, 1977), p. 45. J. Ward Perkins et al., *Pompeii A.D. 79* (Boston, 1978), nos. 156, 171. J. P. C. Kent and K. S. Painter, eds., *Wealth of the Roman World,* exh. cat. (London, 1977), no. 11. H. B. Walters, *Catalogue of the Bronzes, Greek, Roman and Etruscan, in the Department of Greek and Roman Antiquities, British Museum* (London, 1899), no. 833, fig. 23. A. Carandini, *La secchia Doria: una 'storia di Achille' tardo-antica,* Studi miscellanei 9 (1963–64), p. 17, fig. 10. K. Weitzman, ed., *Age of Spirituality,* exh. cat. (New York, 1979). H. Pflug in *Antike Helme* (Mainz, 1988), pp. 365–367, figs. 1, 2, color pl. 7. *Los bronces romanos en España,* exh. cat. (Museo Nacional de Arte Romano de Merida, Madrid, 1990), no. 258.

—J. H.

GOLD AND SILVER

170
Snake Bracelet

Egypto-Roman, first century A.D.
Gold; DIAM: 7.25–6.8 cm; WEIGHT: about 113 g

The bracelet, a relatively late member in the long series of Hellenistic snake bracelets, belongs to an Egypto-Roman type datable in the first century A.D. (see Ogden). It has the form of a snake in a single coil with a full-size head at one end and a smaller, secondary head, known only in examples from Egypt, at the end of the wavy tail section. Behind the main head is a section of chased crosshatched decoration representing scales followed by a linear and dotted design possibly meant to suggest ears of wheat. The wavy tail also has crosshatched scales, incised before the

170

171

bracelet was bent, as they typically are in genuine pieces. The underside of the head end is embellished with S-shaped chased lines to represent the belly. A bracelet from the Schimmel collection, now in the Metropolitan Museum of Art (1988.22: Muscarella), is of the same type as ours.

BIBLIOGRAPHY: J. Ogden, *Independent Art Research, Ltd., Report 89078,* March 14, 1990; J. Ogden, *Ancient Jewellery* (London, 1992), p. 8f., fig. 1.

RELATED REFERENCES: O. W. Muscarella, ed., *Ancient Art: The Norbert Schimmel Collection* (Mainz, 1974), no. 71. F. Landenius, "Two Spiral Snake Armbands," *Medelhavsmuseet Bulletin* 13 (Stockholm, 1978), p. 37ff.

—A. H.

171

Statuette of a Young Satyr Playing the Double Aulos

Roman, probably second century A.D.
Silver with gilding; H: 4.2 cm

The rustic-looking youth strides forward playing a double aulos. The instrument has one straight pipe and one with a flaring, upcurved end (see Wrede). The figure is nude except for a gilded panther-skin worn diagonally across his torso. The Dionysiac panther-skin, the impish facial features, and the stiff, tousled hair seem to characterize this being as a satyr, though the ears, which would be pointed like an animal's, are hidden under the hair and the place where a tail would be is covered by the garment.

The date of the figure is difficult to pin down. The vagueness about traditional pagan iconography, along with the round face and popeyes, might suggest a date in Antonine times, when other small, solid-cast silver sculptures were produced.

BIBLIOGRAPHY: Galerie Nefer, Zurich, *Catalogue 8,* 1990, no. 24.

RELATED REFERENCES: H. Wrede, *RM* 95 (1988), p. 97ff., fig. 1. C. C. Vermeule, *Greek and Roman Sculpture in Gold and Silver* (Boston, 1974). For the bronze statuette of a round-faced satyr boy, see H. Menzel, *Die römischen Bronzen aus Deutschland,* vol. 2, *Trier* (Bonn, 1966), no. 58b, pl. 102. For a togate boy playing the double aulos, clearly a human participant in a ceremony or festival, see E. Babelon and A. Blanchet, *Catalogue des bronzes antiques de la Bibliothèque Nationale* (Paris, 1895), no. 880.

—A. H.

172

Roundel with the Head of Medusa

Roman, probably second–early third century A.D.
Silver with gilding, remains of bronze attachments on the back; DIAM: 8.5–8.7 cm
Condition: Intact except for very slight losses to edge; bronze straps broken off.

The piece is worked in high relief by the repoussé technique. Four silver rivets through the border mark the places where bronze attach-

172

ments were fastened to the back and show that it was threaded onto a vertical strap, probably as a bridle ornament. The head of Medusa, seen in three-quarter view, has regular features and a long, straight nose. Her expression shows a rather subtle pathos, indicated by knitted brows and a tightly closed, down-turned mouth. Her hair blows back in an Alexander-like mane, with wings springing from it at each side of her forehead. A snake emerges from either temple, and two snake tails are knotted beneath her chin. The hair and, startlingly but effectively, the eyes are gilded.

This type of Medusa had entered the repertory at least by early Hellenistic times (see *LIMC,* s.v. "Gorgo, Gorgones") and enjoyed a long popularity (see *LIMC,* s.v. "Gorgones Romanae"). The self-assured but rather insensitive execution suggests that the present piece is a Roman interpretation. The pinched expression of grief, ultimately derived from the Hellenistic figures of the Small Attalid Dedication, recalls the gorgoneia from the frieze of the Trajaneum at Pergamon (see *Pergamon*); however, our piece could equally well be a production of the Antonine-Severan period, which saw a revival of flamboyant Hellenistic styles.

BIBLIOGRAPHY: Sotheby's, New York, *Antiquities,* May 30, 1986, lot 42.

RELATED REFERENCES: *LIMC,* vol. 4, p. 285ff., nos. 116, 135, 190, s.v. "Gorgo, Gorgones" (S.-C. Dahlinger, I. Krauskopf). *LIMC,* vol. 4, p. 345ff., nos. 83, 89, 92, s.v. "Gorgones Romanae" (O. Paoletti). *Pergamon,* exh. cat. (Ingelheim am Rhein, 1972), nos. 31, 32.

—A. H.

173
Necklace

Roman; third century A.D.
Gold with amethyst; L: 40 cm
Condition: Well preserved though with some adjustment to render it wearable; central motif added later, possibly in antiquity.

The necklace consists of seventy hollow double-leaf-shaped motifs made of sheet gold and joined by loop-in-loop links affixed to their backs. At the center is an oval cabochon amethyst surrounded by four of the double-leaf shapes arranged in a quatrefoil.

The piece belongs to a well-known type, the elements of which usually have stylized bird or leaf forms. Several of the published versions, often with double-bird motifs, come from Egypt. Two examples in a find group from Nikolaevo in Bulgaria (see *Thracian Treasures*) were found with coins of the mid-third century A.D. One necklace from the Nikolaevo find (no. 454) differs from ours in having openwork scroll terminals, but the other example (no. 452) has tapered terminals, ultimately derived from earlier terminals in the form of Herakles' club, which resemble those of our piece.

Although the halves of such necklaces are normally symmetrical, a central decorative element is not common. However, the more elaborate of the two from the Nikolaevo find (no. 454) has a central setting containing a rock crystal. In our example the frame of the amethyst is assembled from four of

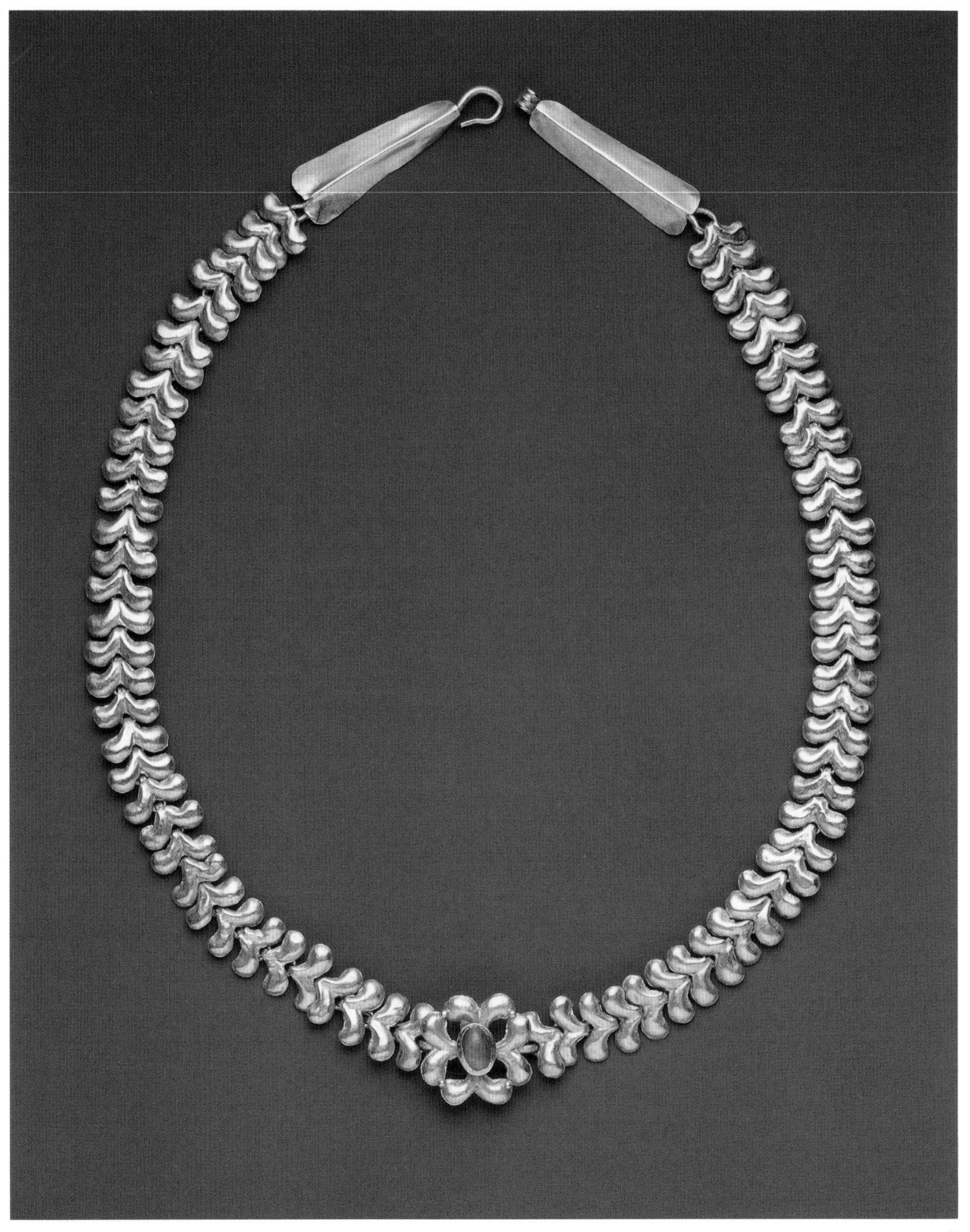

173

the symmetrical sections soldered into a quatrefoil. The remains of wire links can be seen on their backs, showing that they were once part of the rest of the necklace and that the new arrangement is probably a later alteration (see Ogden). From a typological and constructional point of view, there is nothing to indicate that the adaptation is other than ancient.

BIBLIOGRAPHY: J. Ogden, *Independent Art Research, Report 90165*, December 18, 1990.

RELATED REFERENCES: *Thracian Treasures from Bulgaria*, exh. cat. (London, 1976), p. 84f., nos. 437–456. M. C. Ross, *Catalogue of the Byzantine and Early Medieval Antiquities in the Dumbarton Oaks Collection*, vol. 2, *Jewelry, Enamels and the Art of the Migration Period* (Washington, D.C., 1965), p. 13ff., no. 8.

—A. H.

GLASS

174
Flask

Roman, third century A.D.
Glass; H: 14.2 cm; DIAM (OF MOUTH): 3.6 cm; (OF BODY): 11.6 cm
Condition: Surface encrustation on interior and underside; some whitish iridescence on exterior; one crack from shoulder to center of base, through both clear glass and overlaid trails, as well as ring foot; another, shorter crack nearby is present only in the base; small chip on bottom, above ring foot and near the larger crack; the lower blue trail on neck of vessel has two sections missing at opposite points on neck.

174

The free-blown flask of colorless glass has a globular body with a narrow neck and flaring mouth. The neck and base are ornamented with applied blue glass trails. The figural zone contains blue and white trails arranged in foliate patterns. From undulating stems, flattened leaves spread over the curving surface of the flask.

Originally thought to have been created in the Rhineland, "snake-thread" glass is now believed to have been developed first in the eastern Mediterranean. Two stylistic groups existed side by side, one with freely applied trails, and a variety known as the "flower and bird" group because of its figural representations. Shortly after its emergence in the eastern Mediterranean, the style spread to the western provinces of the Roman empire, particularly the city of Cologne (Colonia Claudia Ara Agrippinensium), where many examples have been excavated. The eastern workshops are characterized by their use of colorless glass for both body and trails, the crosshatched patterning found on some of the trails, and the "flower and bird" patterns found on some of the vessels. The western workshops incorporated colored trails into their designs and indented the trails with single lines, rather than crosshatching. The Fleischman flask is an eclectic mix of shape, color, and patterning that successfully blends elements ascribed to both eastern and western manufacture. The globular shape of the flask compares most closely to two vessels excavated in Cologne (see Fremersdorf). One of these is composed of colorless glass, an eastern characteristic, but its trails are marked with single lines, a western trait. The other is decorated with clear and white trails, and the trails are likewise indented with single lines. The decoration of the Fleischman flask, however, belongs to the "flower and bird" style, and its trails are crosshatched. It is these two elements that identify it as a product of an eastern atelier.

BIBLIOGRAPHY: Unpublished.

RELATED REFERENCES: F. Fremersdorf, *Römische Gläser mit Fadenauflage in Köln,* Die Denkmäler des Römischen Köln 5 (Cologne, 1959), p. 42, N 119, pl. 20; p. 49, N 6049, pl. 48. For an overview of "snake-thread" glass, see D. Harden et. al., *Glass of the Caesars* (Milan, 1987), pp. 105–108. For "flower and bird" style, see D. Barag, "'Flower and Bird' and Snake-Thread Glass Vessels," *Annales du 4e Congrès International d'Etude Historique du Verre, Ravenne-Venise, 13–20 mai, 1967* (Liège, 1969), pp. 55–66.

—K. W.

TERRACOTTA

175
Kalathos

Roman, second half of first century B.C.
Glazed terracotta; H: 15.1 cm; DIAM (OF RIM): 16 cm; DIAM (OF FOOT): 9.6 cm
Condition: Reconstructed; two pieces of rim, part of handle, and fragments of vine stem, grapes, and leaves missing.

The large mug has a flaring body with an everted rim. An egg molding (*ovolo*) runs above the base. Its handle is made from two strips of clay with a short added strip used to create the effect of a knot. From each side of the handle runs a branch of grapevine with a cluster of grapes. Above the junction are two birds. A fillet, rendered in very low relief, hangs down from each side of the handle and twists illusionistically in the breeze. Ovoli, birds, tendrils, grapes, and fillets are rendered with streams and dots of liquid clay (the so-called barbotine technique). The vine stems and leaves are modeled by hand and applied. The veins of the leaves are delicately incised. The exterior and underside of the vase is glazed green, while the interior is glazed yellow with an accidental spill of green.

Roman lead-glazed pottery was produced principally at Tarsus and Smyrna. Kalathoi like this one have been attributed by A. Hochuli-Gysel to Smyrna, where this example must also have been made; its dimensions agree almost to the millimeter with the previously published examples (see Hochuli-Gysel; *Aus den Schatzkammern Eurasiens*). This close agreement indicates that the body of the vase was formed in a mold. The decoration of the published examples was made by hand with the greatest care. In both shape and decoration these kalathoi have parallels in the finest silverware of Augustan and late Republican times. This piece is notable for the manner in which its decoration breaks free of the surface in the most naturalistic way possible. The theme of the grapevine is suggested, of course, by the wine that would have been drunk from the cup.

BIBLIOGRAPHY: Unpublished.

RELATED REFERENCES: A. Hochuli-Gysel, *Kleinasiatische glasierte Reliefkeramik* (Bern, 1977), pp. 44–45, 173, pls. 11, 20, 22, 24, 28. *Aus den Schatzkammern Eurasiens,* exh. cat. (Zurich, 1993), no. 86.

—J. H.

175

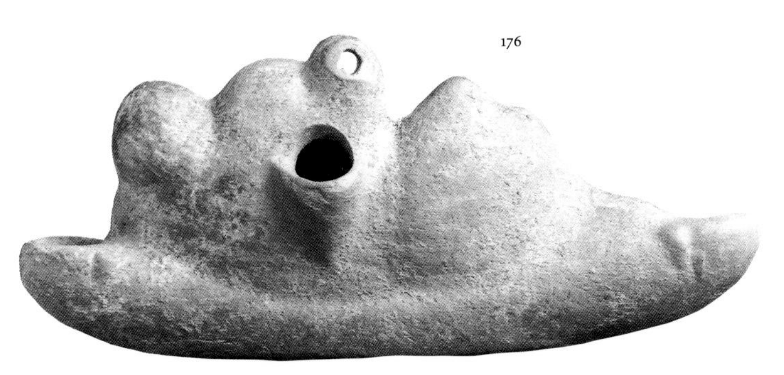

176

176

Lamp in the Form of a Reclining Comic Actor

Roman, second century A.D.
Terracotta; H: 7 cm; L: 16.3 cm
Condition: Intact; tan clay body coated with reddish slip; mold-made with free-hand retouching.

The lamp has a nozzle at either end and a filling hole behind the figure; there is a loop for suspension. The masked and costumed actor reclines on his right side as if in drunken merriment. His legs are crossed and his right hand raised towards his face. He wears a long, close-fitting, textured undersuit and a short draped overgarment, as well as a bearded smiling comic mask.

Another example of this lamp type is in the Metropolitan Museum of Art (17.194.1810). The costume is a possible parallel for the network garments worn by the bronze actor (see cat. no. 153 above) and the comic muse Thalia on sarcophagi. The texture, however, might also represent fleece, as frequently worn by satyr-actors in late Classical vase painting. The head of the figure is small compared with those of earlier comic actor figures, and the treatment of the mask vague, with decreased emphasis on the mouth. These features suggest a relatively late date.

BIBLIOGRAPHY: Unpublished.

—A. H.

STONE

177

Portrait Head of an Old Man

Roman, late first century B.C.–early first century A.D.
Marble; H: 34.9 cm; W: 17.7 cm; DEPTH: 24.7 cm
Condition: Nose broken off.

This superb portrait of an older man is an exceptional example of an early phase of the Roman portrait tradition. At this time, toward the end of the first century B.C. or early first century A.D., the male sitter is usually portrayed as an embodiment of the values dear to the late Republic and early Empire. These include *gravitas* (sober dedication to the duties of a citizen) uncompromised by the dramatic flair of late Hellenistic portraiture. Our sitter's emotionless expression is thus not a reflection of a humorless disposition but is instead the approved demeanor for a commemorative image—for slave and senator alike. The slight rightward turn of the head prevents the portrait from seeming too static.

Notwithstanding the sobriety of the man depicted, his features are described with what we presume to be painstaking accuracy. Although the nose is broken off, enough remains to suggest that it would have been quite pronounced. The close-cropped hair is rendered in short chisel strokes in front and short ridges in back; this mixed technique of rendering the hair is somewhat unusual, and the bathing-cap-like quality of the back of the head is the only dissonant note in this first-rate sculpture. The M-shaped hairline is high on the deeply furrowed forehead. The wrinkles and creases on the neck and face seem at first to be located as precisely as streets on a road map, but closer inspection betrays a formulaic, schematic approach to these indications of maturity. The three wishbone-shaped incisions on the proper left cheek, which are not mirrored on the smooth right cheek, are implausibly regular.

The compact, fine-grained marble probably points to the Italian quarries of Luna, which were not extensively worked until the time of the emperor Augustus. This identification of the stone, along with the smooth lines on the neck, regular patterns of wrinkles on the left cheek, and size of the bust, argue for an Augustan date around the last quarter of the first century B.C. to the first decade of the first century A.D. The portrait was made as a small bust, rather than for insertion into a statue or herm. This might suggest that the sculpture

177

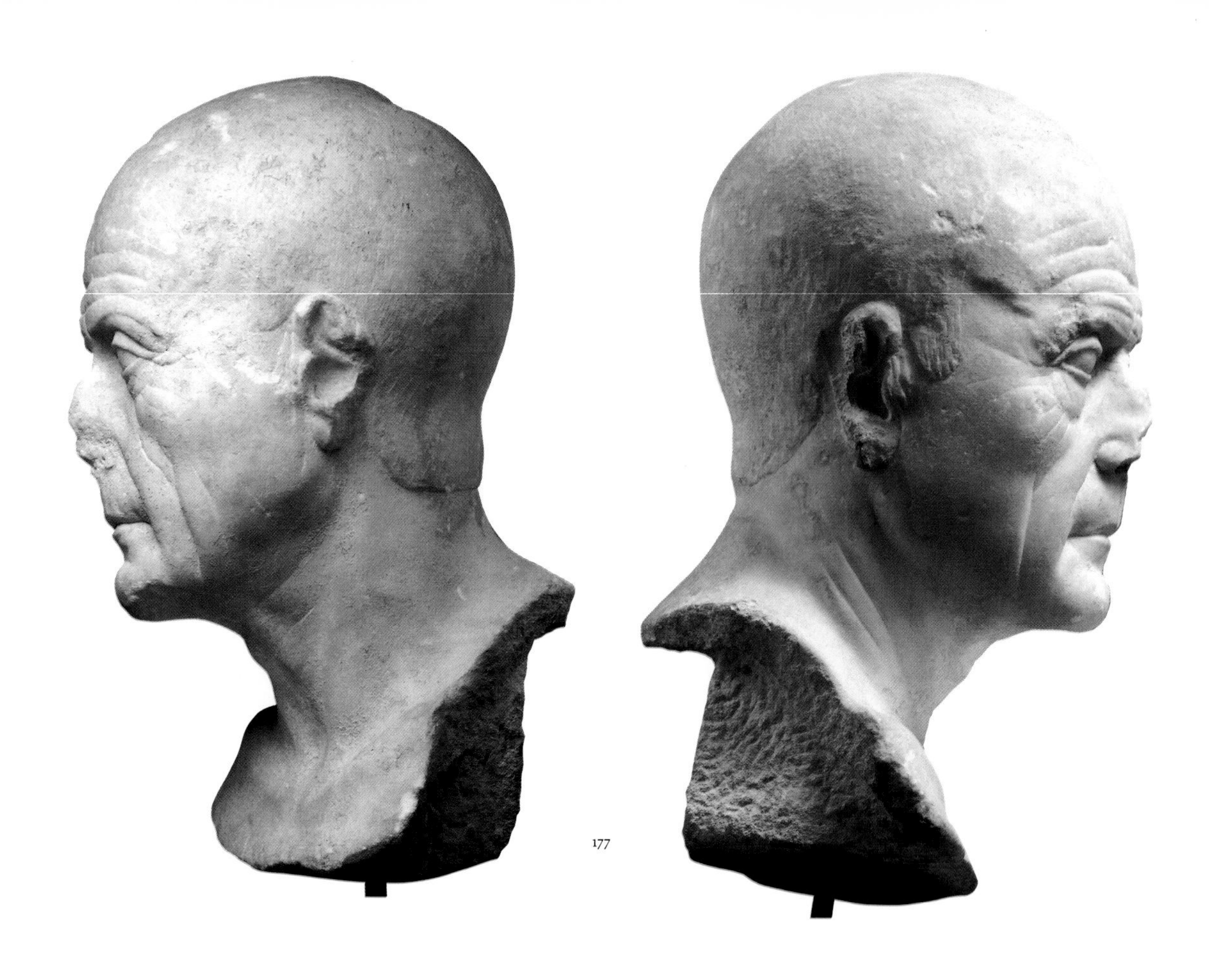

177

was destined for a burial niche in an underground tomb. Alternatively, it might be an image made during the lifetime of a *paterfamilias* of the Augustan period, or even an example of the *imagines maiorum* (recreated portraits of long-dead ancestors intended for a Roman family's display at home). The abstract quality of the cap of hair on the back of the neck may suggest that the head was meant to be seen from the front alone, which would have been achieved by its display in a niche.

The Fleischman portrait is easily among the finest examples of this type of portraiture to be found in any collection, public or private.

BIBLIOGRAPHY: *Burlington Magazine* (December 1977), pl. 1; H. Jucker and D. Willers, *Gesichter: Griechische und römische Bildnisse aus Schweizer Besitz* (Bern, 1982), p. 63, no. 21.

RELATED REFERENCES: For similar heads, see V. Poulsen, *Les Portraits romains*, vol. 1 (Copenhagen, Ny Carlsberg Glyptotek, 1962), p. 55, no. 20, pl. 32; P. Zanker, "Das Bildnis des M. Holconius Rufus," *AA* (1981), pp. 358ff., figs. 17, 18.

—M. A.

178
Grave Relief of a Silversmith

Roman, first quarter of first century A.D.
Marble; H: 79.9 cm; W: 58.5 cm; DEPTH: 31.7 cm
Condition: Upper left corner of relief, figure's nose, and top of tool held in right hand missing.

The relief is inscribed: P·CURTILIUS·PLACAT[US] / FABER·ARGENTARIUS (P[aulus] Curtilius is at rest, maker of silver). The impulse to commemorate the dead as they were in life, without a hint of timelessness, was typical of the

178

Roman temperament. Paulus Curtilius was, as the inscription on this impressive monument tells us, a silversmith. He is shown frontally, framed in a deep marble relief from chest height up. Remnants of lead on the top edges of the thick block of stone were probably points of attachment for the relief, which was very likely set into a large architectural funerary monument. Such tomb-buildings lined the roads out of Ostia, Pompeii, and every other Roman town of note, and allowed freed slaves, guilds, and family groups to commemorate themselves in large-scale commissions.

As a silversmith, Paulus Curtilius would have enjoyed some standing in his community. Paulus's pride in his profession was such that in addition to having himself identified as *argentarius* in his funerary inscription, he had the monument carved in the more costly marble rather than travertine and chose to have the tools of his trade included in the portrait. The sitter rests his left hand on a small vase with a nude male figure running to the right—perhaps a satyr or soldier—while he holds a chasing tool in his left hand and what was probably a mallet in his right.

Paulus's portrait is expressive and accomplished. His full head of hair is combed forward in the manner popular around the time of Augustus—the end of the first century B.C. or beginning of the first century A.D. His prominent cheekbones, somewhat slack jaw, and the rings on his neck are the features of an older man. He sports a ring and wears a tunic with a schematically rendered toga that is pulled open to reveal his hands and tools.

BIBLIOGRAPHY: The Merrin Gallery, *The Majesty of Ancient Egypt and the Classical World* (New York, 1986), p. 27.

RELATED REFERENCES: For background on freedmen's reliefs, see P. Zanker, "Grabreliefs römischer Freigelassener," *Jahrbuch des Deutschen Archäologischen Instituts* 90 (1977), pp. 267–315. D. Kleiner, *Roman Group Portraiture* (New York, 1977). Compare a relief in the British Museum: S. Walker, *Roman Art in the British Museum* (London, 1991), p. 27, fig. 27.

—M. A.

179
Statuette of Dionysos with an Animal

Roman, mid-first century A.D.
Fine-grained white marble; H (WITH PLINTH): 62.3 cm
Condition: Nose tip, right front of wreath, right forearm (separately worked), left wrist, head, right front leg, and possibly part of animal's tail missing. Reassembled from three pieces very evenly broken along horizontal lines.

The figure of a draped Dionysos belongs to a category of highly sophisticated Roman statues made in the first century A.D. and influenced by the rococo Archaism of late Hellenistic Rhodes and Asia Minor (see Fullerton, p. 128ff.; Borrelli). The filmily draped torso of Priapos in the Museo Nuovo of the Musei Capitolini (see Zanker, fig. 191) is the most famous example, but several related statues, all of them headless, have been noted: Apollo figures in the Vatican and the Villa Borghese (see Fullerton, figs. 74, 75), and female images in Verona (see *Arte e civiltà,*) Turin, Madrid (see Fullerton, p. 147f., nos. 1, 3), and on the Rome art market (from the Strickland collection, see de Lachenal), as well as an unpublished replica of the Strickland statue on the New York market. In all these the poses of the figures mimic Archaic stiffness and the arrangement of the drapery has a mannered complexity, but the modeling is refined and naturalistic. Delicately carved, diagonally worn animal skins are combined with clinging, layered, transparent garments in axial compositions.

Although it is a playful small-scale adaptation, our statuette is especially informative because it is the only one of the series with its original head. The god's face is shaded by a luxuriant wreath of vine leaves and grape clusters. His hair hangs in long, wavy ribbonlike strands, two to either side, at the front, and in a mass, caught near the end by a flat barrette in back. His heavy-lidded eyes are lowered and his mysterious smile is emphasized by the long mustache, the ends of which curl upward in a manner most unusual for antiquity. His beard, with an offset upper section, is trimmed to a broad Archaic spade shape. In a recherché detail reminiscent of the "Pausanias"-Pindar type (see Richter/Smith) the tip has been tied off to form a decorative tassel. In type and expression our figure's head resembles a larger head with a spade-shaped beard and secretive smile in the Palazzo dei Conservatori (see Zanker, fig. 192). It has long been hypothesized that the Museo Nuovo Priapos torso may have had a head of this kind (see von Steuben). Our statuette seems to show that the suggestion is correct.

Within the series small rampant animals are favorite accessories. Most members of the category have lost all but fragments of the animals, but in our piece the whole body survives. Dionysos, like an Archaic Master of the Beasts, holds the creature up by one foreleg. The animal's originally back-turned head and its powerful body with coat rendered in rows of snail-shell curls seem borrowed from Achaemenian art, perhaps from a precious metal object such as an amphora-rhyton with animal handles, preserved in a temple treasury. The cloven-hoofed creature is probably a goat, like the one accompanying a Dionysos in the Pushkin Museum, Moscow (see *Antique Sculpture*). The massive chest and developed genitals show that despite its toylike size the creature is an adult male.

Refined decorative sculptures datable to Claudian or Neronian times offer the best comparisons for our piece. The heads of body herms (decorative figures ending below in rectangular shafts) in the British Museum (see Smith), in Florence (see Milani), and in the Lateran collection (see Benndorf/Schöne) have the same facial expressions and the same rippling, ribbonlike locks of hair. The use of pinpoint drill holes to create a delicate foamy effect is seen in several of the mask reliefs from Pompeii (see Bieber) as well as in the Claudian female portraits from Baia (see Tocco Sciarelli) and the Munich Artemis Braschi (see Fuchs), a perhaps slightly later eclectic creation which has many features in common with our series. Dionysos' bizarre tasseled beard and upward-curled mustache have a parallel from the Vesuvian cities on a small double herm from the Contrada Bottaro Villa, now in Boston (see Vermeule/Comstock). Little rampant animals accompany the Artemis Braschi and the nude Ar-

chaistic Apollo from the House of the Menander at Pompeii (see Ward Perkins et al.); a "real" goat climbs on a body herm from the House of Marcus Lucretius at Pompeii (see Spinazzola). Our statuette's low rectangular plinth, made in one with the figure, is also found in works from the Vesuvian sites (e.g., the drunken Hercules from the Casa dei Cervi at Herculaneum: *Rediscovering Pompeii*).

BIBLIOGRAPHY: Unpublished.

RELATED REFERENCES: M. Fullerton, *The Archaistic Style in Roman Statuary,* Mnemosyne Suppl. 110 (Leiden and New York, 1990). L. Borelli, "Una scuola di 'manieristi' dell'ellenismo rodio-asiatico," *Atti dell'Accademia Nazionale dei Lincei. Rendiconti* 8, 4 (1949), p. 336ff. P. Zanker, *Augustus und die Macht der Bilder* (Munich, 1987). *Arte e civiltà romana nell'Italia settentrionale,* exh. cat. (Bologna, 1964), pl. 198. L. de Lachenal, "Su una statua di Menade," *Bollettino d'arte* 65 (1980), pp. 1–6. G. M. A. Richter and R. R. R. Smith, *The Portraits of the Greeks* (Ithaca, 1984), frontis, p. 176ff. H. von Steuben in *Führer durch die öffentlichen Sammlungen classischer Altertümer in Rom,* vol. 2 (Tübingen, 1966), nos. 1512, 1699. L. I. Akimova et al., *Antique Sculpture from the Collection of the Pushkin Museum in Moscow* (Moscow, 1987), p. 85, no. 46. A. H. Smith, *A Catalogue of Sculpture in the Department of Greek and Roman Antiquities of the British Museum* (London, 1904), no. 1745. L. A. Milani, *Il R. Museo Archeologico di Firenze* (Florence, 1912), p. 320, no. 110, pl. CLVII. O. Benndorf and R. Schöne, *Antike Bildwerke des Lateranischen Museums* (Leipzig, 1867), p. 105f., nos. 181, 188. M. Bieber, *The History of the Greek and Roman Theater,* 2nd ed. (Princeton, 1961), figs. 562, 564, 569 (cf. no. 573 from Rome). G. Tocco Sciarelli, ed., *Baia: il ninfeo sommerso del Punto Epitaffio* (Naples, 1983), pls. 126–128, 158–162. M. Fuchs, *Römische Idealplastik* (Munich, 1992), p. 38ff. C. C. Vermeule and M. Comstock, *Sculpture in Stone and Bronze in the Museum of Fine Arts, Boston* (Boston, 1988), nos. 41–202. J. B. Ward Perkins, et al., *Pompeii A.D. 79,* exh. cat. (Boston, 1978), no. 83. V. Spinazzola, *Le arti decorative in Pompeii e nel Museo Nazionale di Napoli* (Milan and Rome, 1928), pl. 65. IBM, Gallery of Science and Art, and IBM-Italia, *Rediscovering Pompeii,* exh. cat. (New York and Rome, 1990), no. 190.

—A. H.

180
Head of the Diadoumenos of Polykleitos

Roman, late first-century A.D. copy, after a Greek original of about 430 B.C.
Fine-grained white marble; H: 22.8 cm; W: 15.2 cm; DEPTH: 17.7 cm
Condition: Generally good; broken at the neck; surface of break weathered with light encrustation; hard brown encrustation on carved surface, especially on proper left side of head from top to bottom.

The head comes from a Roman copy of an original Greek bronze statue of a complete figure of a young athlete who is tying a fillet or band, a mark of victory, around his head. "Polykleitos of Sikyon . . . made a statue of the Diadoumenos or Binding his Hair—a youth, but soft-looking—famous for having cost 100 talents" (Pliny *NH* 34.55 [Loeb ed., trans. H. Rackham]). The continuing popularity of the Diadoumenos in antiquity is attested by Lucian's reference to it (*Philopseudes* 18) as famous for its beauty. The best and earliest copy of the original bronze statue was excavated on the island of Delos in the nineteenth century and is now in the National Archaeological Museum, Athens (1826: Machaira). In the complete statue the athlete is depicted frontally with his weight on his right leg, lifting both hands up to his head to tie on the broad victory ribbon. He has his head turned slightly to his right, giving the whole figure a counterpoised sense of movement.

Current theories on the identity of the subject favor an anonymous victorious athlete, the god Apollo in the guise of a victorious athlete, or as the eponymous hero of a play by Euripides, *Alexander.* All scholars seem to agree that the person portrayed is no ordinary mortal but either a deified human being or a god. In any event, the statue embodies the idea of the perfect athlete.

The sculptor of the head has made a distinction between the differing textures of hair and skin. The abrasive used to achieve a smoother surface on the skin is evidenced in the fine marks preserved on the face. The individual locks of hair are very carefully carved and modeled. The centers of three side locks below the fillet on the proper right side and two side locks on the proper left side are drilled. The proper left ear is larger than the proper right by nearly one centimeter. The inner corners of the eyes are drilled. The mouth is separated by a running drill channel with deeply drilled corners. The top of the head is flattish then squared off where it meets the back of the head. A sharp division is made between the neck and the underside of the chin, which are almost at right angles to one another.

180

180

Among the many replicas, the Fleischman Diadoumenos, though slightly smaller in scale, compares most closely in style, technique of carving, and details such as the slightly downcast eyes and the bordered flat fillet, to one in Kassel (see Bieber).

BIBLIOGRAPHY: Unpublished.

RELATED REFERENCES: V. Machaira in *Mind and Body* (Athens, 1989), no. 220. M. Bieber, *Die antiken Skulpturen und Bronzen des königlichen Museum Fridericianum in Cassel* (Marburg, 1915), pp. 10–11, no. 6, pls. 11–13. D. Kreikenbom, *Bildwerke nach Polyklet* (Berlin, 1990), pp. 109–140, nos. V 1–V 60. P. Bol, "Diadumenos," in *Polyklet: Der Bildhauer der griechischen Klassik* (Liebieghaus, Frankfurt am Main, 1990), pp. 206–212, nos. 68–81. *LIMC*, vol. 2, no. 468, s.v. "Apollo" (V. Lambrinoudakis). *LIMC*, vol. 1, no. 18, s.v. "Alexandros" (R. Hampe).

—J. B. G.

181

Statue of Bes

Italy or Egypt, Roman period, perhaps Hadrianic, circa second century A.D.
Gabbro; H: 45 cm

The god Bes, protector of the home—especially the sleeping quarters, of women pregnant and in childbirth, and of children themselves, was a favorite divinity in Egypt from the New Kingdom onward. His grotesque form was regarded as both amusing and frightening. Traditionally he was closely associated with the goddess of music and love, Hathor, but during the Greco-Roman period, the attributes and associates of this goddess as well as her persona merged with those of Isis. Bes came to be seen in particular as protector of the young sun-god Horus, Isis' son. Bes was also held to be a talented dancer and musician, able to perform both arts simultaneously, and he is depicted as such on temple walls in Roman times, in particular at Philae, an island south of Aswan. His other aspect was as a fierce warrior.

Bes's true identity has been a topic of some interest in recent decades. As with many mythological figures, pinpointing his species is difficult. However, Bes does have a great many specific features symptomatic of certain types of human dwarf, and while he has often been generally called a dwarf, this identification has been questioned at times. Thus it may be useful to outline those features here. It is important to note, however, that even experts in dwarfism find the various types extremely difficult to differentiate, sometimes switching diagnoses after a few years.

Bes was first identified as an achondroplastic dwarf by P. A. Vassal. Several of the hallmarks of this disease are carefully depicted in this statue: normal-length torso with enlarged abdomen and buttocks, rhizomelia (shortness of the upper parts of limbs), permanently flexed elbows and knees, saddle-nose, flattened cheekbones, enlarged mandible, and trident hands with equal-length fingers. Additional signs of achondroplasia are spinal kyphosis (hunchback) and a tilted pelvis, both typical for Bes and evident here. The protruding belly and forward-angled thighs partially obscure normal-sized genitalia. Achondroplasts usually have bulbous foreheads. While this example does not, many Bes figures do. This Bes's arms are shown longer than is the norm either for Bes or for achondroplastics. The muscle power of achondroplastic dwarves is superior, enabling them to perform the feats of strength and acrobatics for which Bes was famous.

As here Bes is usually shown with cauliflower ears, stubby feet, perhaps even clubfeet, and a protruding tongue apparently caused by a misshapen mouth. These are all symptoms of the rarer diastrophic dwarfism, and since varieties of dwarfism are difficult to distinguish, it is possible that these features would be enough to change Vassal's diagnosis. It has been suggested to this writer by Dr. George Thompson of University Hospitals, Cleveland, that Bes's long hair and beard could be the result of short, permanently flexed limbs which make hygiene difficult for dwarves. The leopard-skin often worn by Bes is absent here as is the tail, a feature which, along with the mane of hair and misshapen facial features, has caused some scholars to regard Bes as partly leonine.

Along with Isis, Osiris, Sarapis, and Harpocrates, Bes is one of the deities most frequently represented among Roman terracottas from Egypt, which understandably survive in far greater numbers than stone statuary of the same period.

Among these he almost always occurs in one of two poses: wearing a tall, wide, feathered headdress, brandishing a sword in his right hand, and holding a shield on his left arm; or wearing the same headdress but standing or squatting with his hands pressed on his upper thighs. The Fleischman statue may be a large version of the latter, but certain features suggest another possibility.

Normally, Bes's crown of feathers would have presented a long, rectangular section at its base rather than a square as it does here. Furthermore, the statue's plain, uninscribed back pillar is the same width where it breaks above Bes's head as it is at the statue's bottom, suggesting that a substantial weight continued above Bes's head. Normally back pillars on Egyptian hard-stone statues narrow significantly above the statue's neck, the part which needs the greatest support. Therefore, the structure on top of Bes's head could have been rather large—perhaps a bust of Isis—as on a composite statue in the Ägyptisches Museum, Berlin.

Although the workmanship, both in terms of the method of carving and the glossy polish, is Roman and not Egyptian, it is not known whether this statue was made to adorn a temple in Egypt, perhaps in Alexandria, or a villa like Hadrian's in Italy. Gabbro comes from Wadi Semna in Egypt, but the quarry (called by the Romans *mons ophites* because of the stone's resemblance to snakeskin), according to geologist James Harrell, was exploited only by the Romans, who simply cut and shipped the stone, leaving behind no partially carved statuary, unlike the Egyptians who carved in their quarries. No inscriptions relating to Bes exist in Italy; however, two statuettes and a painting of him were found at the Iseum in Pompeii.

BIBLIOGRAPHY: Unpublished.

RELATED REFERENCES: M. Malaise, *Les Conditions de pénétration et de diffusion des cultes égyptiens en Italie* (Leiden, 1972), pp. 214–215. Idem, "Bes et les croyances solaires," in S. Israelit-Groll, ed., *Studies in Egyptology Presented to Miriam Lichtheim,* vol. 2 (Hebrew University, Jerusalem, 1990), pp. 680–729. J. F. Romano, *The Bes-Image in Pharaonic Egypt,* Ph.D. diss., New York University, 1989 (Ann Arbor, Mich., 1989). P. A. Vassal, "La Physico-pathologie dans le panthéon égyptien: les dieux Bes et Ptah, le nain et l'embryon," *Bulletin de la Société d'Anthropologie* 10, 7 (1956), pp. 168–181. J. A. Bailey, M.D. F.A.C.S., *Disproportionate Short Stature: Diagnosis and Management* (Philadelphia, 1973), pp. 59–89, 100–116. F. Dunand, *Catalogue des terres cuites grèco-romaines d'Egypte* (Paris, 1990), pp. 38–47, nos. 30–59. A. Roullet, *The Egyptian and Egyptianizing Monuments of Imperial Rome* (Leiden, 1972), p. 92, no. 122, pl. XCI.

—A. K.

182

Statue of Aphrodite (Venus Genetrix type)

Roman, second-century A.D. copy, after a Greek original of about 410 B.C.
Fine-grained white marble; H: 97.7 cm; W: 30.5 cm; DEPTH: 31.7 cm
Condition: Head, right arm below the shoulder, left arm below the elbow, pieces of drapery on left side, ends of the first three toes on both feet, parts of base missing; surfaces show modern re-cutting, possibly the result of the removal of old restorations; left shoulder reconstructed (two distinct pieces of marble used); surface of marble on proper left side, more at the back than the front, pitted and weathered as if from water damage.

The under life-size statue is supported by a high, profiled plinth, oval in plan with a concave central zone between upper and lower moldings all carved from one block of marble. The original finished surface of the statue, which is in generally good condition, was polished; the fine lines of the abrasive used are preserved. Traces of a mortar-type substance can be seen in the deepest part of the drapery folds.

Aphrodite, the goddess of sex and sexual love, was represented in antiquity in varying states of dress, from fully clothed in the earlier periods to completely nude by the Hellenistic period. The manner of her dress in this statue, with sheer clinging fabric emphasizing her femininity, especially her genitalia, and one breast bared, identifies the figure as a copy of what is commonly known as the Venus Genetrix or, more properly, Aphrodite Fréjus/Naples, after the best copy of the Greek original (Musée du Louvre MA 525), once thought to be from Fréjus, but actually from near Naples. The original statue is a Roman copy of what must have been a famous and popular Greek original, as many copies, both full-size and reduced, have survived. The profiled base, made in one with the figure, indicates that this copy was made in the second century A.D. The accomplished workmanship and the uniform, fine-grained marble suggest that the piece was produced in a major center.

The identity of the sculptor of the original statue has been much discussed. The figure is similar in character to the sandal-binding Nike on the parapet of the Temple of Athena Nike on the Acropolis. The play of long, finely ridged folds of thin, clinging fabric that reveals the forms of the body beneath as if it were nude, and a serene, slightly remote and languorous aura are typical of the late fifth century B.C. Attempts to identify the artist of the original as Kallimachos or Alkamenes remain inconclusive, as do attempts to identify a specific statue of Aphrodite as the prototype.

Figures similar to the Louvre statue appear on Roman coins with the inscription *Veneri genetrici*. Some think that the coin types reproduce the cult statue of Venus Genetrix by the sculptor Arkesilaos (Pliny *NH* 35.156) for the goddess's temple in the Forum of Julius Caesar, but as figures of various types with the same inscriptions occur on other Roman coins, the identifi-

182

cation is not certain. The majority of scholars now think that it is unlikely that the statue for the temple of Venus Genetrix was the type we see here, variously called the Aphrodite Louvre/Naples, Aphrodite Fréjus, or Aphrodite Fréjus/Genetrix.

BIBLIOGRAPHY: Unpublished.

RELATED REFERENCES: *LIMC*, vol. 2, pp. 33–38, esp. nos. 225–240, s.v. "Aphrodite" (A. Delivorrias, G. Berger-Doer, A. Kossatz-Deissmann). M. Bieber, *Ancient Copies* (New York, 1977), pp. 46–47, figs. 124–135. M. Brinke, *Kopienkritische und typologische Untersuchungen zur statuarischen Überlieferung der Aphrodite Typus Louvre-Neapel* (Hamburg, 1991). E. M. Moormann, "Eine Kopie der Aphrodite von Kallimachos in Amsterdam," *Bulletin Antieke Beschaving* 65 (1990), pp. 45–50. P. Karanastassis, "Untersuchungen zur kaiserzeitlichen Plastik in Griechenland. I: Kopien, Varianten und Umbildungen nach Aphrodite-Typen des 5., Jhs. v. Chr.," *AM* 101 (1986), esp. pp. 207–259.

—J. B. G.

183
Statuette of Nemesis with Portrait Resembling the Empress Faustina I

Roman, mid-second century A.D.
Marble; H: 46.4 cm; W: 20.4 cm;
DEPTH: 12.7 cm
Provenance: Formerly in the Dattari Collection, Alexandria, Egypt.

Roman portraits often took the form of figures from mythology—women frequently favored Aphrodite and men occasionally saw themselves as Mars. The disjunction between an idealized body and an often less than ideal face seems not to have troubled the sitter, but it startles the modern viewer.

183

183

This highly sophisticated statuette has been the subject of much speculation with regard to its portraitlike features. Although the line of the figure's jaw is somewhat more taut and her neck more slender than canonical images of the empress Faustina I (reigned A.D. 138–141), the hairstyle and certain aspects of the facial features strongly resemble those of the empress. Carved from an imperfect block of large-grained marble, probably from the Greek islands or the Greek East, this sensitive sculpture may be plausibly interpreted as a small-scale portrait of Faustina. In that the subject is Nemesis, the personification of stalwart values and revenge, it is difficult to propose a matching mate which might have represented her husband, the emperor Antoninus Pius (reigned A.D. 138–161), since Nemesis is not usually associated with a male being. The possibility cannot be excluded that if this is a portrait of Faustina, it was a freestanding commission, perhaps from a shrine to the imperial cult. The statuette is fully carved in back, and the wings are worked in the round as well.

With her left hand, Nemesis steadies the wheel of fortune, symbolizing her influence over destiny, which rests on the globe of the cosmos; below it is an altar. Nemesis triumphantly rests her right foot on a conquered transgressor (*hubristes*), who lies prone at the foot of the base. His diminutive scale is in keeping with the tradition of second-century iconography, according to which captive and conquered figures shrunk in scale.

Nemesis wears a belted peplos and military boots. Her role as a goddess of destiny meant that she merged well with the goddess Tyche-Fortuna; the military character of this image may link her to the goddess Roma, which would buttress the statuette's identification as Faustina, since the empress would thereby be aptly linked to the protectress of the empire.

The marble is marred by dark inclusions, which, along with its provenance in the Dattari Collection of Alexandria, confirms that the work was made in Lower Egypt. The paucity of marble in Egypt led to its importation from sources throughout the Mediterranean; Egyptian artisans used blocks that would have been discarded elsewhere in the Empire.

BIBLIOGRAPHY: O. Rubensohn, "Griechisch-römishe Funde in Ägypten," *AA* (1905), p. 69; Hôtel Drouot, Paris, *Collection Lambros-Dattari*, auct. cat., 1912, no. 339; S. Reinach, *Répertoire de la statuaire greque et romaine*, vol. 4 (Paris, 1913), p. 235, no. 1; P. Perdrizet, "Némésis," *BCH* 36 (1912), pp. 251ff., pl. 1; N. Rostovtzeff, "Pax Augusta Glaudiana," *Journal of Egyptian Archaeology* 12 (1926), p. 27, pl. 10.7; A. Adriani, *Repertorio d'arte dell'Egitto greco-romano*, ser. AII (1961), p. 66, no. 220, fig. 339; Münzen und Medaillen, A.G., Basel, *Kunstwerke der Antike*, Auktion 56, 1980, no. 169; F. Baratte, " Une statue de Némésis dans les collections du Louvre," *Revue du Louvre* 3 (1981), p. 176, figs. 14, 15.

—M. A.

Checklist of Objects in the Fleischman Collection* Not Included in the Exhibition

* This catalogue and checklist reflect the contents of the collection as of January 1994.

Greece from the Geometric Period to the Late Classical

184. *Statuette of a Ram*
Northern Greek, eighth–seventh century B.C.
Bronze; H: 4 cm

185. *Armlet*
Greek, late eighth–early seventh century B.C.
Bronze; MAX. DIAM: 11 cm

186. *Finial in the Form of a Long-billed Bird on a Stand*
Northern Greek, circa 750–700 B.C.
Bronze; H: 4.8 cm

187. *Goose Pendant*
Northern Greek, Geometric, circa 700 B.C.
Bronze; MAX. H: 4.6 cm; L: 5.2 cm

188. *Phiale*
East Greek or Phrygian, circa seventh century B.C.
Bronze; H: 6 cm; DIAM 18.5 cm
Found with no. 189.

189. *Phiale*
Asia Minor, circa seventh century B.C.
Bronze; H: 4.4 cm; DIAM: 18.7 cm
Found with no. 188.

190. *Fragmentary Statuette of a Warrior*
Greek, second half of sixth century B.C.
Bronze; H: 6.1 cm

191. *Statuette of a Goose*
Greek, circa 500 B.C.
Bronze; H: 3.2 cm

192. *Statuette of a Winged Boar*
Greek, fifth century B.C.
Bronze with lead; H: 4.4 cm; L: 5.3 cm

193. *Tripod Pyxis*
Corinthian, circa 570 B.C.
Terracotta; H: 7.9 cm
Bibliography: Charles Ede Ltd., London, *Corinthian Pottery,* Catalogue 8, 1992, no. 18.

194. *Black-figured Mastoid Cup*
Attic, circa 535–520 B.C.
Terracotta; H: 4.2 cm

195. *Olpe*
Greek, circa 500 B.C.
Terracotta; H: 14 cm

196. *Black-glazed Cup*
Attic, fifth century B.C.
Terracotta; H: 3.9 cm; W (WITH HANDLES): 15.5 cm; DIAM (OF BOWL): 10.2 cm

197. *Red-figured Squat Lekythos*
Attic, circa 430 B.C.
Terracotta; H: 7.1 cm

198. *Epichysis*
Greek, fourth century B.C.
Terracotta; H: 9.8 cm; DIAM (OF BOTTOM): 8 cm

199. *Bird with Two Chicks*
Boeotian, late sixth century B.C.
Terracotta (orange-brown clay with light engobe); H: 7.8 cm; L: 11 cm

200. *Fragmentary Male Figure*
Greek, Archaic, late sixth century B.C.
Terracotta (fine buff clay covered with a cream slip); H: 14.1 cm

201. *Kore*
Greek, late sixth century B.C.
Terracotta (reddish buff clay with large coarse particles added, covered with a lighter slip); H: 25.5 cm

202. *Ram*
Boeotian, late sixth–fifth century B.C.
Terracotta (brownish clay body); H: 11.9 cm; L: 12.7 cm

203. *Lamp*
Attic, fifth century B.C.
Terracotta; L (WITH NOZZLE): 9 cm; H: 2.4 cm

The Western Greek Colonies in South Italy and Sicily

204. *Fragmentary Vessel Attachment in the Form of a Warrior*
West Greek, mid-sixth century B.C.
Bronze; H: 7.5 cm

205. *Three-disk Cuirass*
South Italian, fourth century B.C.
Bronze; H (EACH PIECE): 30.5 cm; W: 27.5 cm

206. *Two Finials*
West Greek, probably fifth century B.C.
Gold with filler; L (EACH): 4 cm

207. *Fibula*
South Italian, fourth century B.C.
Silver; L: 12 cm

208. *Fibula*
South Italian, second half of fourth century B.C.
Silver, gold; L: 11 cm

209. *Guttus*
South Italian, fourth century B.C.
Terracotta; H: 14 cm; DIAM (OF BODY): 12 cm

210. *Black-glazed Stemless Kylix*
South Italian, fourth century B.C.
Terracotta; W (WITH HANDLES): 18 cm

211. *Red-figured Bell Krater*
Apulian, fourth century B.C.
Terracotta; H: 27.5 cm; DIAM (OF MOUTH): 28.7 cm; (OF FOOT): 13.1 cm
Bibliography: W. K. Zewadski, *Ancient Greek Vases from South Italy in Tampa Bay Collections*, Suppl. 3 (Tampa, Fla., 1989), p. 66, k; Charles Ede, Ltd., London, *Greek Pottery from South Italy*, Catalogue 14, 1990, no. 1.

212. *Bell Krater*
Side A: Seated Woman
Side B: Theatrical Mask
Apulian, second half of fourth century B.C.
Terracotta; H: 24 cm
Bibliography [F298]: A. D. Trendall, "An Apulian Bell-Krater Depicting the Mask of a White-Bearded Phlyax," *Studies in Honor of Eric Handley* (forthcoming).

213. *Red-figured Fish Plate*
Paestan, third quarter of fourth century B.C.
Terracotta; H: 6.5 cm; DIAM: 26 cm

214. *Pyxis*
Apulian, late fourth century B.C.
Terracotta; H: 8.5 cm; MAX. DIAM: 9 cm

215. *Plastic Vase in the Form of a Bird*
Campanian, circa 310–280 B.C.
Terracotta; H: 13.1 cm

216. *Figure of a Banqueter*
Tarantine, late sixth–fifth century B.C.
Terracotta; H: 13 cm; L: 15.3 cm

217. *Bearded Head*
Tarantine, fourth century B.C.
Terracotta; H: 11.5 cm
Provenance: Formerly in the Virzi collection.

218. *Relief Fragment: Female Figure*
Tarantine, late fourth century B.C.
Limestone; H: 19.9 cm

219. *Relief Fragment: Female Torso*
Tarantine, late fourth–early third century B.C.
Limestone; H: 13.8 cm

220. *Capital*
Tarantine, late fourth–early third century B.C.
Limestone; H: 20 cm

221. *Statuette of a Dancer Wearing a Himation*
Tarantine, third century B.C.
Terracotta; H (OF FIGURE): 23.5 cm

222. *Comic Mask*
Tarantine, third–second century B.C.
Terracotta with polychromy; H: 9 cm; DEPTH: 9 cm
Provenance: Formerly in the Virzi collection.

Etruria

223. *Horse's Head*
Etruscan, late sixth century B.C.
Bronze; H: 3 cm

224. *Wine Strainer*
Etruscan, fifth century B.C.
Bronze; H: 27 cm

225. *Wine Ladle and Strainer*
Etruscan, fifth century B.C.
Bronze; L (LADLE): 41 cm; (STRAINER): 26.5 cm

226. *Foot of a Thymiaterion*
Etruscan, fifth century B.C.
Bronze; H: 11.8 cm
Bibliography: Münzen und Medaillen, A.G., Basel, *Kunstwerke der Antike*, Auction 18, November 29, 1956, lot 35; Sotheby's, New York, May 22, 1981, lot 109.

227. *Perfume Dipper*
Etruscan, late fifth–fourth century B.C.
Bronze; H (OVERALL): 26.3 cm; (OF FIGURE): 5.2 cm

228. *Thymiaterion*
Etruscan, fourth century B.C.
Bronze; H: 38.5 cm

229. *Foot of a Cista*
Etruscan, circa fourth century B.C.
Bronze; H: 8.1 cm

230. *Shovel* (Batillum)
Etruscan, third century B.C.
Bronze; L: 40 cm

231. *Pin with Finial in the Form of a Dove Sitting on Pomegranates*
Etruscan, late sixth–fifth century B.C.
Gold; H: 7.7 cm

232. *Pair of Chalices*
Etruscan, late seventh–early sixth century B.C.
Bucchero; H (a): 17.5 cm; (b): 17.9 cm

233. *Male Votive Head*
Etruscan, third century B.C.
Terracotta (light brown clay with many black particles); H: 28 cm

The Hellenistic World after the Death of Alexander

234. *Comic Mask*
Greek, circa 100 B.C.
Bronze; H: 2.3 cm; DEPTH: 3 cm

235. *Statuette of a Grotesque Male Figure*
Late Hellenistic, second half of first century B.C.
Bronze; H: 9.2 cm

236. *Statuette of a Running Mime*
Late Hellenistic, second half of first century B.C.
Bronze; H: 4.4 cm

237. *Torque*
Greco-Celtic, Hellenistic, probably late fourth century B.C.
Silver; DIAM: 15 cm

238. *Ring*
Greco-Celtic, Hellenistic
Gold; MAX. DIAM: 2.2 cm; MAX. H: 0.5 cm

239. *Two Masks*
Hellenistic, probably third century B.C.
Terracotta (fine buff clay)
A. Old Man
H: 5 cm; DEPTH: 6 cm
B. Young Woman
H: 9 cm; DEPTH: 5.5 cm

240. *Statuette of Bes*
Late Hellenistic, second half of first century B.C.
Terracotta; H: 15.8 cm

241. *Two Seated Erotes*
Canosan, circa 300 B.C.
Terracotta (light brown clay); H (a): 10.2 cm; (b): 10 cm

242. *Rearing Horse*
Canosan, circa 300 B.C.
Terracotta with polychromy; MAX. H: 23 cm; L: 23.5 cm

243. *Statuette of Apollo*
Canosan, circa 300 B.C.
Terracotta (light brown clay); H: 21.8 cm
Bibliography: Galerie Nina Borowski, Paris, *20eme Anniversaire*, exh. cat., 1986, no. 6.

Rome and the Provinces from the Republic to the Late Antique

244. *Statuette of a Seated Actor*
Roman
Bronze; H: 4.7 cm

245. *Mouse*
Roman?
Bronze; H: 3 cm; L: 4.8 cm

246. *Appliqué*
Roman, late first century B.C.–first century A.D.
Bronze; H: 5 cm

247. *Statuette of Mercury*
Roman, late first century B.C.–first century A.D.
Bronze; H: 6.5 cm

248. *Lid in the Form of a Comic Mask*
Probably Roman, first century B.C.–first century A.D.
Bronze; H: 2.3 cm; DEPTH: 4 cm

249. *Statuette of a Mime*
Roman, first century B.C.–first century A.D.
Bronze; H: 6.4 cm

250. *Handle Attachment Plate with Relief of a Child Fisherman*
Roman, first century B.C.–first century A.D.
Bronze with silver; H: 4.5 cm

251. *Statuette of a Lion*
Roman, perhaps first century B.C.–second century A.D.
Bronze, inlaid silver(?); H (OF FIGURE): 32 cm; L (OVERALL): 39 cm; (OF BASE): 51 cm; W: 29 cm

252. *Lid in the Form of a Comic Mask*
Roman, first century A.D.
Bronze (?); L: 3.9 cm

253. *Statuette of Apollo*
Roman, circa 100 A.D.
Bronze, silver; H: 18.5 cm

254. *Statuette of a Trumpeter*
Roman, first–second century A.D.
Bronze; H: 14.1 cm
Provenance: Formerly in the collection of James Coats.

255. *Statuette of Aphrodite*
Roman, first–second century A.D.
Bronze with silver inlay; H (WITH BASE): 25.7 cm; (OF FIGURE): 23.2 cm
Bibliography [F26]: *Emory Magazine*, June 1989, ill. p. 3; G. Donato and M. Seefried, *The Fragrant Past*, exh. checklist (Rome, 1989), no. 1.

256. *Beam-end Fitting with Head of Athena*
Roman, first–second century A.D.
Bronze, with lead (?) filling; H: 18 cm; W: 14 cm; DEPTH: 17.8 cm
Bibliography: Münzen und Medaillen A.G., Basel, Auction 60, September 21, 1982, lot 134.

257. *Roundel with Lion's Head*
Roman, circa first–second century A.D.
Bronze; DIAM: 14 cm

258. *Statuette of a Gladiator*
Roman, second century A.D.
Bronze; H: 4.1 cm

259. *Bust of Zeus in a Leaf Chalice*
Roman, second century A.D.
Bronze; H: 6.2 cm

260. *Finial with Protome of a Wild Boar*
Roman, second century A.D.
Bronze; H: 15.8 cm

261. *Brooch*
Roman, second century A.D.
Bronze with enamel; H: 5.2 cm

262. *Theatrical Mask*
Roman, circa second century A.D.
Bronze; H: 3.5 cm

263. *Statuette of Helios*
Roman, second–early third century A.D.
Bronze; H: 18.6 cm

264. *Brooch*
Roman, second half of second–early third century A.D.
Bronze with enamel; H: 2.7 cm; L: 5.4 cm

265. *Miniature Tragic Mask*
Roman, circa second–third century A.D.
Bronze; H: 3.8 cm

266. *Pendant with Comic Mask*
Roman, second–third century A.D.
Bronze; H: 5.2 cm

267. *Ring*
Roman, third century A.D.
Bronze with silver; MAX. DIAM: 3.5 cm; MAX. H: 1.6 cm

268. *Ring*
Roman, first century B.C.–first century A.D.
Gold with black jasper; MAX. DIAM: 2.2 cm; MAX. H: 8 cm

269. *Millefiori Patella Bowl*
Roman, first century B.C.–first century A.D.
Glass; H: 4.7 cm; DIAM: 9.7 cm

270. *Cameo*
Roman, end of first century B.C.–early first century A.D.
Glass, white on purple ground; H: 1.6 cm

271. *Intaglio*
Roman, early first century A.D.
Amethyst (in Neoclassical gold mount); H: 3.3 cm; W: 3 cm
Provenance: Formerly in the Behague-Bearn collection.
Bibliography: *Burlington Fine Arts Club: Exhibition of Greek Art* (London, 1904), p. 247 (case O, no. 62, Comtesse de Bearn collection); J. Spier in *Antike Kunst* 2 (1991), p. 94f., figs. 6, 7, pl. 10.

272. *Head of a Julio-Claudian Prince*
Roman, second quarter of first century A.D.
Silver, filled with lead; H: 2.7 cm
Bibliography: Sotheby's, London, *Antiquities*, July 13, 1987, lot 64.

273. *Cameo*
Roman, first half of first century A.D.
Sardonyx in three layers: brownish-black ground, white middle layer, and transparent red-brown top layer (Neoclassical gold and onyx setting); H (OF ANCIENT FRAGMENT): 1.9 cm; W: 1.9 cm

274. *Ring*
Roman, first–second century A.D.
Gold with amethyst; MAX. DIAM: 2.2 cm; MAX. H: 8 cm

275. *Necklace*
Roman, second century A.D.
Gold; L: 50.2 cm

276. *Miniature Theatrical Mask*
Roman, circa second century A.D.
Ivory; H: 1.5 cm

277. *Ring*
Parthian (?), second–third century A.D.
Silver; MAX. DIAM: 3 cm; MAX. H: 1.2 cm

278. *Roundel with Mask*
Roman, second–third century A.D.
Bone; DIAM: 5.4 cm

279. *Cameo*
Parthian or Sassanian (?), third–fourth century A.D. (?)
Agate with dark brown ground and gray-white top layer; H: 2.5 cm; W: 2.1 cm; DEPTH: 0.4–0.5 cm

280. *Fresco Fragment*
Roman, first century A.D.
H: 19 cm; W: 22 cm

281. *Two Fresco Fragments*
Roman, first century A.D.
MAX. H (a): 33 cm; W: 32 cm
MAX. H (b): 46 cm; W: 35.5 cm

282. *Head of Medusa*
Roman, early second century A.D.
Large-crystalled yellow-white marble; H: 30 cm
Provenance: Formerly in the collection of J. and U. Thimme.
Bibliography: *Pergamon*, exh. cat. (Ingolheim am Rhein, 1972), no. 31, cf. also no. 32.

283. *Portrait Head of Faustina the Younger*
Roman, circa 170 A.D.
Fine-grained, slightly translucent white marble; H: 24 cm
Bibliography: Sotheby's, New York, *Antiquities*, March 1–2, 1984, no. 76.

284. *Two Heads of Sea Monsters*
Roman, late first–second century A.D. (?)
Africano marble; L (a): 28 cm; MAX. H: 17 cm; L (b): 26 cm; MAX. H: 17 cm

285. *Nude Male Torso*
Roman, probably second century A.D.
Large-crystalled white marble; H: 42 cm

286. *Portrait Head of a Boy as Mercury*
Roman, second half of second century A.D.
Opaque white marble, probably Luna; H: 16 cm

Addenda

287. *Fresco Fragment with the Heads of Two Women*
Roman, first century A.D.
H: 28.5 cm; W: 21.5 cm

288. *Statuette of Apollo*
Roman, second century A.D.
Silver; H: 3.8 cm

289. *Black-figured Column Krater with Odysseus under the Ram*
Attic, second half of sixth century B.C.
Terracotta; H: 33 cm; W: 35.5 cm; DIAM (OF RIM): 30 cm

290. *Statuette of Apollo*
Hellenistic, circa 100 B.C.
Silver; H: 20 cm

291. *Mask of a Satyr*
Hellenistic, third–second century B.C.
Terracotta with polychromy; H: 12 cm; L: 15.5 cm

292. *Statuette of Apollo*
Hellenistic, second half of second century B.C.
Bronze; H: 29 cm

293. *Key with Handle in the Form of a Dog*
Roman, first century A.D.
Bronze (handle) and iron: L (OVERALL): 10.9 cm; (OF HANDLE): 7.7 cm

294. *Statuette of an Actor Seated on an Altar*
Romano-Egyptian, first century A.D.
Terracotta; H: 13.6 cm

295. *Two-handled Cup with Relief Decoration of Eight Male Figures and an Animal in a Landscape*
Roman, first century A.D.
Silver; H: 12.5 cm; DIAM: 16.3 cm

Glossary of Frequently Cited Terms

amphora
two-handled storage vessel

aryballos
small container for oils or ointments

calyx krater
large vessel with flaring straight walls and a bulbous offset above the foot that resembles the calyx of a flower; used for mixing wine and water

chiton
a dress with or without sleeves, usually made of a thin fabric like linen or silk

cista
basket or box; may be of various materials such as wood, stone, or bronze

emblema
inlaid relief medallion

fillet
flat, narrow architectural molding or a similarly shaped band of cloth

himation
an oblong cloth worn as an outer garment (a shawl or cloak) over a chiton

hoplite
heavily armed foot soldier

hydria
wide-bodied vessel with a narrow mouth and three handles; used for carrying and pouring water

kalathos
a basket, narrow at the base; often used for wool

kalpis
a kind of hydria (see above) with a curving body

kantharos
a special type of drinking cup with two arching vertical handles; a shape closely associated with the god Dionysos and his retinue

keras
drinking horn

kore
Greek for "maiden"; used for sculpted female figures of the Archaic period

kouros
Greek for "youth"; used for sculpted male figures of the Archaic period

krater
a large wide-mouthed vessel, often characterized by some aspect of its shape, i.e. volute krater, calyx krater, etc.; used for mixing

kymation
an architectural molding with a double curvature (a vertical S-shape), usually with the concave part at the top, sometimes decorated with leaves or other designs; also called a cyma

lekanis
flat two-handled bowl with a lid and a foot

lekythos
vessel for oil, cosmetics, or unguents

Lesbian kymation
a kymation with the convex curve at the top; also called a cyma reversa

mural crown
a crown shaped like a miniature city wall; worn by personifications of cities and of good fortune

neck amphora
a type of amphora on which the neck meets the body at a sharp angle

oinochoe
a vessel for pouring and serving liquids

olpe
a kind of oinochoe

ovolo
a convex architectural molding with an egg-shaped profile that has its maximum projection toward the top; often decorated with egg-and-dart pattern

patera
a flat dish with handles

peplos
a dress usually made of a heavy fabric like wool

petasos
a broad-brimmed cap often associated with travelers

phiale
a flat handleless cup used for pouring libations

pilos
a kind of pointed helmet or cap

pithos
a large storage vessel

polos
a cylindrical ceremonial crown

protome
a decorative device in the form of the head or bust of a figure or creature

psykter
a wine cooler

sakkos
a coarse cloth of hair; anything made of such cloth, for example, the hair wrap seen on South Italian vases

situla
a bucket-shaped container

skyphos
a deep two-handled cup with a low foot

sphendone
a sling or a cloth shaped like a sling; a headband broad in the front and narrow in the back

stele
an upright block or slab usually of stone; used as a grave marker, boundary stone, etc.

stephané
a crescent-shaped diadem

thymiaterion
an incense burner

thyrsos
a staff tipped with a pine cone; carried by devotees of Dionysos

torus
a convex architectural molding, semicircular in profile

volute krater
a krater characterized by handles in the shape of volutes that extend above the rim of the vase